Archives of Virology

Supplementum 2

R. I. B. Francki, C. M. Fauquet
D. L. Knudson, F. Brown (eds.)

Classification and Nomenclature of Viruses

Fifth Report of the International Committee on Taxonomy of Viruses

Virology Division of the International Union of Microbiological Societies

Springer-Verlag Wien New York

Dr. R. I. B. Francki †

Department of Crop Protection, Waite Agricultural Research Institute
University of Adelaide, South Australia, Australia

Dr. C. M. Fauquet

Department of Biology, ORSTOM
Washington University, St. Louis, Missouri, U.S.A.

Dr. D. L. Knudson

Department of Entomology
Colorado State University, Fort Collins, Colorado, U.S.A.

Dr. F. Brown

Department of Microbiology
University of Surrey, Guilford, Surrey, England

Printed in Austria by Ferdinand Berger & Söhne, A-3580 Horn

Printed on acid-free paper

ISSN 0939-1983
ISBN 3-211-82286-0 Springer-Verlag Wien NewYork
ISBN 0-387-82286-0 Springer-Verlag NewYork Wien

Contents

Contributors 3
In memory of Richard Francki 5
Preface 7

Officers and Members of the ICTV 1984-1990
- Executive Committee 8
- Life Members 8
- Bacterial Virus Subcommittee 10
- Coordination Subcommittee 16
- Fungal Virus Subcommittee 18
- Invertebrate Virus Subcommittee 20
- Plant Virus Subcommittee 22
- Vertebrate Virus Subcommittee 24
- Virus Data Subcommittee 36
- National Representatives 36

President's Report 1987-1990 38
The Format for Submission of New Taxonomic Proposals 43
The Rules of Virus Nomenclature 46
The Statutes of the ICTV 48
References 52

The Viruses 53
- Glossary of Abbreviations and Virological Terms 55
- Virus Diagrams 57
- Virus Families and Groups in Order of Presentation 63
- Listing of Virus Families and Groups
  - Table I : Alphabetical listing 80
  - Table II : Listing by host 82
  - Table III : Listing by nucleic acid type 84
- Key to Identification of Virus Families and Groups 86

Descriptions of Virus Families and Groups 91
- Some Unclassified Viruses and Virus-Like Agents 400

Author Index 407
Virus Index 421
Order, Families, Groups and Genera Index 447

Classification and Nomenclature of Viruses

Fifth Report of the International Committee on Taxonomy of Viruses

Editors

R.I.B. Francki (deceased)
Department of Crop Protection,
University of Adelaide,
Waite Agricultural Research Institute,
Adelaide, South Australia 5064

C.M. Fauquet
Department of Biology,
ORSTOM/Washington University,
CB 1137,
St Louis, MO 63130 USA

D.L. Knudson
Department. of Entomology,
Colorado State University,
Fort Collins, CO 80523 USA

F. Brown
Department of Microbiology,
University of Surrey,
Guilford, Surrey GU2 5XH United Kingdom

Contributors

H.-W. Ackermann
A.M. Aubertin
M. Bar-Joseph
O.W. Barnett
R.W. Briddon
A.A. Brunt
M.J. Buchmeier
K.W. Buck
C.H. Calisher
D. Cavanagh
J.M. Coffin
P. Dobos
J.J. Esposito
R. Frisque
S.A. Ghabrial
R. Goldbach
R.I. Hamilton
I.H. Holmes
M.C. Horzinek
C.R. Howard
R. Hull
J.E. Johnson
H.D. Klenk
R. Koenig
B.E.L. Lockhart
H. Lot
J. Maniloff
G.P. Martelli
M.A. Mayo
J.B. McCormick
R. Milne
P. Minor
T.J. Morris
A.F. Murant
D. Peters
C.R. Pringle
J.W. Randles
M.A. Rezaian
D.J. Robinson
B. Roizman
R.R. Rueckert
W.C. Russell
E.P. Rybicki
S. Salquero
F.L. Schaffer
G. Siegl
J. Stanley
J.H. Strauss
D. Stoltz
K. Tomaru
J. van Duin
J.E. van Etten
M.H.V. van Regenmortel
C.C. Wang
M. Wilson
G. Wengler
W.H. Wunner
W. Zillig

In Memory of Richard Francki, 1930-1990

Richard Ignacy Bartolomiej Francki was born in Warsaw, Poland on 10 September 1930. He attended primary school in Gdynia from 1936-1939. Near the beginning of World War II Richard's father, an officer in the Polish navy, moved with his family to England. Richard continued his primary education in Hereford, and secondary schooling at Kelly College in Tavistock, Devon. It was here that he developed his excellent command of English. His family migrated to New Zealand in 1948. In 1956 he married Zofia Bozenna Surynt.

Richard continued his education in New Zealand at Auckland University College graduating with a Masters degree in Botany in 1958. In 1959 at a somewhat older age than most research students, he enrolled for a PhD degree under my supervision. After only a few months I recognized that Richard had a natural talent for research. Seven papers in Nature, Virology, and Biochimica Biophysica Acta arose from his PhD studies.

In 1961 Richard took up an appointment as Lecturer in the Department of Plant Pathology, Waite Agricultural Research Institute, of the University of Adelaide. Apart from visits abroad he remained in Adelaide, being promoted to Senior Lecturer in 1967 and Reader in 1972. He spent four periods of up to 12 months in other laboratories: 1964-65 in the Department of Botany and Plant Biochemistry, University of California, Los Angeles; 1970 in the Department of Agricultural Biochemistry, University of Arizona, Tucson; 1977 in the Department of Virology, Agricultural University, Wageningen; and 1985-86 in the Department of Plant Pathology, Cornell University. These visits widened the base of his experience and interest in the field of plant virology. His 135 research papers dealt with many different viruses and covered a range of topics from molecular biology to applied field work. In addition he was author, coauthor or editor of some 38 books, reviews or chapters dealing with a variety of topics.

Space does not allow justice to be done here to Richard's research contributions in plant virology. Sufficient to say that his wide practical experience in research provided a sound foundation for his interest in viral taxonomy. He was Chairman of the Plant Virus Subcommittee of the ICTV from 1976-1981, the same period when I was President. He was always a hardworking and reliable contributor to the work of the organisation. During meetings of the Executive Committee, discussions frequently became quite heated. Richard's contributions were always clear, to the point, and above all, put forward in a gentlemanly fashion. Important developments in plant virus taxonomy took place during his chairmanship of the Subcommittee. In 1987 Richard was elected President of the ICTV. He was deeply concerned that no updated report from the ICTV had been published since 1982. He worked extremely hard to ensure that a Fifth Report would be produced as soon as possible after the August 1990 Berlin Virology Congress. He had nearly completed this task when he became terminally ill after a courageous battle with cancer lasting many years. He died at home in Adelaide on November 14th. Richard's zest for life and for his research, his caring for others, particularly his students will long be remembered by his many friends and colleagues around the world. The sympathy of all of us goes to his wife Zofia and their two sons and three daughters.

R. E. F. Matthews
Auckland, New Zealand
11 December 1990

Preface

The Fifth Report of the International Committee on Taxonomy of Viruses (ICTV), summarizes the proceedings and decisions reached by the ICTV at its meetings held at the International Congresses of Virology in Sendai (1984), Edmonton (1987) and Berlin (1990). This report has been organized in the same way as the previous ones (Wildy, 1971; Fenner, 1976; Matthews, 1979; 1982), yet it encompasses many more families and groups of viruses than previous reports, and it includes new tables, diagrams and keys. The officers and members of the ICTV study groups from 1984 to 1990 are listed, as the current ICTV statutes and rules of nomenclature. Information on the format for submission of new taxonomic proposals to the ICTV is also provided.

Since the Fourth Report of the ICTV (1982), 19 new virus families and groups have been described. This report includes 2,430 viruses belonging to 73 families or groups, as well as virus satellites and viroids descriptions, but it does not include descriptions not approved by the ICTV. It now will be possible to publish such preliminary, and in some cases controversial, descriptions in the Virology Division pages of the *Archives of Virology* -- this will allow virologists to carry on the kind of interim dialogue that is necessary for arriving at broad agreement on taxonomic matters. Similarly, a listing of acronyms of plant viruses, developed by the members of the Plant Virus Subcommittee, will soon be published informally in the *Archives of Virology*, but it is hoped that in the next three years a universal acronym listing for all viruses will be approved by the ICTV and be included in the Sixth Report of the ICTV.

The names of virologists who provided initial and revised compilations of virus family and group descriptions are indicated at the beginning of each description. For clarity, the term '*Reported by*' is issued to indicate the chair of the concerned ICTV study-group; '*Revised by*' is used to indicate the person providing a revised compilation; and '*Compiled by*' is used to indicate the person providing a new description. In all cases these named virologists have worked with the many members of the various subcommittees and study-groups of the ICTV -- it is only by the combined work of all these virologists that this report has been completed.

The editors would like to express their gratitude to R.E.F. Mathews, who edited the Third and Fourth ICTV Reports which form the backbone of this report. The editors would also like to express their gratitude to F.A. Murphy, the incoming President of the ICTV, who helped in the editing of this report. Finally, the editors would like to express their gratitude to all the persons who contributed to this report, and particularly to C.J. Grivell, E.G. Cabot, and J.W. Randles of the University of Adelaide, and B. Delannay of Washington University, St Louis.

The editing of the ICTV reports has in the past been done by the President of the ICTV; however, this time, the President, Richard Francki, was not able to complete this task because of ill health. The editors, on behalf of all virologists, dedicate this report to Richard Francki's memory.

C.M. Fauquet
April 1991
St Louis, Missouri, USA

Officers and Members of the International Committee on Taxonomy of Viruses

EXECUTIVE COMMITTEE 1987-1990

President	R.I.B. FRANCKI	AUSTRALIA
Vice-President	H.W. ACKERMANN	CANADA
Secretary	K.W. BUCK	UNITED KINGDOM
Secretary	C.M. FAUQUET	FRANCE
Subcommittee Chairs		
Bacterial Virus	H.W. ACKERMANN	CANADA
Virus data	D.L. KNUDSON	USA
Fungal Virus	S.A. GHABRIAL	USA
Invertebrate Virus	M.D. SUMMERS	USA
Plant Virus	G.P. MARTELLI	ITALY
Vertebrate Virus	D.H.L. BISHOP	UNITED KINGDOM
Elected members		
Member	P. AHLQUIST	USA
Member	L. BERTHIAUME	CANADA
Member	Y. GHENDON	SWITZERLAND
Member	B.M. GORMAN	AUSTRALIA
Member	R. GRANADOS	USA
Member	M.A. MAYO	UNITED KINGDOM
Member	D. PETERS	THE NETHERLANDS
Member	J.M. VLAK	THE NETHERLANDS

LIFE MEMBERS 1987-1990

Life Member	F. BROWN	UNITED KINGDOM
Life Member	F.J. FENNER	AUSTRALIA
Life Member	F.O. HOLMES	USA
Life Member	A. LWOFF	FRANCE
Life Member	R.E.F. MATTHEWS	NEW ZEALAND
Life Member	H.G. PEREIRA	BRAZIL

Officers and Members of the International Committee on Taxonomy of Viruses

EXECUTIVE COMMITTEE 1984-1987

President	F. BROWN	UNITED KINGDOM
Vice-President	J.P.H. van der WANT	THE NETHERLANDS
Secretary	J. MAURIN	FRANCE
Secretary	K. W. BUCK	UNITED KINGDOM
Subcommittee Chairs		
Bacterial Virus	A. EISENSTARK	USA
Virus data	J.C. ATHERTON	AUSTRALIA
Fungal Virus	K.W. BUCK	UNITED KINGDOM
Invertebrate Virus	J.F. LONGWORTH	NEW ZEALAND
Plant Virus	R.I. HAMILTON	CANADA
Vertebrate Virus	D.W. KINGSBURY	USA
Elected members		
Member	M. BERGOIN	FRANCE
Member	T. GRAF	FRG
Member	D.C. KELLY	UNITED KINGDOM
Member	M.B. KOROLEV	USSR
Member	S. MATSUMOTO	JAPAN
Member	A.F. MURANT	UNITED KINGDOM
Member	L. van VLOTEN DOTING	THE NETHERLANDS
Member	D.W. VERWOERD	SOUTH AFRICA

LIFE MEMBERS 1984-1987

Life Member	C. ANDREWES	UNITED KINGDOM
Life Member	F.J. FENNER	AUSTRALIA
Life Member	F.O. HOLMES	USA
Life Member	A. LWOFF	FRANCE
Life Member	R.E.F. MATTHEWS	NEW ZEALAND
Life Member	H.G. PEREIRA	BRAZIL

BACTERIAL VIRUS SUBCOMMITTEE 1984-1987

Chair	J.N. COETZEE	SOUTH AFRICA
Vice-Chair	H.-W. ACKERMANN	CANADA
Member	S.N. CHATTERJEE	INDIA
Member	T.S. DHILLON	HONG KONG
Member	R.W. HEDGES	BELGIUM
Member	A.W. JARVIS	NEW ZEALAND
Mmeber	L.A. JONES	USA
Member	V.N. KRYLOV	USSR
Member	J. MANILOFF	USA
Member	S. OGATA	JAPAN
Member	S. SAFFERMAN	USA
Member	M. SHILO	ISRAEL
Member	T. SOZZI	SWITZERLAND

Bacillus Phage Study Group

Clostridium Phage Study Group

Cubic, Filamentous & Pleomorphic Phages Study Group

BACTERIAL VIRUS SUBCOMMITTEE 1987-1990

Chair	H.-W. ACKERMANN	CANADA
Vice-Chair	A.W. JARVIS	NEW ZEALAND
Member	S.N. CHATTERJEE	INDIA
Member	K. FURUSE (until 1989)	JAPAN
Member	L.A. JONES	USA
Member	V.N. KRYLOV	USSR
Member	J. MANILOFF	USA
Member	S. OGATA	JAPAN
Member	J.J. PATEL	NEW ZEALAND
Member	J. ROCOURT	FRANCE
Member	R.S. SAFFERMAN	USA
Member	L. SELDIN	BRAZIL
Member	T. SOZZI	ITALY
Member	P.R. STEWART	AUSTRALIA
Member	J. van DUIN	THE NETHERLANDS
Member	L. WUNSCHE	DDR

Bacillus Phage Study Group

Chair	L. SELDIN	BRAZIL
Member	D.H. DEAN	USA
Member	P.S. LOVETT	USA
Member	E. NAGY	HUNGARY
Member	A.S. TIKHONENKO	USSR
Member	T.A. TRAUTNER	FRG
Member	P.S. VARY	USA

Clostridium Phage Study Group

Chair	S. OGATA	JAPAN
Member	M.W. EKLUND	USA
Member	D.T. JONES	NEW ZEALAND
Member	D.E. MAHONY	CANADA
Member	K. OGUMA	JAPAN
Member	G. SCHALLEHN	FRG

Cubic, Filamentous & Pleomorphic Phages Study Group

Chair until 1989	K. FURUSE	JAPAN
Chair since 1989	J. van DUIN	THE NETHERLANDS
Member	D. BAMFORD	FINLAND
Member	D.T. DENHARDT	CANADA
Member	A.H. HAVELAAR	THE NETHERLANDS
Member	K. KODAIRA	JAPAN
Member	J. MANILOFF	USA

1984-1987

Cyanophage Study Group

Lactobacillus Phage Study Group

Lactococcus & Streptococcus Phage Study Group

Listeria & Coryneform Phages Study Group

Pseudomonas Phage Study Group

1987-1990

Cyanophage Study Group

Chair	R.S. SAFFERMAN	USA
Member	R. CANNON	USA
Member	P.R. DESJARDINS	USA
Member	B.V. GROMOV	USSR
Member	R. HASELKORN	USA
Member	L. SHERMAN	USA

Lactobacillus Phage Study Group

Chair	T. SOZZI	ITALY
Member	J.-P. ACCOLAS	FRANCE
Member	T. ALATOSSOVA	FINLAND
Member	M. MATA	FRANCE
Member	P. RITZENTHALER	FRANCE
Member	K. TREVORS	CANADA
Member	K. WATANABE	JAPAN

Lactococcus & Streptococcus Phage Study Group

Chair	A.W. JARVIS	NEW ZEALAND
Member	G. FITZGERALD	IRELAND
Member	M. MATA	FRANCE
Member	A. MERCENIER	FRANCE
Member	I.B. POWELL	AUSTRALIA
Member	C. RONDA	SPAIN
Member	M.-L. SAXELIN	FINLAND
Member	M. TEUBER	FRG

Listeria & Coryneform Phages Study Group

Chair	J. ROCOURT	FRANCE
Member	N. GROMAN	USA
Member	S. ORTEL	GDR
Member	R. RAPPUOLI	ITALY
Member	A. TRAUTWETTER	FRANCE

Pseudomonas Phage Study Group

Chair	V.N. KRYLOV	USSR
Member	A.B. JUSUPOVA	USSR
Member	A.M. KROPINSKI	CANADA
Member	L.A. KULAKOV	USSR

1984-1987

Rhizobium Phage Study Group

Staphylococcus Phage Study Group

Tailed Phage of Enterobacteria Study Group

Vibrio Phage Study Group

1987-1990

Rhizobium Phage Study Group

Chair	J.J. PATEL	NEW ZEALAND
Member	M. KOWALSKI	POLAND
Member	M. WERQUIN	FRANCE

Staphylococcus Phage Study Group

Chair	P. STEWART	AUSTRALIA
Member	M. BES	FRANCE
Member	Y. DUVAL-IFLAH	FRANCE

Tailed Phage of Enterobacteria Study Group

Chair	H.W. ACKERMANN	CANADA
Member	T.S. DHILLON	HONG KONG
Member	M.S. DUBOW	CANADA
Member	F. GRIMONT	FRANCE
Member	B. KARSHA-WYSOCKI	CANADA
Member	B. KWIATKOWSKI	POLAND
Member	M. MAMET	CANADA
Member	A. O'BRIEN	USA
Member	J.-F. VIEU	FRANCE

Vibrio Phage Study Group

Chair	S.N. CHATTERJEE	INDIA
Member	S.A. AMAD	INDIA
Member	M.Q. ANSARI	INDIA
Member	S.C. BHATTACHARYA	INDIA
Member	J. DAS	INDIA
Member	S.S. KASATIYA	BANGLADESH
Member	T. KAWATA	JAPAN
Member	T. KOGA	JAPAN
Member	M. MAITI	INDIA
Member	S.C. PAL	INDIA

COORDINATION SUBCOMMITTEE 1984-1987

Chair	F. BROWN	UNITED KINGDOM
Member	D. BAXBY	UNITED KINGDOM
Member	K.W. BUCK	UNITED KINGDOM
Member	P. DOBOS	CANADA
Member	A. EISENSTARK	USA
Member	R. I. HAMILTON	CANADA
Member	K. JOKLIK	USA
Member	D.W. KINGSBURY	USA
Member	J. F. LONGWORTH	NEW ZEALAND
Member	P. SCOTTI	NEW ZEALAND
Member	R.E. SHOPE	USA
Member	G. SIEGL	SWITZERLAND
Member	D. WILLIS	USA

Birnaviridae Study Group

Iridoviridae Study Group

Chair	D. WILLIS	USA
Member	A.M. AUBERTIN	FRANCE
Member	G. DEVAUCHELLE	FRANCE
Member	E. VINUELA	SPAIN
Member	D. KELLEY	UNITED KINGDOM
Member	R. FLÜGEL	FRG
Member	A. GRANOFF	USA
Member	D. STOLTZ	USA

Parvoviridae Study Group

Chair	G. SIEGL	SWITZERLAND
Member	T.S. BATES	USA
Member	K.I. BERNS	USA
Member	B.J. CARTER	USA
Member	D.C. KELLY	UNITED KINGDOM
Member	E. KURSTAK	CANADA
Member	P. TATTERSALL	USA

COORDINATION SUBCOMMITTEE 1987-1990

Chair	R.I.B. FRANCKI	AUSTRALIA
Member	A.M. AUBERTIN	FRANCE
Member	D.H.L. BISHOP	UNITED KINGDOM
Member	P. DOBOS	CANADA
Member	S.A. GHABRIAL	USA
Member	I.H. HOLMES	AUSTRALIA
Member	D.L. KNUDSON	USA
Member	G.P. MARTELLI	ITALY
Member	G. SIEGL	SWITZERLAND
Member	M.D. SUMMERS	USA
Member	W.H. WUNNER	USA

Birnaviridae Study Group

Chair	P. DOBOS	CANADA
Member	A. AZAD	AUSTRALIA
Member	H. BECHT	FRG
Member	K. FAHEY	AUSTRALIA
Member	N. IKAMOTO	JAPAN
Member	T.-A. LEONG	USA
Member	H. MULLER	FRG
Member	B. NICHOLSON	USA
Member	R. RENO	USA

Iridoviridae Study Group

Chair	A.M. AUBERTIN	FRANCE
Member	V.G. CHINCHAR	USA
Member	G. DARAI	FRG
Member	V.L. SELIGY	CANADA

Parvoviridae Study Group

Chair	G. SIEGL	SWITZERLAND
Member	K.I. BERNS	USA
Member	M. BLOOM	USA
Member	B.J. CARTER	USA
Member	M. LEDERMAN	USA
Member	T. TAL	ISRAEL
Member	P. TATTERSALL	USA
Member	P. TIJSSEN	CANADA

1984-1987

Reoviridae Study Group

Chair	W.K. JOKLIK	USA
Member	T.H. FLEWETT	UNITED KINGDOM
Member	R.I.B.FRANCKI	AUSTRALIA
Member	I.H. HOLMES	AUSTRALIA
Member	H. HUISMAN	SOUTH AFRICA
Member	T. HUKUHARA	JAPAN
Member	A.Z. KAPIKIAN	USA
Member	S. KAWASE	JAPAN
Member	O. LOVISOLO	ITALY
Member	D.A. MEBUS	USA
Member	R.G. MILNE	ITALY
Member	E.L. PALMER	USA
Member	C.C. PAYNE	UNITED KINGDOM
Member	D.V.R. REDDY	INDIA
Member	C. VAGO	FRANCE
Member	D.W. VERWOED	SOUTH AFRICA
Member	H.A. WOOD	USA

Rhabdoviridae Study Group

Chair	R.E. SHOPE	USA
Member	J. CRICK	UNITED KINGDOM
Member	R.I.B. FRANCKI	AUSTRALIA
Member	D.L. KNUDSON	USA
Member	D. PETERS	THE NETHERLANDS
Member	L. SCHNEIDER	FRG

FUNGAL VIRUS SUBCOMMITTEE 1984-1987

Chair	K.W. BUCK	UNITED KINGDOM
Member	R.F. BOZARTH	USA
Member	J.A. BRUENN	USA
Member	S.A. GHABRIAL	USA
Member	Y. KOLTIN	ISRAEL
Member	C.J. RAWLINSON	UNITED KINGDOM
Member	H.A. WOOD	USA
Member	S. YAMASHITA	JAPAN

Algal & Protozoal Viruses

1987-1990

Reoviridae Study Group

Chair	I.H. HOLMES	AUSTRALIA
Member	S.M. ELEY	UNITED KINGDOM
Member	M.K. ESTES	USA
Member	B.W. FALK	USA
Member	B.M. GORMAN	AUSTRALIA
Member	Y. HOSHINO	JAPAN
Member	W.K. JOKLIK	USA
Member	K.S. KIM	USA
Member	D. KNUDSON	USA
Member	M. McCRAE	UNITED KINGDOM
Member	R.G. MILNE	ITALY
Member	D. NUSS	USA
Member	C.C. PAYNE	UNITED KINGDOM
Member	N. PLUS	FRANCE
Member	E. SHIKATA	JAPAN
Member	J.R. WINTON	USA

Rhabdoviridae Study Group

Chair	W.H. WUNNER	USA
Member	C.H. CALISHER	USA
Member	A.O. JACKSON	USA
Member	E.W. KITAJIMA	BRASIL
Member	M. LAFON	FRANCE
Member	J.C. LEONG	USA
Member	S.T. NICOL	USA
Member	D. PETERS	THE NETHERLANDS
Member	J.S. SMITH	USA

FUNGAL VIRUS SUBCOMMITTEE 1987-1990

Chair	S.A. GHABRIAL	USA
Member	K.W. BUCK	UNITED KINGDOM
Member	R.F. BOZARTH	USA
Member	J.A. BRUENN	USA
Member	Y. KOLTIN	ISRAEL
Member	C.P. ROMAINE	USA
Member	J. van ETTEN	USA
Member	R.B. WICKNER	USA
Member	S. YAMASHITA	JAPAN

Algal & Protozoal Viruses

Chair	J. van ETTEN	USA
Member	J.A. DODDS	USA
Member	A.J. GIBBS	AUSTRALIA
Member	C.C. WANG	USA

INVERTEBRATE VIRUS SUBCOMMITTEE 1984-1987

Chair	J.F. LONGWORTH	NEW ZEALAND
Member	M. BERGOIN	FRANCE
Member	P. FAULKNER	CANADA
Member	R.R. GRANADOS	USA
Member	C. IGNOFFO	USA
Member	D.C. KELLY	USA
Member	D.L. KNUDSON	USA
Member	E. KURSTAK	CANADA
Member	N. MOORE	UNITED KINGDOM
Member	C.C. PAYNE	UNITED KINGDOM
Member	M.D. SUMMERS	USA

Baculoviridae Study Group

Chair	M.D. SUMMERS	USA
Member	E. CARSTENS	CANADA
Member	G. CROIZIER	FRANCE
Member	D.C. KELLY	UNITED KINGDOM
Member	L.K. MILLER	USA
Member	G.F. ROHRMANN	ARGENTINA
Member	D.B. STOLTZ	CANADA
Member	J.M. VLAK	THE NETHERLANDS

Nodaviridae Study Group

Chair	R.R. RUECKERT	USA
Member	J.F. LONGWORTH	NEW ZEALAND
Member	N.F. MOORE	UNITED KINGDOM
Member	T.J. MORRIS	USA
Member	J.F. NEWMAN	SOUTH AFRICA
Member	C. REINGANUM	AUSTRALIA

Polydnaviridae Study Group

INVERTEBRATE VIRUS SUBCOMMITTEE 1987-1990

Chair	M.D. SUMMERS	USA
Member	E. CARSTENS	CANADA
Member	C.C. PAYNE	UNITED KINGDOM
Member	R.R. RUECKERT	USA
Member	D.B. STOLTZ	CANADA
Member	J.M. VLAK	THE NETHERLANDS
Member	M. WILSON	USA

Baculoviridae Study Group

Chair	M. WILSON	USA
Member	E. CARSTENS	CANADA
Member	J. COUCH	USA
Member	W. DOERFLER	FRG
Member	L. VOLKMAN	USA

Nodaviridae/Tetraviridae Study Group

Chair	R.R. RUECKERT	USA
Member	D.A. HENDRY	SOUTH AFRICA
Member	J. JOHNSON	USA
Member	P. SCOTTI	NEW ZEALAND

Polydnaviridae Study Group

Chair	D.B. STOLTZ	CANADA
Member	N. BECKAGE	USA
Member	P. DUNN	USA
Member	J.A. FLEMING	USA
Member	P. KRELL	CANADA
Member	M.D. SUMMERS	USA

PLANT VIRUS SUBCOMMITTEE 1984-1987

Chair	R.I. HAMILTON	CANADA
Member	M. BAR-JOSEPH	ISRAEL
Member	A.A. BRUNT	UNITED KINGDOM
Member	J.R. EDWARDSON	USA
Member	H.M. GARNETT	SOUTH AFRICA
Member	R.M. GOODMAN	USA
Member	H.T. HSU	USA
Member	R. HULL	UNITED KINGDOM
Member	R. KOENIG	FRG
Member	G. P. MARTELLI	ITALY
Member	R.G. MILNE	ITALY
Member	A.F. MURANT	UNITED KINGDOM
Member	J.W. RANDLES	AUSTRALIA
Member	E. SHIKATA	JAPAN
Member	J.H. TREMAINE	CANADA
Member	M.H.V. van REGENMORTEL	FRANCE
Member	L. van VLOTEN DOTING	THE NETHERLANDS

Potyvirus Study Group

PLANT VIRUS SUBCOMMITTEE 1987-1990

Chair	G.P. MARTELLI	ITALY
Member	O.W. BARNETT	USA
Member	R. GOLDBACH	THE NETHERLANDS
Member	R.I. HAMILTON	CANADA
Member	R. KOENIG	FRG
Member	H. LOT	FRANCE
Member	K. MAKKOCHAIR	SYRIA
Member	R.G. MILNE	ITALY
Member	T.J. MORRIS	USA
Member	A.F. MURANT	UNITED KINGDOM
Member	J.W. RANDLES	AUSTRALIA
Member	E. RYBICKI	SOUTH AFRICA
Member	L.F. SALAZAR	PERU
Member	K. TOMARU	JAPAN
Member	A. VARMA	INDIA

Potyvirus Study Group

Chair	O.W. BARNETT	USA
Member	A.A. BRUNT	USA
Member	J. DIJKSTRA	THE NETHERLANDS
Member	W.G.DOUGHERTY	USA
Member	J.R. EDWARDSON	USA
Member	R. GOLDBACH	THE NETHERLANDS
Member	J. HAMMOND	USA
Member	J.H. HILL	USA
Member	R. JORDAN	USA
Member	K. MAKKOUK	SYRIA
Member	F. MORALES	COLOMBIA
Member	S.T. OHKI	JAPAN
Member	D. PURCIFULL	USA
Member	E. SHIKATA	JAPAN
Member	D.D. SHUKLA	AUSTRALIA
Member	I. UYEDA	JAPAN

VERTEBRATE VIRUS SUBCOMMITTEE 1984-1987

Chair	D.W. KINGSBURY	USA
Vice-Chair	D.H.L. BISHOP	UNITED KINGDOM
Member	J.J. ESPOSITO	USA
Member	S.D. GARDNER	UNITED KINGDOM
Member	I. GUST	AUSTRALIA
Member	M.C. HORZINEK	THE NETHERLANDS
Member	A.P. KENDAL	USA
Member	M.P. KILEY	USA
Member	W.E. RAWLS	CANADA
Member	B. ROIZMAN	USA
Member	R.R. RUECKERT	USA
Member	F.L. SCHAFFER	USA
Member	S. SIDDELL	FRG
Member	V. ter MEULEN	FRG
Member	H. VARMUS	USA
Member	G. WADELL	SWEDEN
Member	E.G. WESTAWAY	AUSTRALIA

Adenoviridae Study Group

Chair	G. WADELL	SWEDEN
Member	A. BARTHA	HUNGARY
Member	T.H. BROKER	USA
Member	R. DREZIN	USSR
Member	M. GREEN	USA
Member	H. GINSBERG	USA
Member	C. HIERHOLZER	USA
Member	S.S. KALTER	USA
Member	I. MAICHE-LAUPPE	FRG
Member	U. PETTERSON	SWEDEN
Member	W.C. RUSSELL	UNITED KINGDOM
Member	H. van ORMONDT	THE NETHERLANDS
Member	R. WIGAND	FRG

Arenaviridae Study Group

Chair	W.E. RAWLS	CANADA
Member	D.H.L. BISHOP	UNITED KINGDOM
Member	M.J. BUCHMEIER	USA
Member	R.W. COMPANS	USA
Member	C.E. COTO	ARGENTINA
Member	K.M. JOHNSON	USA
Member	F. LEHMAN-GRUBE	FRG
Member	F.A. MURPHY	USA
Member	I.R. PEDERSEN	DENMARK
Member	C.J. PFAU	USA
Member	M.C. WEISSENBACHER	ARGENTINA

VERTEBRATE VIRUS SUBCOMMITTEE 1987-1990

Chair	D.H.L. BISHOP	UNITED KINGDOM
Vice-Chair		
Member	M.J. BUCHMEIER	USA
Member	C.H. CALISHER	USA
Member	D. CAVANAGH	UNITED KINGDOM
Member	J.M. COFFIN	USA
Member	J.J. ESPOSITO	USA
Member	R. FRISQUE	USA
Member	M.C. HORZINEK	THE NETHERLANDS
Member	C.R. HOWARD	UNITED KINGDOM
Member	H.-D. KLENK	FRG
Member	J.B. McCORMICK	USA
Member	P. MINOR	UNITED KINGDOM
Member	C.R. PRINGLE	UNITED KINGDOM
Member	B. ROIZMAN	USA
Member	W.C. RUSSELL	UNITED KINGDOM
Member	F.L. SCHAFFER	USA
Member	J.H. STRAUSS	USA
Member	G. WENGLER	FRG

Adenoviridae Study Group

Chair	W.C. RUSSELL	UNITED KINGDOM
Member	A. BARTHA	HUNGARY
Member	J.C. DE JONG	THE NETHERLANDS
Member	K. FUJINAGA	JAPAN
Member	H. GINSBERG	USA
Member	C. HIERHOLZER	USA
Member	Q.G. LI	CHINA
Member	V. MAUTNER	UNITED KINGDOM
Member	I. NASZ	HUNGARY
Member	G. WADELL	SWEDEN

Arenaviridae Study Group

Chair	M.J. BUCHMEIER	USA
Member	D.D. AUPERIN	USA
Member	M.T. FRANZE-FERNANDEZ	ARGENTINA
Member	J.-P. GONZALEZ	SENEGAL
Member	C.R. HOWARD	UNITED KINGDOM
Member	F. LEHMAN-GRUBE	FRG
Member	J.B. McCORMICK	USA
Member	C.J. PETERS	USA
Member	V. ROMANOWSKI	ARGENTINA
Member	P.J. SOUTHERN	USA

1984-1987

Bunyaviridae Study Group

Chair	D.H.L. BISHOP	UNITED KINGDOM
Member	C.H. CALISHER	USA
Member	C. CHASTEL	FRANCE
Member	M.P. CHUMAKOV	USSR
Member	J.M. DALRYMPLE	USA
Member	C. HANNOUN	FRANCE
Member	D.K. LVOV	USSR
Member	I. MARSHALL	AUSTRALIA
Member	R. PETTERSSON	SWEDEN
Member	J. POTERFIELD	UNITED KINGDOM
Member	R.E. SHOPE	USA
Member	E.G. WESTAWAY	AUSTRALIA

Caliciviridae Study Group

Chair	F.L. SCHAFFER	USA
Member	H.L. BACHRACH	USA
Member	D. BLACK	UNITED KINGDOM
Member	J.N. BURROUGHS	UNITED KINGDOM
Member	C.R. MADELEY	UNITED KINGDOM
Member	S.H. MADIN	USA
Member	R.C. POVEY	CANADA
Member	F. SCOTT	USA
Member	A.W. SMITH	USA
Member	M.J. STUDDERT	AUSTRALIA

Coronaviridae Study Group

Chair	S. SIDDELL	FRG
Member	R. ANDERSON	CANADA
Member	D. CAVANAGH	UNITED KINGDOM
Member	K. FUJIWARA	JAPAN
Member	H.-D. KLENK	FRG
Member	J. LAPORTE	FRANCE
Member	M.R. MACNAUGHTON	UNITED KINGDOM
Member	M. PENSAERT	BELGIUM
Member	S.A. STOHLMAN	USA
Member	L. STURMAN	USA
Member	B. van der ZEIJST	THE NETHERLANDS

1987-1990

Bunyaviridae Study Group

Chair	C.H. CALISHER	USA
Member	B.J. BEATY	USA
Member	J.M. DALRYMPLE	USA
Member	R.M. ELLIOTT	UNITED KINGDOM
Member	N. KARABATSOS	USA
Member	H.W. LEE	KOREA
Member	D.K. LVOV	USSR
Member	P.A. NUTTALL	UNITED KINGDOM
Member	D. PETERS	THE NETHERLANDS
Member	R. PETTERSSON	SWEDEN
Member	C. SCHMALJOHN	USA
Member	R.E. SHOPE	USA

Caliciviridae Study Group

Chair	F.L. SCHAFFER	USA
Member	D. BLACK	UNITED KINGDOM
Member	S. CHIBA	JAPAN
Member	D. CUBITT	UNITED KINGDOM
Member	A.W. SMITH	USA
Member	M.J. STUDDERT	AUSTRALIA

Coronaviridae Study Group

Chair	D. CAVANAGH	UNITED KINGDOM
Member	D.A. BRIAN	USA
Member	L. ENJUANES	SPAIN
Member	K.V. HOLMES	USA
Member	M.M.C. LAI	USA
Member	H. LAUDE	FRANCE
Member	S. SIDDELL	FRG
Member	W.J.M. SPAAN	THE NETHERLANDS
Member	F. TAGUCHI	JAPAN
Member	P.J. TALBOT	CANADA

1984-1987

Filoviridae Study Group

Chair	M. KILEY	USA
Member	T.W. BOWEN	UNITED KINGDOM
Member	M. ISAACSON	JAPAN
Member	A.O. JACKSON	USA
Member	K.M. JOHNSON	USA
Member	S.R. PATTYN	BELGIUM
Member	D.I.H. SIMPSON	UNITED KINGDOM
Member	P. SUREAU	FRANCE
Member	R. SWANEPOEL	SOUTH AFRICA
Member	G. van der GROEN	BELGIUM
Member	R.R. WAGNER	USA
Member	A. WEBB	USA

Flaviviridae Study Group

Hepadnaviridae Study Group

Chair	I. GUST	AUSTRALIA
Member	C.J. BURELL	AUSTRALIA
Member	A.G. COULEPIS	AUSTRALIA
Member	W.S. ROBINSON	USA
Member	A.J. ZUCKERMAN	UNITED KINGDOM

1987-1990

Filoviridae Study Group

Chair	J.B. McCORMICK	USA
Member	M. KILEY	USA
Member	D.W. KINGSBURY	USA
Member	H.-D. KLENK	FRG
Member	G.W. WERTZ	USA

Flaviviridae Study Group

Chair	G. WENGLER	FRG
Member	D.W. BRADLEY	USA
Member	M.S. COLLETT	USA
Member	F.X. HEINZ	AUSTRIA
Member	R.W. SCHLESINGER	USA
Member	J.H. STRAUSS	USA

Hepadnaviridae Study Group

Chair	C.R. HOWARD	UNITED KINGDOM
Member	C.J. BURELL	AUSTRALIA
Member	J.L. GERIN	USA
Member	W.H. GERLICH	FRG
Member	I. GUST	AUSTRALIA
Member	K. KOIKE	JAPAN
Member	P.L. MARION	USA
Member	W. MASON	USA
Member	J. NENBOLD	USA
Member	A.R. NEURATH	USA
Member	W. ROBINSON	USA
Member	H. SCHALLER	FRG
Member	P. TIOLLAIS	FRANCE
Member	H. WILL	FRG
Member	Y.-M WEN	PEOPLE'S REPUBLIC OF CHINA

1984-1987

Herpesviridae Study Group

Chair	B. ROIZMAN	USA
Member	L.E. CARMICHAEL	USA
Member	F. DEINHARDT	FRG
Member	T. HUNG	USA
Member	H. LUDWIG	FRG
Member	H. NAHMIAS	USA
Member	W. PLOWRIGHT	UNITED KINGDOM
Member	F. RAPP	USA
Member	P. SHELDRICK	FRANCE
Member	M. TAKAHASHI	JAPAN
Member	G. de THE	FRANCE
Member	K.E. WOLF	USA

Orthomyxoviridae Study Group

Chair	A.P. KENDAL	USA
Member	Y. GHENDON	USSR
Member	B.W.J. MAHY	UNITED KINGDOM
Member	C. SCHOLTISSEK	FRG
Member	A. SUGUIRA	JAPAN
Member	R.G. WEBSTER	USA

Papovaviridae Study Group

Chair	S.D. GARDNER	UNITED KINGDOM
Member	G. BARBANTI-BRODANO	ITALY
Member	L.V. CRAWFORD	UNITED KINGDOM
Member	P.M. HOWLEY	USA
Member	W.F.H. JARRETT	UNITED KINGDOM
Member	K.V. SHAH	USA
Member	K.K. TAKEMOTO	USA
Member	J. van der NOORDAA	THE NETHERLANDS
Member	D.L. WALKER	USA
Member	H. ZUR HAUSEN	FRG

1984-1987

Herpesviridae Study Group

Chair	B. ROIZMAN	USA
Member	R.C. DESROSIERS	USA
Member	C. LOPEZ	USA
Member	A.C. MINSON	UNITED KINGDOM
Member	M. STUDDERT	AUSTRALIA

Orthomyxoviridae Study Group

Chair	H.-D. KLENK	FRG
Member	N. COX	USA
Member	R.A. LAMB	USA
Member	D.K. LVOV	USSR
Member	B. MAHY	USA
Member	K. NAKAMURA	JAPAN
Member	P. PALESE	USA
Member	R. ROTT	FRG

Papovaviridae Study Group

Chair	R.J. FRISQUE	USA
Member	G. BARBANTI-BRODANO	ITALY
Member	L.V. CRAWFORD	UNITED KINGDOM
Member	S.D. GARDNER	UNITED KINGDOM
Member	P.M. HOWLEY	USA
Member	W.F.H. JARRETT	UNITED KINGDOM
Member	G. ORTH	FRANCE
Member	K.V. SHAH	USA
Member	J. van der NOORDAA	THE NETHERLANDS
Member	H. ZUR HAUSEN	FRG

1984-1987

Paramyxoviridae Study Group

Chair	V. ter MEULEN	FRG
Member	D.J. ALEXANDER	UNITED KINGDOM
Member	M. BRATT	USA
Member	P.W. CHOPPIN	USA
Member	R.P. HANSON	USA
Member	Y. HOSAKA	JAPAN
Member	S.J. MARTIN	UNITED KINGDOM
Member	E. NORRBY	SWEDEN
Member	M. PONS	USA
Member	R. ROTT	FRG

Picornaviridae Study Group

Chair	R.R. RUECKERT	USA
Member	V.I. AGOL	USSR
Member	R. CROWELL	USA
Member	H.J. EGGERS	FRG
Member	M. GRUBMAN	USA
Member	O.L. KEW	USA
Member	J. LONGWORTH	NEW ZEALAND
Member	J. MELNICK	USA
Member	N. MOORE	UNITED KINGDOM
Member	J. MORRIS	USA
Member	P.J. PROVOST	USA
Member	G. SIEGL	SWITZERLAND
Member	E. WIMMER	USA

Poxviridae Study Group

Chair	J.J. ESPOSITO	USA
Member	D. BAXBY	UNITED KINGDOM
Member	D. BLACK	UNITED KINGDOM
Member	S. DALES	CANADA
Member	K. DUMBELL	SOUTH AFRICA
Member	F. FENNER	AUSTRALIA
Member	R. GRANADOS	USA
Member	J. HOLOWCZAK	USA
Member	W.K. JOKLIK	USA
Member	G. MCFADDEN	CANADA
Member	M. MACKETT	UNITED KINGDOM
Member	B. MOSS	USA
Member	J. NAKANO	USA
Member	D. PICKUP	USA
Member	A. ROBINSON	NEW ZEALAND
Member	D. TRIPATHY	USA

1984-1987

Paramyxoviridae Study Group

Chair	C.R. PRINGLE	UNITED KINGDOM
Member	D.J. ALEXANDER	UNITED KINGDOM
Member	M.A. BILLETER	SWITZERLAND
Member	P.L. COLLINS	USA
Member	Y. HOSAKA	JAPAN
Member	D.W. KINGSBURY	USA
Member	M.A. LIPKIND	ISRAEL
Member	C. ORVELL	SWEDEN
Member	B. RIMA	UNITED KINGDOM
Member	R. ROTT	FRG
Member	V. ter MEULEN	FRG

Picornaviridae Study Group

Chair	P. MINOR	UNITED KINGDOM
Member	F. BROWN	UNITED KINGDOM
Member	A. KING	UNITED KINGDOM
Member	N. KNOWLES	UNITED KINGDOM
Member	S. LEMON	USA
Member	S. MARTIN	UNITED KINGDOM
Member	J. MELNICK	USA
Member	N. MOORE	UNITED KINGDOM
Member	A. PALMENBERG	USA
Member	R.R. RUECKERT	USA
Member	M. YIN MURPHY	MALAYSIA

Poxviridae Study Group

Chair	J.J. ESPOSITO	USA
Member	D. BAXBY	UNITED KINGDOM
Member	D. BLACK	UNITED KINGDOM
Member	S. DALES	CANADA
Member	G. DARAI	FRG
Member	K. DUMBELL	SOUTH AFRICA
Member	R. GRANADOS	USA
Member	W.K. JOKLIK	USA
Member	G. MCFADDEN	CANADA
Member	B. MOSS	USA
Member	R. MOYER	USA
Member	D. PICKUP	USA
Member	A. ROBINSON	NEW ZEALAND
Member	H. ROUHANDEH	USA
Member	D. TRIPATHY	USA

1984-1987

Retroviridae Study Group

Chair	H. VARMUS	USA
Member	P. BIGGS	UNITED KINGDOM
Member	J.M. COFFIN	USA
Member	M. ESSEX	USA
Member	R. GALLO	USA
Member	T.M. GRAF	FRG
Member	Y. HINUMA	JAPAN
Member	R. JAENISCH	USA
Member	R. NUSSE	THE NETHERLANDS
Member	S. OROSZLAN	USA
Member	J. SVOBODA	CSECHOSLOVAKIA
Member	N. TEICH	UNITED KINGDOM
Member	K. TOYOSHIMA	JAPAN

Togaviridae Study Group

Chair	E.G. WESTAWAY	AUSTRALIA
Member	M.A. BRINTON	USA
Member	S.Y. GAIDAMOVITCH	USSR
Member	M.C. HORZINEK	THE NETHERLANDS
Member	A. IGARASHI	JAPAN
Member	L. KAARIAINEN	FINLAND
Member	D.K. LVOV	USSR
Member	J.S. POTERFIELD	UNITED KINGDOM
Member	P.K. RUSSELL	UNITED KINGDOM
Member	D.V. TRENT	USA

Torovirus Study Group

Chair	M.C. HORZINEK	THE NETHERLANDS
Member	T.H. FLEWETT	UNITED KINGDOM
Member	L. SAIF	USA
Member	W.J.M. SPAAN	THE NETHERLANDS
Member	M. WEISS	SWITZERLAND
Member	G. WOODE	USA

1987-1990

Retroviridae Study Group

Chair	J.M. COFFIN	USA
Member	M. ESSEX	USA
Member	R. GALLO	USA
Member	T.M. GRAF	FRG
Member	Y. HINUMA	JAPAN
Member	E. HUNTER	USA
Member	R. JAENISCH	USA
Member	R. NUSSE	THE NETHERLANDS
Member	S. OROSZLAN	USA
Member	J. SVOBODA	CZECHOSLOVAKIA
Member	N. TEICH	UNITED KINGDOM
Member	K. TOYOSHIMA	JAPAN
Member	H. VARMUS	USA

Togaviridae Study Group

Chair	J.H. STRAUSS	USA
Member	C.H. CALISHER	USA
Member	L. DALGARNO	AUSTRALIA
Member	J. DALRYMPLE	USA
Member	R.F. PETTERSSON	SWEDEN
Member	C.M. RICE	USA
Member	W.J.M. SPAAN	THE NETHERLANDS

Torovirus Study Group

Chair	M.C. HORZINEK	THE NETHERLANDS
Member	T.H. FLEWETT	UNITED KINGDOM
Member	L. SAIF	USA
Member	W.J.M. SPAAN	THE NETHERLANDS
Member	M. WEISS	SWITZERLAND
Member	G. WOODE	USA

VIRUS DATA SUBCOMMITTEE 1984-1987

Chair	J.C. ATHERTON	AUSTRALIA
Member	H.-W. ACKERMANN	CANADA
Member	A.J. GIBBS	AUSTRALIA
Member	N. KARABATSOS	USA
Member	D.L. KNUDSON	USA

NATIONAL REPRESENTATIVES 1984-1987

Member	P. COOPER	AUSTRALIA
Member	W. FRISCH-NEGGEMEYER	AUSTRIA
Member	O.A. de CARVALHO PEREIRA	BRAZIL
Member	P. ANDONOV	BULGARIA
Member	G. CONTRERAS	CHILE
Member	M. OUF	EGYPT
Member	E. TAPIO	FINLAND
Member	A. KIRN	FRANCE
Member	H.J. EGGERS	FRG
Member	A. TSOTSOS	GREECE
Member	K. BANERJEE	INDIA
Member	M. LA PLACA	ITALIA
Member	S. KONISHI	JAPAN
Member	C. MATSUI	JAPAN
Member	H. UETAKE	JAPAN
Member	Y.T. YANG	KOREA (SOUTH)
Member	C. FERNANDEZ-TOMAS	MEXICO
Member	A. CHABAUD	MOROCCO
Member	A. FABIYI	NIGERIA
Member	R. MENDES	PERU
Member	R. BOZEMANN-RODRIGUEZ	PHILIPPINES
Member	M. MORZYCKA	POLAND
Member	N. CAJAL	ROUMANIA
Member	E. NORRBY	SWEDEN
Member	M.C. HORZINEK	THE NETHERLANDS
Member	E.T. CETIN	TURKEY
Member	R. SOMMA-MOREIRA	URUGUAY
Member	H.S. GINSBERG	USA
Member	S.Y. GAIDAMOVICH	USSR
Member	J. ESPARZA	VENEZUELA
Member	D. SUTIC	YUGOSLAVIA

VIRUS DATA SUBCOMMITTEE 1987-1990

Chair	D.L. KNUDSON	USA
Member	J.C. ATHERTON	AUSTRALIA
Member	N. KARABATSOS	USA

NATIONAL REPRESENTATIVES 1987-1990

Member	A. BOUGUERMOUH	ALGERIA
Member	M. WEISSENBACHER	ARGENTINA
Member	I.H. HOLMES	AUSTRALIA
Member	A.J. SARKER	BANGLADESH
Member	V. ter MEULEN	FRG
Member	M.A. CHERNESKY	CANADA
Member	Q.F. PANG	CHINA
Member	B. KORYCH	CZECHOSLOVAKIA
Member	A. KIRN	FRANCE
Member	B.F. VESTERGAARD	DENMARK
Member	E. ALLAM	EGYPT
Member	T. HOVI	FINLAND
Member	G.L. FRENCH	HONG KONG
Member	I. NASZ	HUNGARY
Member	N. RISHI	INDIA
Member	Y. BECKER	ISRAEL
Member	A. OYA	JAPAN
Member	M. TAKAHASHI	JAPAN
Member	A.R. BELLAMY	NEW ZEALAND
Member	C. IROEGBU	NIGERIA
Member	G. HAUKENES	NORWAY
Member	D.W. VERWOERD	SOUTH AFRICA
Member	R. NAJERA	SPAIN
Member	G. WADELL	SWEDEN
Member	O. HALLER	SWITZERLAND
Member	D. PETERS	THE NETHERLANDS
Member	A.A. BRUNT	UNITED KINGDOM
Member	D.H. WATSON	UNITED KINGDOM
Member	O.W. BARNETT	USA
Member	S. CVETNIC	YUGOSLAVIA

President's Report 1987-1990

R. I. B. FRANCKI
President of the International Committee on Taxonomy of Viruses 1987-1990

The International Committee on Taxonomy of Viruses (ICTV) and its Executive Committee held a series of meetings before and during the Eighth International Congress of Virology in Berlin during August 1990. The following summarizes decisions made by the ICTV during those meetings:
(i) Changes in the Rules of the ICTV.
(ii) Changes of membership of the Executive Committee.
(iii) Details of the new taxonomic proposals approved by the ICTV.

CHANGES IN THE ICTV RULES

Rules 4 and 13, as detailed in the Fourth Report of the ICTV (Matthews, 1982), have been abolished and rules 5, 12, 13, 14 and 20 have been modified as follows:

(i) Rule 5 which stated that "existing latinized names shall be retained whenever feasible" has been changed to "existing names shall be retained whenever feasible".
(ii) Rule 12 which stated that "the genus name and species epithet, together with the strain designation, must give an unambiguous identification of the virus" has been changed to read "a virus name, together with a strain designation, must provide an unambiguous identification and need not include the genus or group name".
(iii) Rule 14 which stated that "A species epithet should consist of a single word, or, if essential, a hyphenated word. The word may be followed by numbers or letters". It has now been changed to read "A virus name should be meaningful and consist of as few words as possible".
(iv) Rule 20 which stated that "The ending of the name of a viral genus is . . . virus" has been changed to read "The genus name should be a single meaningful word ending in . . . virus".

The full current set of ICTV Rules of Nomenclatures are found on page 9.

ELECTION OF THE EXECUTIVE COMMITTEE OF THE ICTV FOR THE TERM 1990-1993

Following elections in Berlin the membership of the Executive Committee is as follows:

President	F. Murphy	USA
Vice President	K. W. Buck	United Kingdom
Secretaries	C. Fauquet	USA
	C. Pringle	United Kingdom

Elected Members	H. W. Ackermann	Canada
	P. Ahlquist	USA
	L. Berthiaume	Canada
	C. Calisher	USA
	R. Goldbach	The Netherlands
	J. Maniloff	USA
	M. A. Mayo	United Kingdom
	G. Rohrmann	USA

Subcommittee Chairs

Coordination	F. Murphy (ex officio)	USA
Bacterial Virus	A. Jarvis	New Zealand
Fungal Virus	S. A. Ghabrial	USA
Invertebrate Virus	M. D. Summers	USA
Plant Virus	G. P. Martelli	Italy
Vertebrate Virus	D. H. L. Bishop	United Kingdom
Virus Data	A. J. Gibbs	Australia

NEWLY APPROVED TAXONOMIC PROPOSALS

A. Coordination Subcommittee

a. Reoviridae Study Group

1. The genus name *Cypovirus* is established for the cytoplasmic polyhedrosis virus group.
2. A new genus, *Coltivirus*, is established in the family ***Reoviridae*** with Colorado tick fever virus as the type species.
3. A new genus, *Aquareovirus*, is established in the family ***Reoviridae*** with the golden shiner virus as the type species.

B. Bacterial Virus Subcommittee

1. The family of viruses consisting of the F3 phage group has been named ***Lipothrixviridae*** with a single genus *Lipothrixvirus*.
2. A genus, *Spiromicrovirus*, has been established within the family ***Microviridae*** and *Spiroplasma* virus SpV4 as the type species.
3. A genus, *Levivirus*, has been established within the family ***Leviviridae*** (earlier known as supergroup A) with the MS2 phage group as the type species.
4. Another second genus, *Allolevivirus*, has been established within the family ***Leviviridae*** with the Qβ phage group as the type species.
5. A monogeneric family, yet un-named, has been established to include virus-like particles or archaebacteria with SSV1 phage as the type species.
6. Acholeplasma phage group L51 has been designated as the type species *Plectrovirus* (family ***Inoviridae***).

7. Phage fd has been designated as the type species of the genus *Inovirus* (family ***Inoviridae***).

C. Fungal Virus Subcommittee

Algal and Protozoal Virus Study Group

1. A family, ***Phycodnaviridae***, has been established to include dsDNA viruses with polyhedral particles which infect *Chlorella*-like green algae including a single genus, *Phycodnavirus*, with *Paramecium bursaria chlorella* virus-1 as the type species.
2. A genus, *Giardiavirus*, has been established to include viruses of parasitic protozoa with dsRNA and isometric particles and *Giardia lamblia* (strain Pastland 1) has been designated as the type species.

D. Invertebrate Virus Subcommittee

Baculovirus Study Group

1. A subfamily, the ***Nudibaculovirinae***, comprising the non-occluded baculoviruses has been established within the family ***Baculoviridae***.
2. Two genera, the nuclear polyhedrosis viruses (NPV) and the granulosis viruses (GV), have been established within the ***Eubaculovirinae***.
3. Two subgenera have been established within the NPV genus, one comprising viruses with multiple nucleocapsids per envelope (MNPV) and the other comprising viruses with a single nucleocapsid per envelope (SNPV).
4. *Autographa californica* multiple nuclear polyhedrosis virus has been designated as the type species of the MNPV subgenus.
5. *Bombyx mori* nuclear polyhedrosis virus has been designated as the type species of the SNPV subgenus.
6. *Trichoplusia ni* granulosis virus has been designated as the type species of the GV genus.
7. A genus to include the non-occluded baculoviruses (NOB) has been established within the subfamily ***Nudibaculovirinae***.
8. *Heliothis zea* non-occluded baculovirus has been designated as the type species of the NOB genus.

Polydnavirus Study Group

1. A genus, *Ichnovirus*, has been established within the family ***Polydnaviridae*** to include polydnaviruses with individual nucleocapsids in the form of a prolate ellipsoid surrounded by two envelopes.
2. *Compoletis sonovensii* virus has been designated as the type species *Ichnovirus* genus.

3. A genus, *Bracovirus*, has been established within the family ***Polydnaviridae*** to include polydnaviruses within cylindrical nucleocapsids of variable length and a single envelope.
4. *Cotesia melanoscela* virus has been designated as the type species of the *Bracovirus* genus.

E. Plant Virus Subcommittee

1. A new group of plant viruses, as yet un-named, with bacilliform particles and dsDNA is established with Commelina yellow mottle virus as the type member.
2. The ***geminivirus*** group has been divided into 3 subgroups with the following type members:

Subgroup I	-	maize streak virus
Subgroup II	-	beet curly top virus
Subgroup III	-	bean golden mosaic virus.

F. Vertebrate Virus Subcommittee

Hepadnavirus Study Group

1. A family, ***Hepadnaviridae***, has been established to include hepatotropic and similar DNA viruses that replicate via reverse transcription.

Paramyxovirus Study Group

1. An order, ***Mononegavirales***, has been established to include the families ***Filoviridae***, ***Paramyxoviridae*** and ***Rhabdoviridae***.
2. The sub-families, ***Paramyxovirinae*** and ***Pneumovirinae*** have been established within the family ***Paramyxoviridae*** to include the existing genera *Paramyxovirus* and *Morbillivirus*, and the genus *Pneumovirus*, respectively.

Poxvirus Study Group

1. A genus, *Molluscipoxvirus*, has been established within the subfamily ***Chordopoxvirinae*** of the family ***Poxviridae*** with *Molluscum contagiosm* virus as the type species.
2. A genus, *Yatapoxvirus*, has been established within the subfamily ***Chordopoxvirinae*** of the family ***Poxviridae*** with Yaba monkey tumour virus as the type species.

Torovirus Study Group

1. A genus, *Torovirus*, with possible affinities with members of the ***Coronaviridae*** family, has been established and Berne virus has been designated as the type species.

Togavirus and Flavivirus Study Group

1. The genus *Pestivirus* has been transferred from the ***Togaviridae*** to the ***Flaviviridae*** family.

Bunyaviridae Study Group and Plant Virus Subcommittee

1. A genus, *Tospovirus*, which infects plants and is transmitted by thrips, has been established within the family ***Bunyaviridae*** with tomato spotted wilt as the type species.

Retrovirus Study Group

1. The three sub-families, ***Oncovirinae***, ***Lentivirinae*** and ***Spumavirinae*** have been eliminated from the family ***Retroviridae*** and members of the family have been divided into seven genera as follows:

   The type B retroviruses
   The mammalian type C retroviruses
   The avian retroviruses
   The type D retroviruses
   Spumavirus (foamy viruses)
   The HTLV-BLV viruses
   Lentivirus.

The Format for Submission of New Taxonomic Proposals

Contents

I. Initiation of New Proposals
II. Processing of New Proposals
III. Publication of New Proposals
IV. Timing of Events in the Period 1990-1993
V. Standard Format for Presenting New Taxonomic Proposals

Over the last years the Executive Committee of ICTV has evolved procedures and rules to facilitate the processing and assessment of new taxonomic proposals for viruses. This section, which summarizes the present position, is provided to assist virologists wishing to make a contribution to the work of ICTV.

I. Initiation of New Proposals

The key units in the organization of the ICTV are the host-oriented subcommittees. Most of these subcommittees are organized into study groups of working virologists. New taxonomic proposals are usually initiated by these study groups, and less commonly by the subcommittees themselves.

It should be emphasized that, apart from the formal organization, it is perfectly in order for any individual virologist to initiate a new taxonomic proposal. Any such proposal should be in the format outlined below, and should be sent to the Chairperson of the appropriate subcommittee for consideration.

II. Processing of New Proposals

A taxonomic proposal originating in a study group or favorably considered by a study group after receipt from an individual virologist is forwarded to the appropriate subcommittee. If it is approved by the subcommittee, the proposal is then considered by the Executive Committee of ICTV. The Executive Committee of ICTV may approve a proposal, decline to approve, or send it back to the subcommittee for suggested changes.

Proposals approved by the Executive Committee go forward every 3 years to the plenary meeting of the full ICTV membership for final ratification.

III. Publication of New Proposals

Some new proposals pass through the ICTV and are approved without any prior publication. Such proposals then appear first in an official ICTV triennial report. Other proposals are published at an earlier stage in the *Archives of Virology*, which is the official journal of the Virology Division of the International Union of Microbiological Societies.

These publications may be enlarged presentations of taxonomic proposals being formally submitted by ICTV study groups. Two examples of this sort, published in *Intervirology* concern the family ***Caliciviridae*** (Schaffer et al., 1980) and the family ***Bunyaviridae*** (Bishop et al., 1980). In the near future a proposal for establishing the family ***Potyviridae*** family, comprising three genera will be published in *Archives of Virology* (Barnett, 1991). Another proposal for an order to encompasses all the tailed phages, is also in preparation (Ackermann, pers. com.).

Such publications allow individual virologists to scrutinize proposals and to make their views known to the appropriate ICTV subcommittee. It should be emphasized, however, that publication in itself does not give the proposals any status as far as ICTV is concerned.

IV. Timing of Events in the Period 1990-1993

There is a plenary session of the ICTV held every three years at the International Congress of Virology. The next plenary session will be held at the IXth International Congress in Glasgow, Scotland in August 1993.

There is no deadline for submitting proposals to the Executive Committee of the ICTV. Subcommittee chairs can send proposals to the ICTV Secretary for circulation to members before any Executive Committee meeting. New taxonomic proposals should be in the hands of the secretary before May 1993, so that the proposals can be circulated to the members before the Executive Committee of the ICTV during the Virology Congress of 1993.

V. Standard Format for Presenting New Taxonomic Proposals

Chairs of study groups and subcommittees should use the following guidelines and format in preparing new taxonomic proposals.

Guidelines:

1. Each individual taxonomic proposal should be submitted as a separate item (not mixed with explanatory or historical details). For example, a proposal to form a new genus must be separate from a proposal genus and separate from a proposal designating the type species for the genus.

2. Attention is drawn to rule N°20, which requires that approval of a new family must be linked with approval of a type genus and that approval of a new genus must be linked with approval of a type species.

3. Each proposal should contain information in the following format:

Date.........

From the... ... ... Subcommittee or Study group
Taxonomic Proposal N°.:
1. *Proposal:* The taxonomic proposal in its essence, in a form suitable for presentation to ICTV for voting.

2. *Purpose:* A summary of the reasons for the proposal, with any explanatory and historical notes.

3. A summary of the new taxonomic situation within the family, group or genus (e.g. for a new genus- 'The family would now consist of the following genera:... ... ... ')

4. Derivation of any names proposed.

5. New literature references, if appropriate.

The Rules of Virus Nomenclature 1990

Rule 1 The code of bacterial nomenclature shall not be applied to viruses.

Rule 2 Nomenclature shall be international.

Rule 3 Nomeclature shall be universally applied to all viruses.

Rule 4 Existing names shall be retained whenever feasible.

Rule 5 The law of priority shall not be observed.

Rule 6 Sigla may be accepted as names of viruses or virus groups, provided that they are meaningful to workers in the field and are recommended by international study-groups.

Rule 7 No person's name should be used.

Rule 8 Names should have international meaning.

Rule 9 The rules of orthography of names and epithets are listed in Chapter 3, Section 6 of the proposed international code of nomenclature of viruses [Appendix D; Minutes of 1966 (Moscow) meeting].

Rule 10 A virus species is a concept that will normally be represented by a cluster of strains from a variety of sources, or a population of strains from a particular source, which have in common a set of pattern of correlating stable properties that separates the cluster from other clusters of strains.

Rule 11 A virus name, together with a strain designation, must provide an unambiguous identification and need not include the genus or group name.

Rule 12 A virus name should be meaningful and consist of as few words as possible.

Rule 13 Numbers, letters, or combinations thereof may be used as an official species epithet where such numbers and letters already have wide usage for a particular virus.

Rule 14 Newly designated serial numbers, letters or combinations thereof are not acceptable alone as species epithets.

Rule 15 Artificially created laboratory hybrids between different viruses will not be given taxonomic consideration.

Rule 16 Approval by ICTV of newly proposed species, species names and type species will proceed in two steges. In the first stage, provisional approval may be given. Provisionally approved proposals will be published in an ICTV report. In the second stage, after a 3-year waiting period, the proposals may receive the definitive approval of ICTV.

Rule 17 The genus is a group of species sharing certain common caracters.

Rule 18 The genus name should be a single meaningful word ending in "...**virus**".

Rule 19 A family is a group of genera with common characters, and the ending of the name of a viral family is "...**viridae**".

Rule 20 Approval of a new family must be linked to approval of a type genus; approval of a new genus must be linked to approval of a type species.

Guidelines for the Delineation and Naming of Species

1. Criteria for delineation species may vary in different families of viruses.
2. Wherever possible, duplication of an already approved virus species name should be avoid.
3. When a change in the type species is desirable, this should be put forward to ICTV in the standard format for a taxonomic proposal.
4. Subscripts, superscripts, hyphens, oblique bars, or Greek letters should be avoided in future virus nomenclature.
5. When designating new virus names, study groups should recognize national sensitivities with regard to language. If a name is universally used by virologists (those who publish in scientific journals), that name or a derivative of it should be used regardless of national origin. If different names are used by virologists of different national origin, the study group should evaluate relative international usage and recommend the name that will be acceptable to the majority and which will not be offensive in any language.
6. ICTV is not concerned with the classification and naming of strains, variants or serotypes. This is the responsibility of specialist groups.

The Statutes of the I C T V

Article 1

Official name
International Committee on Taxonomy of Viruses (ICTV).

Article 2

Status
The ICTV is a Committee of the Virology Division of the International Union of Microbiology Societies (IUMS).

Article 3

Objectives
1. To develop an internationally agreed taxonomy for viruses.
2. To establish internationally agreed names for taxonomic groups of viruses.
3. To communicate the latest results on the classification and nomenclature of viruses to virologists by holding meetings and publishing reports.

Article 4

Membership

Membership of the ICTV shall be comprised as follows.

A. President and Vice-President
These shall be nominated and seconded by any members of the ICTV and elected at a plenary meeting of the full ICTV membership. They shall be elected for a term of three years and may not serve for more than two consecutive terms of three years.

B. Secretaries
Two permanent secretaries shall be nominated by the Executive Committee and elected at a plenary meeting of the full ICTV membership.

C. Members of the Executive Committee (EC)
The President, Vice-President and Secretaries
Chairs of the Subcommittees (SC)
    Bacterial Virus SC
    Co-ordination Virus SC (The President *ex officio*)
    Fungal Virus SC
    Invertebrate Virus SC
    Plant Virus SC
    Vertebrate Virus SC
    Virus Data SC
Eight elected members.

The Chairs of the Subcommittees shall be elected by the Executive Committee at its mid-term meeting preceding the next plenary meeting of the full ICTV membership for a term of three years and may not serve more than two consecutive terms of three years each.

The eight elected members shall be nominated and seconded by any ICTV member and elected at a plenary meeting of the ICTV for a term of three years and may not serve for more than two consecutive terms of three years each. Generally four of the elected members shall be replaced every three years.

D. National Members
National members shall be nominated by Member Societies of the Virology Division of the IUMS. Societies belonging to the IUMS are considered to be Member Societies of the Division if they have members actively interested in virology. Wherever practicable, each country shall be represented by at least one National Member and no country by more than five National Members. Nominated National Members shall not require further approval by the ICTV.

E. Life Members
Life members shall be nominated by the Executive Committee on account of their outstanding service to virus taxonomy. They shall be elected by the full ICTV.

F. Members of the Bacterial Virus, Co-ordination, Fungal Virus, Invertebrate Virus, Plant Virus, Vertebrate Virus, and Virus Data Subcommittees
These shall be appointed by the Chairs of the Subcommittees and shall not require further approval by the ICTV.

G. Status of Study Group Members
Study Groups may be formed to examine the taxonomy of specialized groups of viruses. A Chair of a Study Group shall be appointed by the Chair of the appropriate Subcommittee and shall be a member of that Subcommittee ex officio and hence also a member of the ICTV.

Chairs of Study Groups shall appoint the members of their Study Groups. Members of Study Groups, other than Chairs, shall not be members of the ICTV, but their names shall be published in the minutes and reports of the ICTV to recognize their valuable contribution to the taxonomy of viruses.

Article 5

Meetings
Plenary meetings of the full ICTV membership shall be held in conjunction with the International Congresses of Virology.
Meetings of the ICTV Executive Committee shall be held in conjunction with the International Congresses of Virology. In addition, a mid-term meeting shall be held between Congresses.

Article 6

Taxonomic Proposals
Taxonomic proposals may be initiated by an individual member of the ICTV, by a Study Group or by a Subcommittee member by sending it to the Chair of the appropriate subcommittee for consideration by that subcommittee. Taxonomic proposals approved by a subcommittee shall be submitted by its chair for consideration by the Executive Committee. Proposals approved by the Executive Committee shall be presented to the next plenary meeting of the full ICTV membership for ratification.
Separate proposals shall be required to establish a new taxonomic group, to name a taxonomic group, to designate the type species and the members of a taxonomic group.

Article 7

Voting
Decisions will be made on the following basis.

(i) At meetings, or postal votes, of the Executive Committee
A simple majority of the votes of those present, or those replying within two months of a questionnaire being sent out.

(ii) At plenary meetings, or postal votes, of the full ICTV membership
A simple majority of the votes of those present, or those replying within two months of a questionnaire being sent out. A quorum consisting of the President or Vice-President together with 15 voting members will be required.

In the event of a tie in (i) or (ii), the President shall have an additional casting vote.

Article 8

The Rules of Nomenclature of Viruses
The rules of nomenclature of viruses, and any subsequent changes, shall be approved by the Executive Committee and at a plenary meeting of the full ICTV membership.

Article 9

Duties of Officers

A. Duties of the President shall be:

1. To preside at meetings of the Executive Committee and plenary meetings of the full ICTV membership.
2. To prepare with the Secretaries the agendas for meetings of the Executive Committee and the plenary meetings of the full ICTV membership.

3. To act as editor for ICTV reports to be published after each plenary meeting of the ICTV.

B. Duties of the Vice-President shall be:

1. To carry out the duties of the President in the absence of the President.
2. To attend meetings of the Executive Committee and plenary meetings of the ICTV.

C. Duties of the Secretaries shall be:

1. To attend meetings of the Executive Committee and plenary meetings of the ICTV.
2. To prepare with the President the agendas for meetings of the Executive Committee and the plenary meetings of the ICTV.
3. To prepare the Minutes of meetings of the Executive Committee and plenary meetings of the ICTV and circulate them to all ICTV members.
4. To act as Treasurer of the ICTV. To handle any funds that may be allocated to the ICTV by the Virology Division of the IUMS or other sources.
5. To keep an up-to-date record of ICTV membership.

Article 10

Publications
No publication of the ICTV shall bear any indication of sponsorship by a commercial agency, or institution connected in any way with a commercial company, except as an acceptable acknowledgment of financial assistance. Furthermore, any publication containing material not authorized, prepared, or edited by the ICTV, or a committee or subcommittee of the ICTV, may not bear the name of the ICTV or the IUMS.

Article 11

ICTV Statutes
The Statutes of the ICTV, and any subsequent changes, shall be approved by the ICTV Executive Committee, by a plenary meeting of the full ICTV membership and by the Virology Division of the IUMS.

Article 12

Disposition of Funds
In the event of dissolution of the ICTV, any remaining funds shall be turned back to the Secretary-Treasurer of the Virology Division of the IUMS.

References

Baltimore, D.: Expression of animal virus genomes. Bact. Rev. *35*: 235-241(1971).

Barnett, O.W.: *Potyviridae*, a proposed family of plant viruses. Arch. Virol. (submitted)

Bishop, D.H.L.; Calisher, C.H.; Casals, J.; Chumakov, M.P.; Gaidamovich, S. Ya.; Hannoun, C.; Lvov, D.K.; Marshall, I.D.; Oker-Blom, N.; Pettersson, R.F.; Porterfield, J.S.; Russell, P.K.; Shope, R.E.; Westaway, E.G.: *Bunyaviridae*. Intervirology *14*: 125-143 (1980).

Cooper, P.D.: Towards a more profound basis for the classification of viruses. Intervirology *4*: 317-319 (1974).

Fenner, F.: Classification and nomenclature of viruses. Second report of the International Committee on Taxonomy of Viruses. Intervirology 7: 1-116 (1976).

Matthews, R.E.F.: Classification and nomenclature of viruses. Summary of results of meetings of he International Committee on Taxonomy of Viruses in Strasbourg, August 1981. Intervirology *16*: 53-60 (1981).

Matthews, R.E.F.: Classification and nomenclature of viruses. Fourth report of the International Committee on Taxonomy of Viruses. Intervirology *17*: 1-199 (1982).

Schaffer, F.L.; Bachrach, H.L.; Brown, F.; Gillespie, J.H.; Burroughs, J.N.; Madin, S.H.; Madeley, C.R.; Povey, R.C.; Scott, F.; Smith, A.W.; Studdert, M.J.: *Caliciviridae*. Intervirology *14*: 1-6 (1980).

Wildy, P.: Classification and nomenclature of viruses. First report of the International Committee on Nomenclature of Viruses. Monogr. Virol., *Vol.5* (Karger, Basel, 1971).

The Viruses

Presentation

This report contains a listing of the virus taxa approved by ICTV between 1970 and 1981. Descriptions of the important characteristics of these taxa are provided, together with a list of members and selected references giving a guide to recent literature. The detailed information has been provided from the work of the subcommittees of ICTV and their various study groups, and from individual virologists.

Names for Viruses, Genera, and Families

In the formal descriptions the order, family, subfamily, genus and species names approved by ICTV are listed under 'International name'. All names of taxa approved by ICTV are printed in italic type.

Names that have not been officially approved are printed in standard type face. The heading 'English vernacular name' is used, even though for a few viruses a name in some other language has been adopted into English usage. Where there is a widely used vernacular synonym, this is included within parentheses. In the virus diagrams, approved names for all taxa are in bold type. For those plant viruses that have been included in the CMI/AAB Descriptions of Plant Viruses the description number is given in parentheses following the name.

Main Characteristics

The 'Main characteristics' section has been further expanded for most taxa. The order of listing of data is standardized for ease of reference. As would be expected, the amount of relevant information available varies quite widely for different families, genera and groups, Since all known plant viruses can be transmitted by grafting and vegetative propagation, these two methods of transmission have been omitted in the descriptions.

List of Members

The lists of members for genera and groups have been updated. In these lists the word 'virus' has been omitted for the sake of brevity, unless it forms part of a single word in the name or unless the plural 'viruses' is required. Three categories of members have been defined as follows:

Other members:

Those viruses, besides the type member, which definitely belong in the family, genus or group.

Probable members:

Those viruses for which information known to study group members strongly indicates membership in the family, genus or group.

Possible members:

Viruses for which taxonomically useful data must be regarded as more tenuous.

To assist readers, fairly extensive lists of names have been included for many of the taxa. It should be remembered, however, that these lists may contain described and named isolates which, on further examination, will be shown to be closely related strains or even indistinguishable isolates of a single virus.

Arrangement of the approved Families and Groups

Seventy-three families and groups of viruses have now been approved by ICTV. Since a taxonomic structure above the level of family has not yet been developed (with the exception of the newly approved Order ***Mononegavirales***), any sequences of listing must be arbitrary. Many virologists consider the kind, and strandedness, of the nucleic acid making up the viral genome and the presence or absence of a lipoprotein envelope to be basically important virus properties. Using these three properties, the 73 families and groups are described in order in the section entitled "The Virus Families and Groups" There are no known ssDNA viruses with envelopes, so these three virus properties give rise to seven clusters of families and groups.

Within two of these clusters, the families can be usefully arranged on other criteria as follows: (i) for the enveloped ssRNA viruses, on the basis of genome strategy (Baltimore, 1971; Cooper, 1974); and (ii) for the non-enveloped ssRNA viruses infecting primarily plants, on the basis of particle morphology and on the number of pieces of RNA comprising the genome. In addition, to save repetition, a general description is given to cover the three families of tailed phages a possible Order in the future. These arrangements remain unchanged from the Third Report. These clusters are not intended to anticipate higher taxa, this subject has not yet been considered by ICTV.

Other pathogens related to viruses

Though not strictly viruses by definition, descriptions of virus satellites and viroids are included.

Index

Following the virus descriptions, there is an index containing all the virus names used in the text. Family, genus and group names approved by ICTV are given in italics. In addition to the main index, page numbers for the approved families and groups are given in the table of content and in the five pages of line drawings for the vertebrate, invertebrate, plant, and bacterial viruses.

Glossary of Abbreviations and Virological Terms

Note: These terms were approved by the Coordination Subcommittee of ICTV for use in ICTV Report but have no official status.

(i) Abbreviations

bp	= base pair
CF	= complement fixing
CPE	= cytopathic effect
D	= diffusion coefficient
DI	= defective interfering
ds	= double-stranded
HI	= hemagglutination inhibition
kbp	= kilo base pair
kDa	= kilo Dalton
MW	= molecular weight
ORF	= open reading frame
RF	= replicative form
RI	= replicative intermediate
RNP	= ribonucleoprotein
ss	= single-stranded

(ii) RNA Replicases, Transcriptases and Polymerases

In the synthesis of viral RNA, the term polymerase has been replaced in general by two somewhat more specific terms: RNA replicase and RNA transcriptase. The term transcriptase has become associated with the enzyme involved in messenger RNA synthesis, most recently with those polymerases which are virion-associated. However, it should be borne in mind that for some viruses it has yet to be established whether or not the replicase and transcriptase activities reflect distinct enzymes rather than alternate activities of a single enzyme. Confusion also arises in the case of the small positive-sense RNA viruses where the term replicase (e.g., Qβ replicase) has been used for the enzyme capable both of transcribing the genome into messenger RNA via an intermediate negative-sense strand and of synthesizing the genome strand from the same template. In the text, the term replicase will be restricted as far as possible to the enzyme synthesizing progeny viral strands of either polarity. The term transcriptase is restricted to those RNA polymerases that are virion-associated and synthesize mRNA. The generalized term RNA polymerase (i.e., RNA-dependent RNA polymerase) is applied where no distinction between replication and transcription enzymes can be drawn (e.g., Qβ, R 17, poliovirus and many plant viruses).

(iii) Other Definitions

Enveloped: possessing an outer (bounding) lipoprotein bilayer membrane

Negative-sense (= minus strand); for RNA or DNA, the strand with a strand: base sequence complementary to the positive-sense strand.

Positive-sense (= plus strand, message strand); for RNA, the strand strand: that contains the coding triplets which can be translated on ribosomes. For DNA, the strand that contains the same base sequence as the mRNA. However, in some dsDNA viruses mRNAs are transcribed from both strands and the transcribed regions may overlap. For such viruses this definition is inappropriate.

Pseudotypes Enveloped virus particles in which the envelope is derived from one virus and the internal constituents from another.

Reverse transcriptase: Virus-encoded RNA-dependent DNA polymerase found as part of the virus particle in *Retroviridae*.

Surface projections: (= spikes, peplomers, knobs); morphological features, usually consisting of glycoproteins, that protrude from the lipoprotein envelope of many enveloped viruses.

Virion: Morphologically complete virus particle.

Viroplasm: (= virus factory, virus inclusion, X-body); a modified region within the infected cell in which virus replication occurs, or is thought to occur.

Virus Diagrams

Revised by C. Fauquet & M.A. Mayo

Virus Diagrams

The following pages provide line drawings for the virus families and groups according to their given major host; bacteria, algae, fungi and protozoæ, invertebrates, vertebrates and plants. All the diagrams have been drawn similarly: there are vertical lines to separate enveloped and non-enveloped viruses and horizontal lines to separate DNA and RNA viruses. Within each of the resulting four separate boxes the viruses having single-stranded (ss) and double-stranded (ds) genomes are indicated. The diagrams do not reflect the importance and/or number of viruses present in each category. When no virus has been identified in a box, it has been left empty or not shown.

All the diagrams have been drawn approximatively to the same scale to provide an indication of the relative sizes of the viruses; but this cannot be taken as definitive for the following reasons: (i) Different viruses within a family or group may vary somewhat in size and shape. In general the size and shape were taken from the type member of the taxon. (ii) Dimensions of some viruses are difficult to determine or only approximatively known. (iii) Some viruses, particularly the larger enveloped ones, are pleomorphic. Only the outlines of most of the smallest viruses are given, with an indication of the icosahedral structure whenever appropriate. The large viruses are given schematically in surface outline, in section, or both, as seems most appropriate to display major morphological characteristics.

Most of the diagrams are reproduced from the Fourth ICTV Report (Matthews, 1982), updated according to the suggestions of the chairmen of the sub-committees or/and of the study-groups as well as of virologists who were kind enough to provide their available drawings. I would like to thank all the persons having contributed to help me to draw these virus diagrams.

FAMILIES OF VIRUSES INFECTING INVERTEBRATES

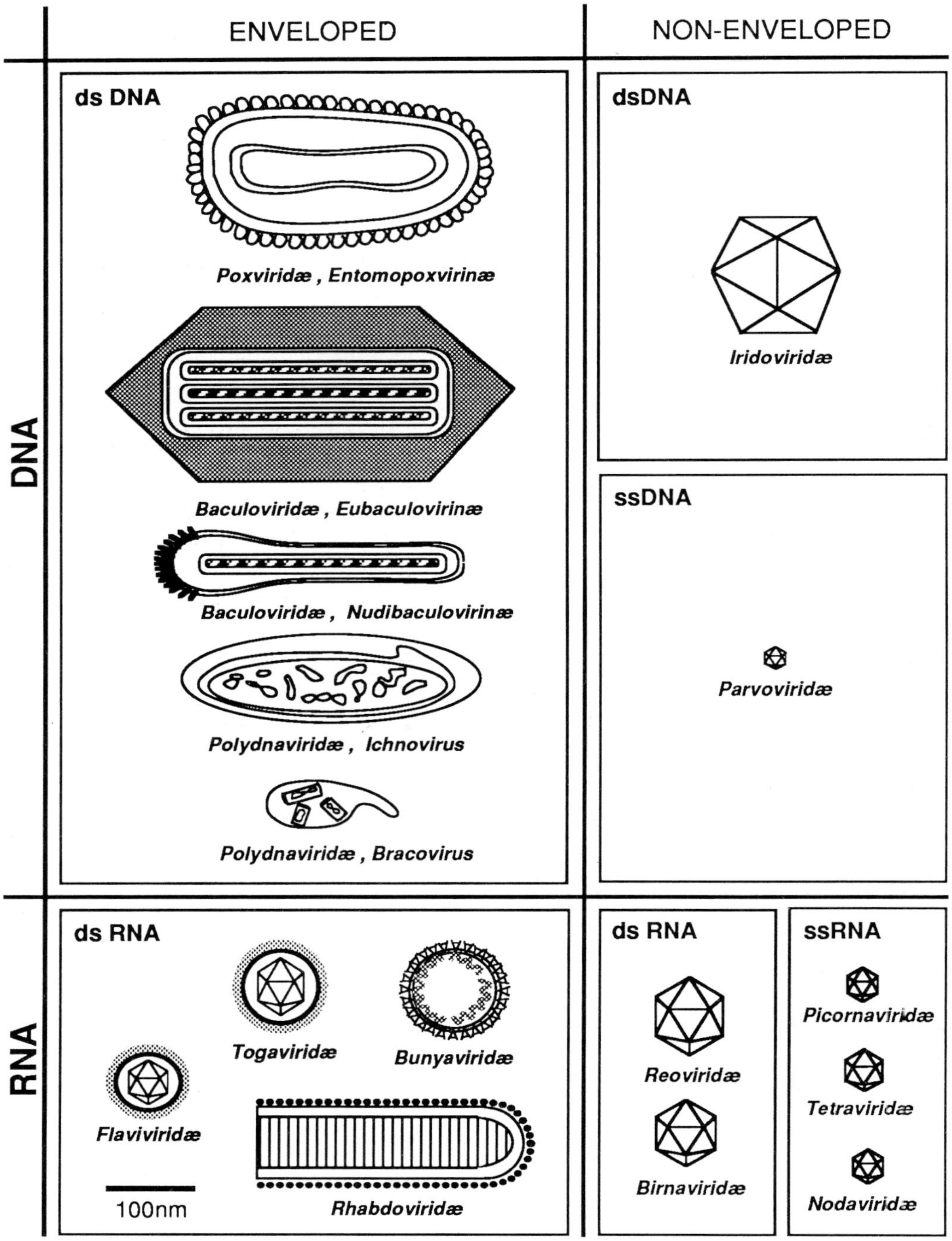

FAMILIES OF VIRUSES INFECTING BACTERIA

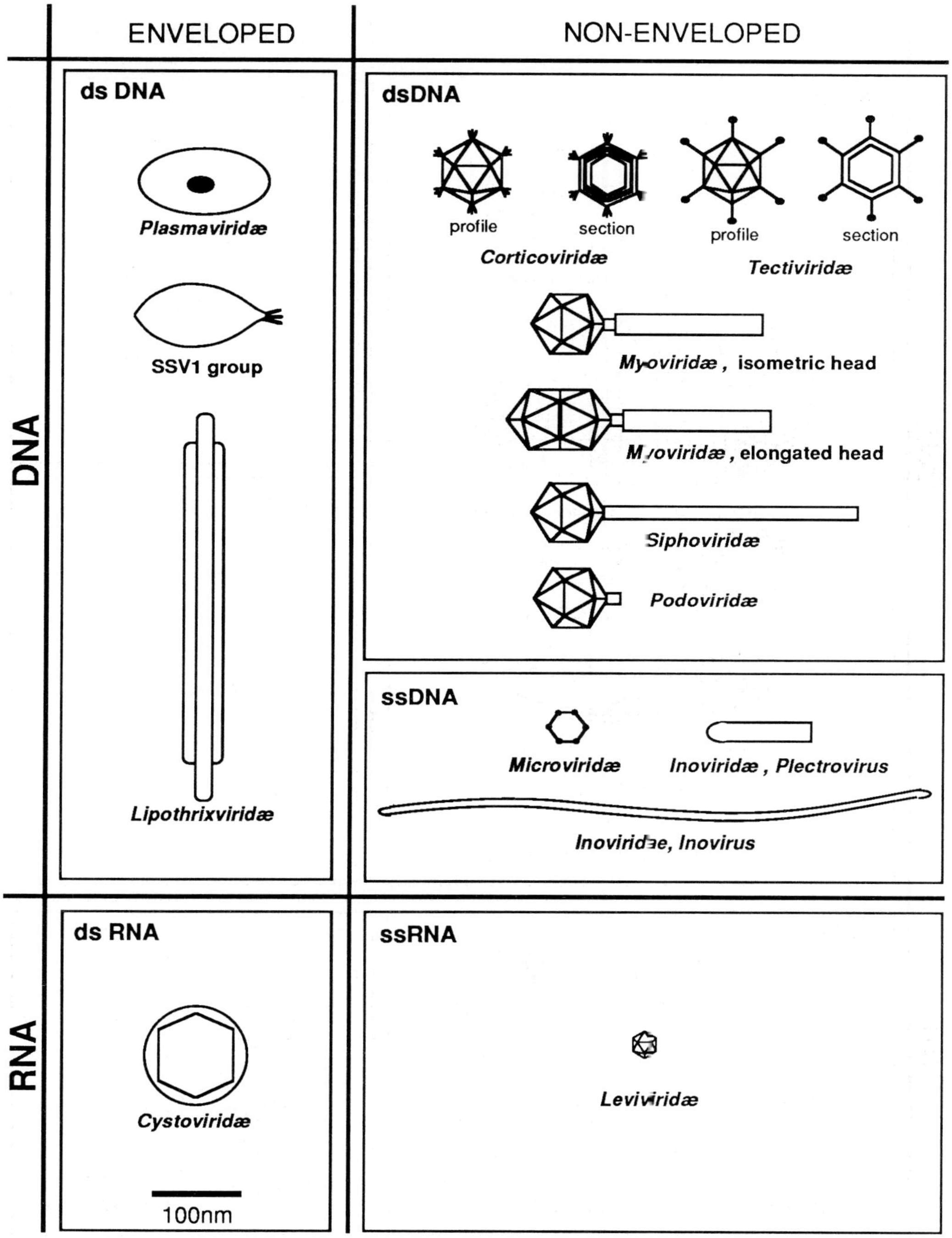

FAMILIES OF VIRUSES INFECTING ALGÆ, FUNGI AND PROTOZOÆ

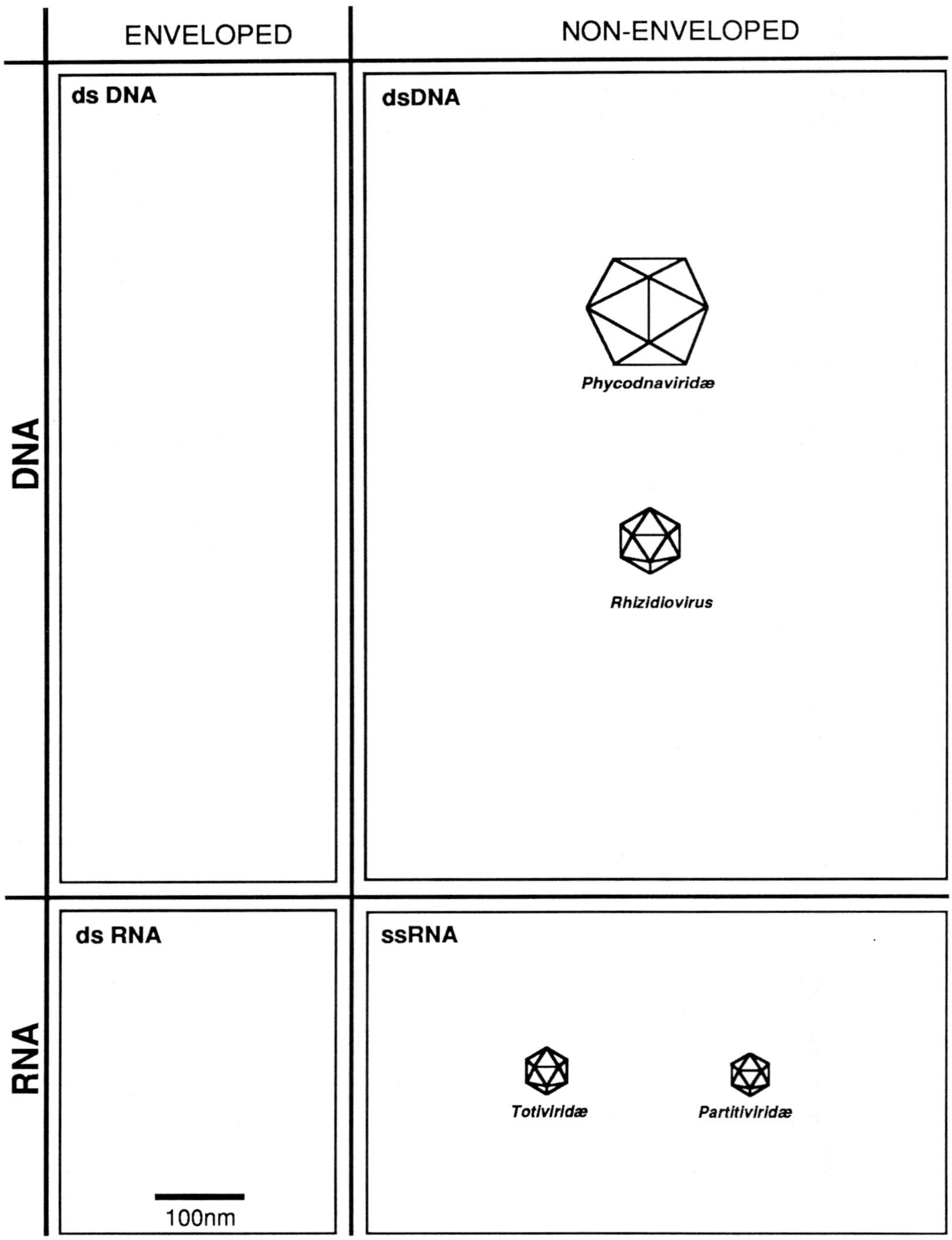

FAMILIES OF VIRUSES INFECTING VERTEBRATES

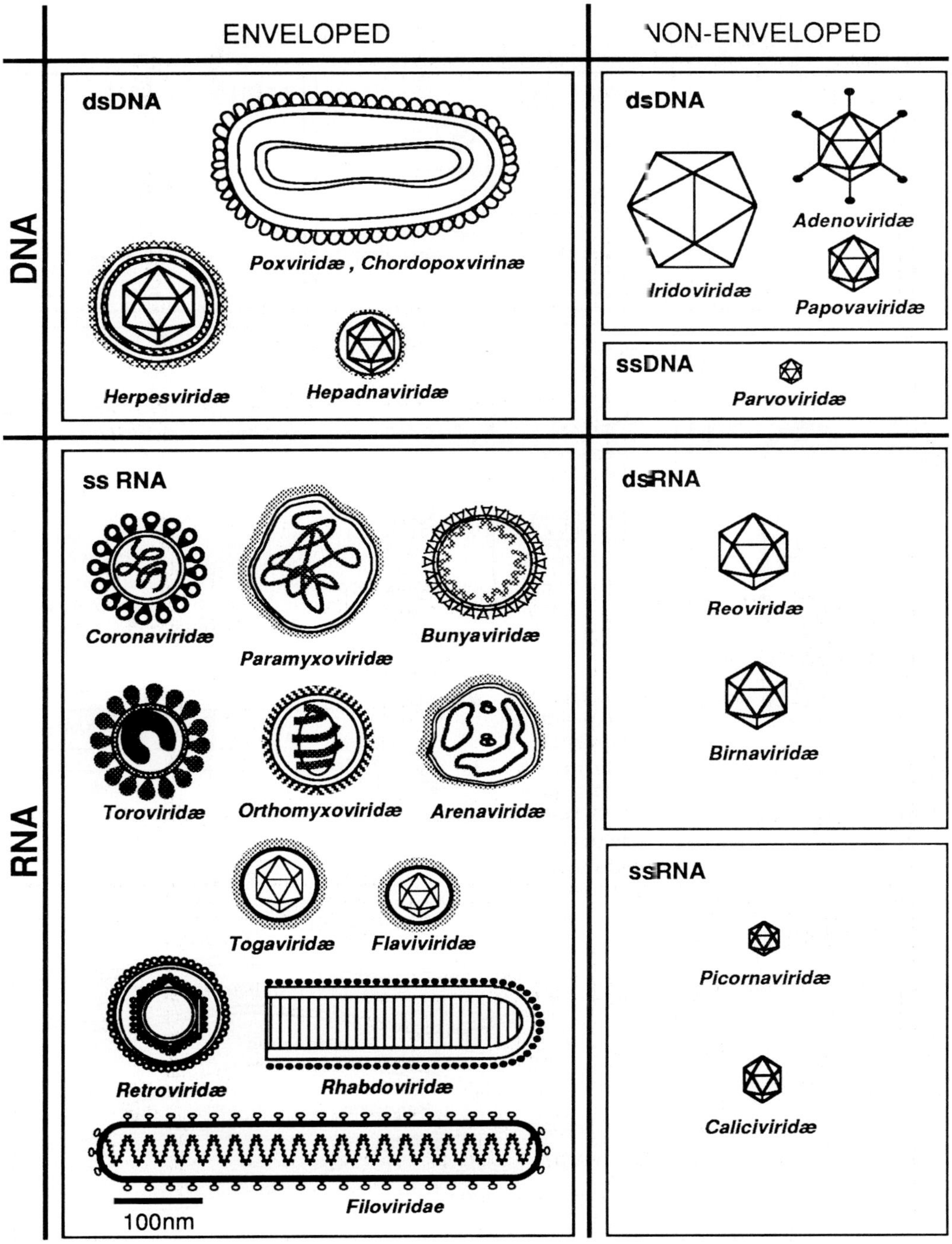

FAMILIES AND GROUPS OF VIRUSES INFECTING PLANTS

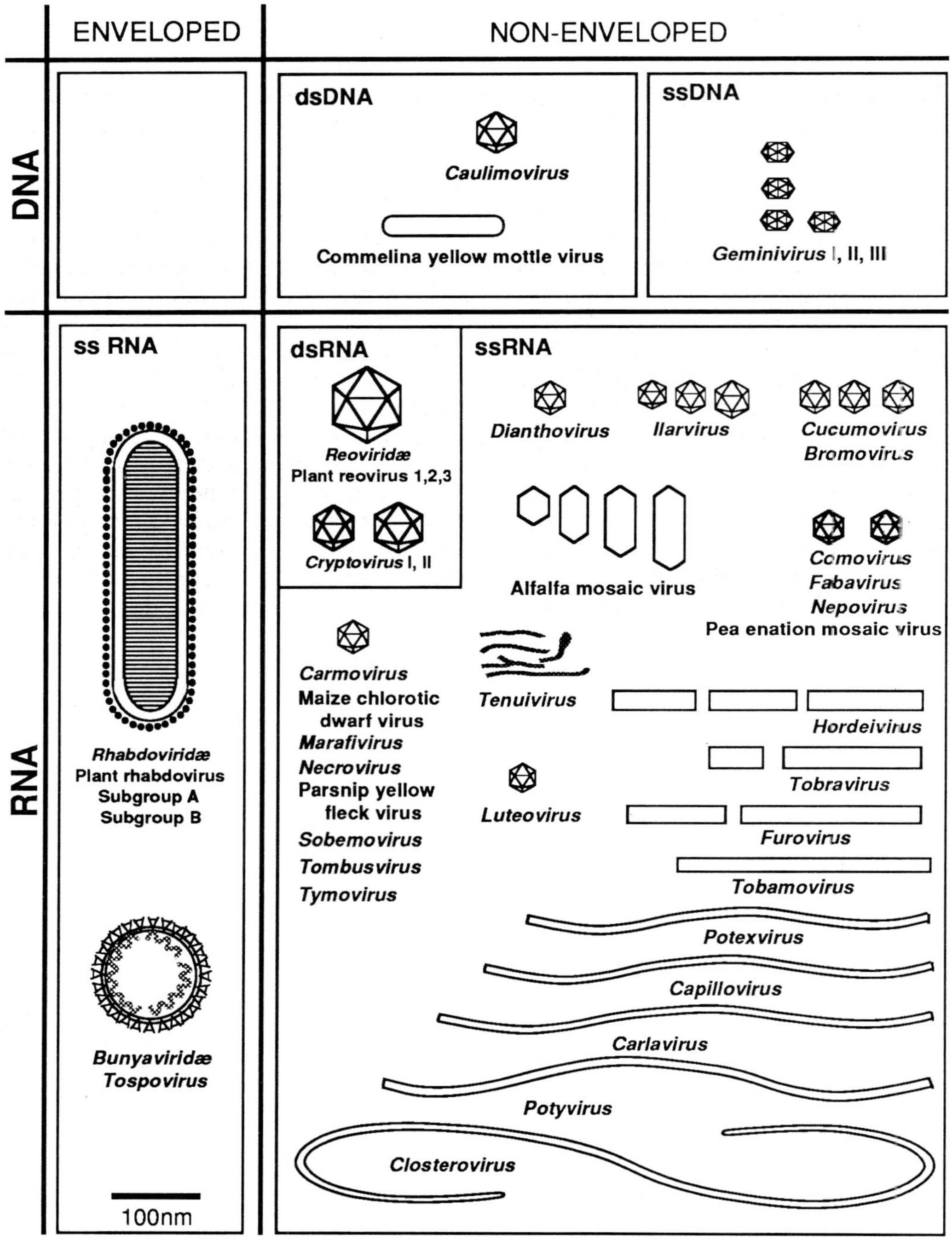

Virus Families and Groups

Revised by C. Fauquet & M.A. Mayo

The following list of virus families and groups has been complemented with virus diagrams to provide an overview of the universal system of virus classification.

The order of presentation of virus families and groups does not reflect any hierarchical or phylogenetic classification but only a convenient order of presentation. Since a taxonomic structure above the level of family or group (with the exception of the order ***Mononegavirales***) has not been developed, any sequence of listing must be arbitrary. The order of presentation is generally the same as in the Fourth ICTV Report (Matthews, 1982). The order of presentation of virus families and groups follows three criteria: (i) the nature of the viral genome, (ii) the strandedness of the viral genome, (iii) the presence or absence of a lipoprotein envelope. There are no known ssDNA viruses with envelopes, so these three criteria give rise to seven clusters comprising the 74 families and groups of viruses. Within two of these clusters, the ssRNA enveloped and non-enveloped viruses, the families have been arranged as follows: (i) the ssRNA enveloped viruses are arranged on the basis of genome strategy (Baltimore, 1971; Cooper, 1974) and (ii) the ssRNA non-enveloped viruses are arranged on the basis of number of pieces of RNA comprising their genome and their virion morphology.

Characterization - Order		Family	Subfamily	Genus/Group	Subgenus/Subgroup
dsDNA	**Enveloped**	***Poxviridæ***	***Chordopoxvirinæ***	***Orthopoxvirus***	
				Parapoxvirus	
				Avipoxvirus	
				Capripoxvirus	
				Leporipoxvirus	
				Suipoxvirus	
				Molluscipoxvirus	
				Yatapoxvirus	
			Entomopoxvirinæ	***Entomopoxvirus* A**	
				***Entomopoxvirus* B**	
				***Entomopoxvirus* C**	
		Herpesviridæ	***Alphaherpesvirinæ***	***Simplexvirus***	
				Varicellovirus	
			Betaherpesvirinæ	***Cytomegalovirus***	
				Muromegalovirus	
			Gammaherpesvirinæ	***Lymphocryptovirus***	
				Rhadinovirus	
		Hepadnaviridæ		***Orthohepadnavirus***	
				Avihepadnavirus	
		Baculoviridæ	***Eubaculovirinæ***	**Nuclear polyhedrosis virus**	
					Multiple nuclear polyhedrosis virus
					Single nuclear polyhedrosis virus
				Granulosis virus	
			Nudibaculovirinæ	**Non-occluded baculovirus**	
		Plasmaviridæ		***Plasmavirus***	
		SSV 1 group			

Type member	Shape	Host	Page
Vaccinia virus	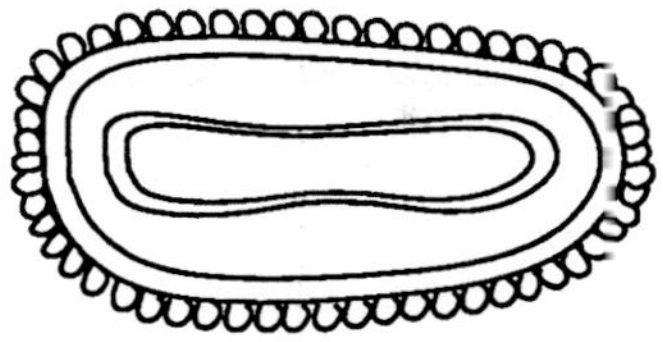	Vertebrate	92
Orf virus			94
Fowlpox virus			94
Sheep pox virus			95
Myxoma virus			96
Swinepox virus			97
Molluscum contagiosum virus			97
Yaba monkey tumor virus			98
Melolontha melolontha entomopoxvirus	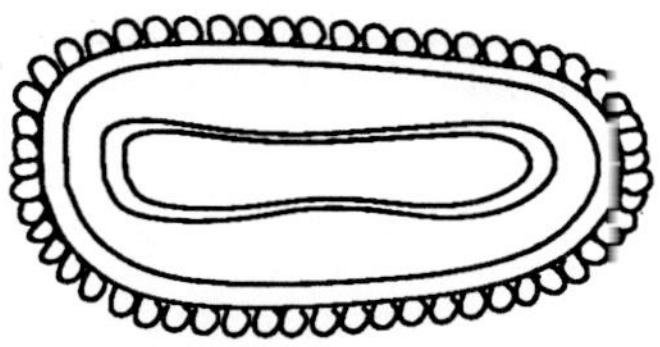	Invertebrate	99
Amsacta moorei entomopoxvirus			99
Chironomus luridus entomopoxvirus			100
Human (alpha) herpesvirus 1		Vertebrate	105
Human (alpha) herpesvirus 3			106
Human (beta) herpesvirus 5			107
Murid (beta) herpesvirus 1			108
Human (gamma) herpesvirus 4			109
Ateline herpesvirus 2			109
Hepatitis B virus		Vertebrate	115
Duck hepatitis B virus			115
Autographa californica multiple nuclear polyedrosis virus	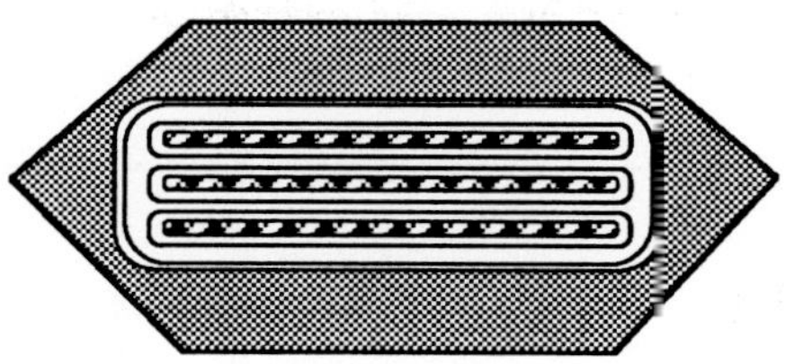	Invertebrate	119
Bombyx mori single nuclear polyedrosis virus			119
Trichoplusia ni granulosis virus			120
Oryctes rhinoceros virus	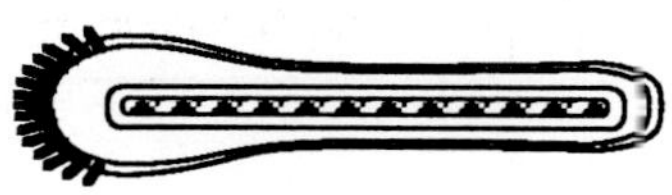	Invertebrate	121
Acholeplasma phage L2	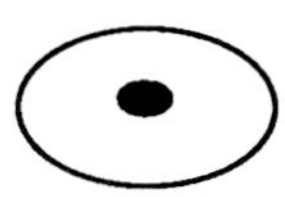	Mycoplasma	125
SSV-1	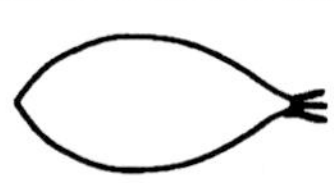	Bacteria	126

Characterization - Order		Family	Subfamily	Genus/Group	Subgenus/Subgroup
dsDNA	**Enveloped**	***Lipothrixviridæ***		***Lipothrixvirus***	
		Polydnaviridæ		***Ichnovirus***	
				Bracovirus	
dsDNA	**Nonenveloped**	***Iridoviridæ***		***Iridovirus***	
				Chloriridovirus	
				Ranavirus	
				Lymphocystivirus	
				Goldfish virus group	
		Phycodnaviridæ		***Phycodnavirus***	
		Adenoviridæ		***Mastadenovirus***	
				Aviadenovirus	
				Rhizidiovirus	
		Papovaviridæ		***Papillomavirus***	
				Polyomavirus	
				Caulimovirus	
				Commelina yellow mottle virus	
		Tectiviridæ		***Tectivirus***	
		Corticoviridæ		***Corticovirus***	

Type member	Shape	Host	Page
Thermoproteus phage TTV1	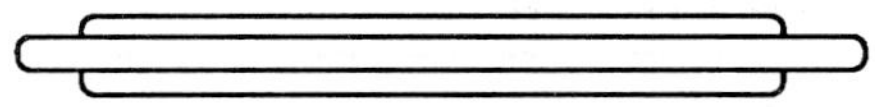	Bacteria	127
Campoletis sonovensis virus		Invertebrate	130
Cotesia melanoscela virus			130
Chilo iridescent virus	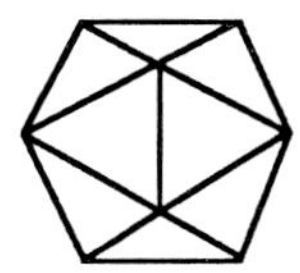		133
Mosquito iridescent virus		Vertebrate	134
Frog virus 3		Invertebrate	134
Flounder isolate			135
Goldfish virus 1			135
Paramecium bursaria Chlorella virus-1	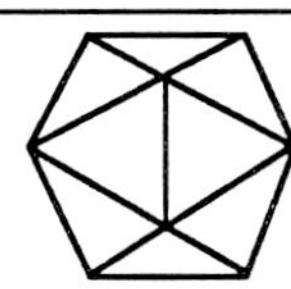	Algæ	137
Human adenovirus 2	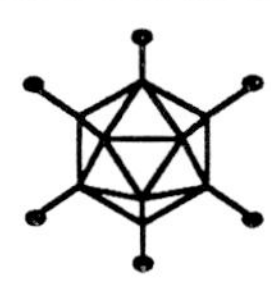		142
Fowl adenovirus 1		Vertebrate	143
Rhizidiomyces virus		Fungus	145
Rabbit (Shope) papilloma virus		Vertebrate	147
Polyoma virus (mouse)			148
Cauliflower mosaic virus		Plant	150
Commelina yellow mottle virus	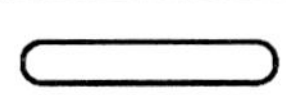	Plant	153
Phage PRD1	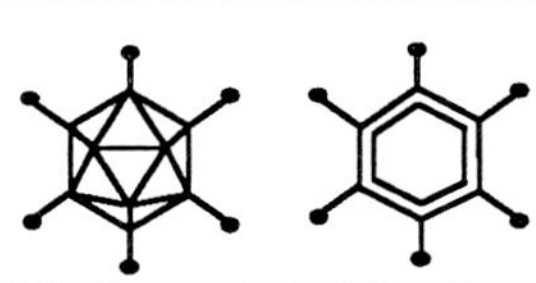	Bacteria	155
Alteromonas Phage PM2	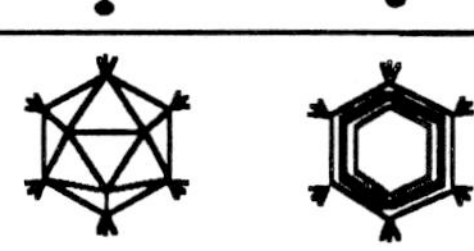	Bacteria	157

Characterization - Order	Family	Subfamily	Genus/Group	Subgenus/Subgroup
dsRNA Enveloped	***Cystoviridæ***		***Cystovirus***	
dsRNA Nonenveloped	***Reoviridæ***		***Orthoreovirus***	
			Orbivirus	
			Coltivirus	
			Rotavirus	
			Aquareovirus	
			Cypovirus	
			Plant reovirus 1	
			Fijivirus (Plant reovirus 2)	
			Plant reovirus 3	
	Birnaviridæ		***Birnavirus***	
	Totiviridæ		***Totivirus***	
			Giardiavirus	
	Partitiviridæ		***Partitivirus***	
			Penicillium chrysogenum **virus**	
			Cryptovirus	
				White clover cryptic virus I
				White clover cryptic virus II
ssRNA Enveloped				
a - No DNA step (i) Positive-sense genome	***Togaviridæ***		***Alphavirus***	
			Rubivirus	
			Arterivirus	
	Flaviviridæ		***Flavivirus***	
			Pestivirus	
			Hepatitis C virus	
	Coronaviridæ		***Coronavirus***	
			Torovirus	

Type member	Shape	Host	Page
Pseudomonas Phage φ6	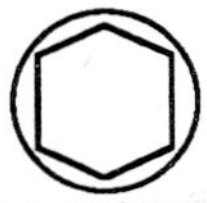	Bacteria	184
Reovirus type 1		Invertebrate	187
Bluetongue virus			188
Colorado tick fever virus		Vertebrate	189
Human rotavirus			190
Golden shiner virus			192
Cytoplasmic polyhedrosis virus from *Bombyx mori*			193
Wound tumor virus		Plant	194
Fiji disease virus			195
Rice ragged stunt virus			197
Infectious pancreatic necrosis virus		Vertebrate Invertebrate	200
Saccharomyces cerevisiæ virus L1		Fungus	203
Giardia lamblia virus		Protozoa	206
Gaeumannomyces graminis virus 019/6-A		Fungus	208
Pennicilium chrysogenum virus			209
White clover cryptic virus I		Plant	213
White clover cryptic virus II		Plant	214
Sindbis virus		Vertebrate	217
Rubella virus		Invertebrate	219
Equine arteritis virus			220
Yellow fever virus		Vertebrate	223
Bovine viral diarrhea virus		Invertebrate	228
Hepatitis C virus			230
Avian infectious bronchitis virus		Vertebrate	234
Berne virus		Vertebrate	237

Characterization - Order	Family	Subfamily	Genus/Group	Subgenus/Subgroup
ssRNA Enveloped				
Mononegavirales (ii) Negative-sense genome single stranded	***Paramyxoviridæ***	***Paramyxovirinæ***	***Paramyxovirus***	
			Morbillivirus	
		Pneumovirinæ	***Pneumovirus***	
	Filoviridæ		***Filovirus***	
	Rhabdoviridæ		***Vesiculovirus***	
			Lyssavirus	
			Plant rhabdovirus	
				Subgroup A
				Subgroup B
(iii) Negative-sense genome multiple stranded	***Orthomyxoviridæ***		***Influenzavirus* A & B**	
			***Influenzavirus* C**	
	Bunyaviridæ		***Bunyavirus***	
			Phlebovirus	
			Nairovirus	
			Hantavirus	
			Tospovirus	
	Arenaviridæ		***Arenavirus***	
b - DNA step	***Retroviridæ***		**Mammalian type B oncovirus**	
			Mammalian type C retrovirus	
			Type D retrovirus	
			Spumavirus	
			HTLV - BLV group	
			Lentivirus	

Type member	Shape	Host	Page
Newcastle disease virus			244
Measles virus		Vertebrate	244
Human respiratory syncytial virus			245
Marburg virus	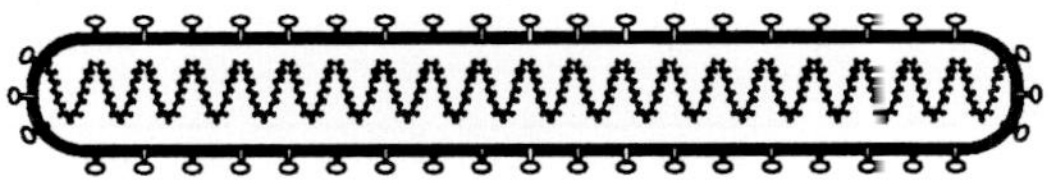	Vertebrate	247
Vesicular stomatitis Indiana virus	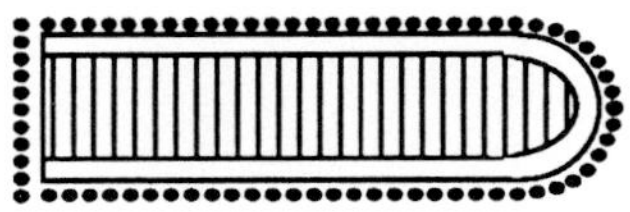	Vertebrate	252
Rabies virus			254
		Invertebrate	
Lettuce necrotic yellow virus		Plant	258
Potato yellow dwarf virus			259
Influenzavirus A/PR/8/34			263
Influenzavirus C/Taylor/1233/47		Vertebrate	270
Bunyamwera virus			274
Sandfly fever Sicilian virus		Vertebrate	277
Crimean-Congo hemorrhagic fever virus		Invertebrate	279
Hantaan virus			280
Tomato spotted wilt virus		Plant	281
Lymphocytic choriomeningitis virus		Vertebrate	284
Mouse mammary tumor virus			293
Murine leukemia virus			294
Mason-Pfizer monkey virus			295
Avian leukosis virus		Vertebrate	295
Human foamy virus			296
Human T-cell lymphotropic virus type 1			297
Human immunodeficiency virus			297

Characterization - Order	Family	Subfamily	Genus/Group	Subgenus/Subgroup
ssRNA Nonenveloped				
Monopartite genomes Isometric particles	***Caliciviridæ***		***Calicivirus***	
			Carmovirus	
	Leviviridæ		***Levivirus*** ***Allolevivirus***	
			Luteovirus	
			Maize chlorotic dwarf virus	
			Marafivirus	
			Necrovirus	
			Parsnip yellow fleck virus	
	Picornaviridæ		***Enterovirus*** ***Hepatovirus*** ***Cardiovirus*** ***Rhinovirus*** ***Aphtovirus***	
			Sobemovirus	
	Tetraviridæ			
			Tombusvirus	
			Tymovirus	

Type member	Shape	Host	Page
Vesicular exanthema of swine virus		Vertebrate	300
Carnation mottle virus		Plant	303
Phage MS2 Phage Qβ		Bacteria	307 307
Barley yellow dwarf virus		Plant	309
Maize chlorotic dwarf virus		Plant	312
Maize rayado fino virus		Plant	314
Tobacco necrosis virus		Plant	316
Parsnip yellow fleck virus		Plant	318
Human poliovirus 1 Human hepatitis A virus Encephalomyocarditis virus Human rhinovirus 1A Aphtovirus O		Vertebrate Invertebrate	322 322 323 323 324
Southern bean mosaic virus		Plant	327
Nudaurelia β virus		Invertebrate	330
Tomato bushy stunt virus		Plant	332
Turnip yellow mosaic virus		Plant	336

Characterization - Order	Family	Subfamily	Genus/Group	Subgenus/Subgroup
ssRNA Nonenveloped				
Monopartite genomes Rod-shaped particles			***Capillovirus***	
			Carlavirus	
			Closterovirus	
			Potexvirus	
			Potyvirus	
			Tobamovirus	
Bipartite genomes Isometric particles			***Comovirus***	
			Dianthovirus	
			Fabavirus	
			Nepovirus	
	Nodaviridæ		***Nodavirus***	
			Pea enation mosaic virus	

Type member	Shape	Host	Page
Apple stem grooving virus		Plant	339
Carnation latent virus		Plant	341
Sugar beet yellows virus		Plant	345
Potato virus X		Plant	348
Potato virus Y	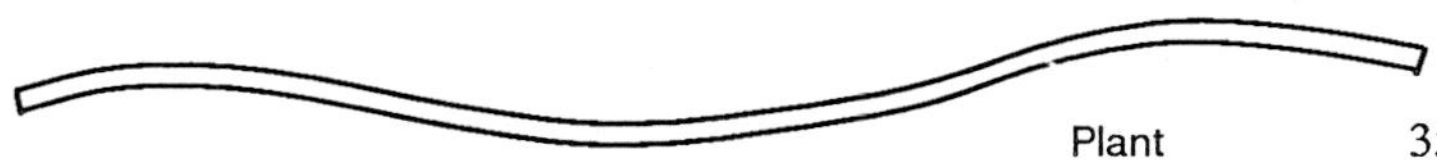	Plant	351
Tobacco mosaic virus		Plant	357
Cowpea mosaic virus		Plant	360
Carnation ringspot virus		Plant	364
Broad bean wilt virus		Plant	366
Tobacco ringspot virus		Plant	368
Nodamura virus		Invertebrate	372
Pea enation mosaic virus		Plant	375

Characterization - Order	Family	Subfamily	Genus/Group	Subgenus/Subgroup
ssRNA	**Nonenveloped**			
Bipartite genomes Rod-shaped particles			***Furovirus***	
			Tobravirus	
Tripartite genomes Isometric particles			***Bromovirus***	
			Cucumovirus	
			Ilarvirus	
Isometric and bacilliform particles			**Alfalfa mosaic virus**	
Rod-shaped particles			***Hordeivirus***	
Tetrapartite genomes			***Tenuivirus***	

Type member	Shape	Host	Page
Soil-borne wheat mosaic virus		Plant	377
Tobacco rattle virus		Plant	380
Brome mosaic virus		Plant	382
Cucumber mosaic virus		Plant	386
Tobacco streak virus		Plant	389
Alfalfa mosaic virus	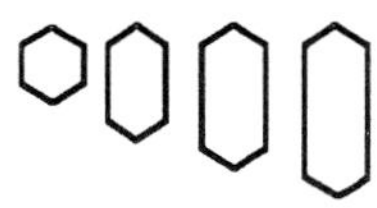	Plant	392
Barley stripe mosaic virus		Plant	395
Rice stripe virus		Plant	398

Listing of Virus Families and Groups

Compiled by R.I.B. Francki & C. Fauquet

TABLE I. Alphabetical Listing of Families and Groups

FAMILIES OR GROUP	MORPHOLOGY	ENVELOPE	NUCLEIC ACID TYPE	NUCLEIC ACID CONFIGURATION	HOST
Adenoviridae	icosahedral	-	dsDNA	linear	V
Alfalfa mosaic	bacilliform	-	ssRNA	3 + strands	P
Arenaviridae	spherical	+	ssRNA	2 - strands	V
Baculoviridae	bacilliform	+	dsDNA	supercoiled	I
Birnaviridae	icosahedral	-	dsRNA	2 segments	V, I
Bromovirus	icosahedral	-	ssRNA	3 + strands	P
Bunyaviridae	spherical	+	ssRNA	3 - strands	V
Caliciviridae	icosahedral	-	ssRNA	1 + strand	V
Capillovirus	rod	-	ssRNA	1 + strand	P
Carlavirus	rod	-	ssRNA	1 + strand	P
Carmovirus	isometric	-	ssRNA	1 + strand	P
Caulimovirus	isometric	-	dsDNA	circular	P
Closterovirus	rod	-	ssRNA	1 + strand	P
Commelina yellow mottle	bacilliform	-	dsDNA	1 circular	P
Comovirus	isometric	-	ssRNA	1 + strand	P
Coronaviridae	pleomorphic	+	ssRNA	1 + strand	V
Corticoviridae	isometric	-	dsDNA	supercoiled	B
Cryptovirus	isometric	-	dsRNA	2 segments	P
Cucumovirus	isometric	-	ssRNA	3 + strands	P
Cystoviridae	isometric	+	dsRNA	3 segments	B
Dianthovirus	isometric	-	ssRNA	2 + strands	P
Fabavirus	isometric	-	ssRNA	2 + strands	P
Filoviridae	bacilliform	+	ssRNA	1 - strand	V
Flaviviridae	spherical	+	ssRNA	1 + strand	V, I
Furovirus	rod	-	ssRNA	2 + strands	P
Geminivirus	isometric	-	ssDNA	1 or 2 circular	P
Hepadnaviridae	isometric	+	dsDNA	circular	V
Herpesviridae	isometric	+	dsDNA	linear	V
Hordeivirus	helical	-	ssRNA	3 + strands	P
Ilarvirus	isometric	-	ssRNA	3 + strands	P
Inoviridae	rod	-	ssDNA	circular	B, M
Iridoviridae	icosahedral	+	dsDNA	linear	V, I
Leviviridae	icosahedral	-	ssRNA	1 + strand	B
Lipothrixviridae	rod	+	dsDNA	linear	B
Luteovirus	isometric	-	ssRNA	1 + strand	P
Maize chlorotic dwarf	isometric	-	ssRNA	1 + strand	P
Marafivirus	isometric	-	ssRNA	1 + strand	P
Microviridae	icosahedral	-	ssDNA	circular	B
Myoviridae	tailed phage	-	dsDNA	linear	B
Necrovirus	isometric	-	ssRNA	1 + strand	P
Nepovirus	isometric	-	ssRNA	2 + strands	P
Nodaviridae	isometric	-	ssRNA	2 + strands	I

FAMILIES OR GROUP	MORPHOLOGY	ENVELOPE	NUCLEIC ACID TYPE	NUCLEIC ACID CONFIGURATION	HOST
Orthomyxoviridae	helical	+	ssRNA	8 - strands	V
Papovaviridae	icosahedral	-	dsDNA	circular	V
Paramyxoviridae	helical	+	ssRNA	1 - strand	V
Parsnip yellow fleck	isometric	-	ssRNA	1 + strand	P
Partitiviridae	isometric	-	dsRNA	2 segments	F
Parvoviridae	icosahedral	-	ssDNA	1 - strand	V, I
Pea enation mosaic	isometric	-	ssRNA	2 + strands	P
Phycodnaviridae	icosahedral	-	dsDNA	1 + linear	A
Picornaviridae	icosahedral	-	ssRNA	1 + strand	V, I
Plasmaviridae	pleomorphic	+	dsDNA	1 circular	B
Podoviridae	tailed phage	-	dsDNA	linear	B
Polydnaviridae	rod, fusiform	+	dsDNA	supercoiled	I
Potexvirus	rod	-	ssRNA	1 + strand	P
Potyvirus	rod	-	ssRNA	1 + strand	P
Poxviridae	oviod	+	dsDNA	linear	V, I
Reoviridae	icosahedral	-	dsRNA	10-12 segments	V, I, P
Retroviridae	spherical	+	ssRNA	dimer 1 + strand	V
Rhabdoviridae	bacilliform	+	ssRNA	1 - strand	V, I, P
Siphoviridae	tailed phage	-	dsDNA	1 linear	B
Sobemovirus	icosahedral	-	ssRNA	1 + strand	P
SSV-1	lemon-shape	+	dsDNA	1+ circular	B
Tectiviridae	icosahedral	-	dsDNA	linear	B
Tenuivirus	rod	-	ssRNA	4 -? strands	P
Tetraviridae	icosahedral	-	ssRNA	1 + strand	I
Tobamovirus	rod	-	ssRNA	1 + strand	P
Tobravirus	rod	-	ssRNA	2 + strands	P
Togaviridae	spherical	+	ssRNA	1 + strand	V, I
Tombusvirus	isometric	-	ssRNA	1 + strand	P
Totiviridae	isometric	-	dsRNA	1 segment	F
Tymovirus	icosahedral	-	ssRNA	1 + strand	P

TABLE II. Families and Groups Listed by Host

FAMILIES OR GROUP	MORPHOLOGY	ENVELOPE	NUCLEIC ACID TYPE	NUCLEIC ACID CONFIGURATION	HOST
Phycodnaviridae	icosahedral	-	dsDNA	1 + linear	A
Corticoviridae	isometric	-	dsDNA	supercoiled	B
Cystoviridae	isometric	+	dsRNA	3 segments	B
Leviviridae	icosahedral	-	ssRNA	1 + strand	B
Lipothrixviridae	rod	+	dsDNA	linear	B
Microviridae	icosahedral	-	ssDNA	circular	B
Myoviridae	tailed phage	-	dsDNA	linear	B
Plasmaviridae	pleomorphic	+	dsDNA	1 circular	B
Podoviridae	tailed phage	-	dsDNA	linear	B
Siphoviridae	tailed phage	-	dsDNA	1 linear	B
SSV-1	lemon-shape	+	dsDNA	1+ circular	B
Tectiviridae	icosahedral	-	dsDNA	linear	B
Inoviridae	rod	-	ssDNA	circular	B, M
Partitiviridae	isometric	-	dsRNA	2 segments	F
Totiviridae	isometric	-	dsRNA	1 segment	F
Baculoviridae	bacilliform	+	dsDNA	supercoiled	I
Nodaviridae	isometric	-	ssRNA	2 + strands	I
Polydnaviridae	rod, fusiform	+	dsDNA	supercoiled	I
Tetraviridae	icosahedral	-	ssRNA	1 + strand	I
Alfalfa mosaic	bacilliform	-	ssRNA	3 + strands	P
Bromovirus	icosahedral	-	ssRNA	3 + strands	P
Capillovirus	rod	-	ssRNA	1 + strand	P
Carlavirus	rod	-	ssRNA	1 + strand	P
Carmovirus	isometric	-	ssRNA	1 + strand	P
Caulimovirus	isometric	-	dsDNA	circular	P
Closterovirus	rod	-	ssRNA	1 + strand	P
Commelina yellow mottle	bacilliform	-	dsDNA	1 circular	P
Comovirus	isometric	-	ssRNA	1 + strand	P
Cryptovirus	isometric	-	dsRNA	2 segments	P
Cucumovirus	isometric	-	ssRNA	3 + strands	P
Dianthovirus	isometric	-	ssRNA	2 + strands	P
Fabavirus	isometric	-	ssRNA	2 + strands	P
Furovirus	rod	-	ssRNA	2 + strands	P
Geminivirus	isometric	-	ssDNA	1 or 2 circular	P
Hordeivirus	helical	-	ssRNA	3 + strands	P
Ilarvirus	isometric	-	ssRNA	3 + strands	P
Luteovirus	isometric	-	ssRNA	1 + strand	P
Maize chlorotic dwarf	isometric	-	ssRNA	1 + strand	P
Marafivirus	isometric	-	ssRNA	1 + strand	P
Necrovirus	isometric	-	ssRNA	1 + strand	P
Nepovirus	isometric	-	ssRNA	2 + strands	P
Parsnip yellow fleck	isometric	-	ssRNA	1 + strand	P
Pea enation mosaic	isometric	-	ssRNA	2 + strands	P
Potexvirus	rod	-	ssRNA	1 + strand	P
Potyvirus	rod	-	ssRNA	1 + strand	P
Sobemovirus	icosahedral	-	ssRNA	1 + strand	P
Tenuivirus	rod	-	ssRNA	4 -? strands	P

FAMILIES OR GROUP	MORPHOLOGY	ENVELOPE	NUCLEIC ACID TYPE	NUCLEIC ACID CONFIGURATION	HOST
Tobamovirus	rod	-	ssRNA	1 + strand	P
Tobravirus	rod	-	ssRNA	2 + strands	P
Tombusvirus	isometric	-	ssRNA	1 + strand	P
Tymovirus	icosahedral	-	ssRNA	1 + strand	P
Adenoviridae	icosahedral	-	dsDNA	linear	V
Arenaviridae	spherical	+	ssRNA	2 - strands	V
Bunyaviridae	spherical	+	ssRNA	3 - strands	V
Caliciviridae	icosahedral	-	ssRNA	1 + strand	V
Coronaviridae	pleomorphic	+	ssRNA	1 + strand	V
Filoviridae	bacilliform	+	ssRNA	1 - strand	V
Hepadnaviridae	isometric	+	dsDNA	circular	V
Herpesviridae	isometric	+	dsDNA	linear	V
Orthomyxoviridae	helical	+	ssRNA	8 - strands	V
Papovaviridae	icosahedral	-	dsDNA	circular	V
Paramyxoviridae	helical	+	ssRNA	1 - strand	V
Retroviridae	spherical	+	ssRNA	dimer 1 + strand	V
Birnaviridae	icosahedral	-	dsRNA	2 segments	V, I
Flaviviridae	spherical	+	ssRNA	1 + strand	V, I
Iridoviridae	icosahedral	+	dsDNA	linear	V, I
Parvoviridae	icosahedral	-	ssDNA	1 - strand	V, I
Picornaviridae	icosahedral	-	ssRNA	1 + strand	V, I
Poxviridae	oviod	+	dsDNA	linear	V, I
Togaviridae	spherical	+	ssRNA	1 + strand	V, I
Reoviridae	icosahedral	-	dsRNA	10-12 segments	V, I, P
Rhabdoviridae	bacilliform	+	ssRNA	1 - strand	V, I, P

TABLE III. Families and Groups Listed by Nucleic Acid Type and Configuration

FAMILIES OR GROUP	MORPHOLOGY	ENVELOPE	NUCLEIC ACID TYPE	NUCLEIC ACID CONFIGURATION	HOST
Commelina yellow mottle	bacilliform	-	dsDNA	1 circular	P
Plasmaviridae	pleomorphic	+	dsDNA	1 circular	B
SSV-1	lemon-shape	+	dsDNA	1 circular	B
Caulimovirus	isometric	-	dsDNA	circular	P
Hepadnaviridae	isometric	+	dsDNA	circular	V
Papovaviridae	icosahedral	-	dsDNA	circular	V
Phycodnaviridae	icosahedral	-	dsDNA	1 linear	A
Siphoviridae	tailed phage	-	dsDNA	1 linear	B
Adenoviridae	icosahedral	-	dsDNA	linear	V
Herpesviridae	isometric	+	dsDNA	linear	V
Iridoviridae	icosahedral	+	dsDNA	linear	V, I
Lipothrixviridae	rod	+	dsDNA	linear	B
Myoviridae	tailed phage	-	dsDNA	linear	B
Podoviridae	tailed phage	-	dsDNA	linear	B
Poxviridae	oviod	+	dsDNA	linear	V, I
Tectiviridae	icosahedral	-	dsDNA	linear	B
Baculoviridae	bacilliform	+	dsDNA	supercoiled	I
Corticoviridae	isometric	-	dsDNA	supercoiled	B
Polydnaviridae	rod, fusiform	+	dsDNA	supercoiled	I
Parvoviridae	icosahedral	-	ssDNA	1 - strand	V, I
Geminivirus	isometric	-	ssDNA	1 or 2 circular	P
Inoviridae	rod	-	ssDNA	circular	B, M
Microviridae	icosahedral	-	ssDNA	circular	B
Totiviridae	isometric	-	dsRNA	1 segment	F
Birnaviridae	icosahedral	-	dsRNA	2 segments	V, I
Cryptovirus	isometric	-	dsRNA	2 segments	P
Partitiviridae	isometric	-	dsRNA	2 segments	F
Cystoviridae	isometric	+	dsRNA	3 segments	B
Reoviridae	icosahedral	-	dsRNA	10-12 segments	V, I, P
Caliciviridae	icosahedral	-	ssRNA	1 + strand	V
Capillovirus	rod	-	ssRNA	1 + strand	P
Carlavirus	rod	-	ssRNA	1 + strand	P
Carmovirus	isometric	-	ssRNA	1 + strand	P
Closterovirus	rod	-	ssRNA	1 + strand	P
Comovirus	isometric	-	ssRNA	1 + strand	P
Coronaviridae	pleomorphic	+	ssRNA	1 + strand	V
Flaviviridae	spherical	+	ssRNA	1 + strand	V, I
Leviviridae	icosahedral	-	ssRNA	1 + strand	B
Luteovirus	isometric	-	ssRNA	1 + strand	P
Maize chlorotic dwarf	isometric	-	ssRNA	1 + strand	P
Marafivirus	isometric	-	ssRNA	1 + strand	P
Necrovirus	isometric	-	ssRNA	1 + strand	P
Parsnip yellow fleck	isometric	-	ssRNA	1 + strand	P
Picornaviridae	icosahedral	-	ssRNA	1 + strand	V, I
Potexvirus	rod	-	ssRNA	1 + strand	P
Potyvirus	rod	-	ssRNA	1 + strand	P

FAMILIES OR GROUP	MORPHOLOGY	ENVELOPE	NUCLEIC ACID TYPE	NUCLEIC ACID CONFIGURATION	HOST
Sobemovirus	icosahedral	-	ssRNA	1 + strand	P
Tetraviridae	icosahedral	-	ssRNA	1 + strand	I
Tobamovirus	rod	-	ssRNA	1 + strand	P
Togaviridae	spherical	+	ssRNA	1 + strand	V, I
Tombusvirus	isometric	-	ssRNA	1 + strand	P
Tymovirus	icosahedral	-	ssRNA	1 + strand	P
Filoviridae	bacilliform	+	ssRNA	1 - strand	V
Paramyxoviridae	helical	+	ssRNA	1 - strand	V
Rhabdoviridae	bacilliform	+	ssRNA	1 - strand	V, I, P
Dianthovirus	isometric	-	ssRNA	2 + strands	P
Fabavirus	isometric	-	ssRNA	2 + strands	P
Furovirus	rod	-	ssRNA	2 + strands	P
Nepovirus	isometric	-	ssRNA	2 + strands	P
Nodaviridae	isometric	-	ssRNA	2 + strands	I
Pea enation mosaic	isometric	-	ssRNA	2 + strands	P
Tobravirus	rod	-	ssRNA	2 + strands	P
Arenaviridae	spherical	+	ssRNA	2 - strands	V
Alfalfa mosaic	bacilliform	-	ssRNA	3 + strands	P
Bromovirus	icosahedral	-	ssRNA	3 + strands	P
Cucumovirus	isometric	-	ssRNA	3 + strands	P
Hordeivirus	helical	-	ssRNA	3 + strands	P
Ilarvirus	isometric	-	ssRNA	3 + strands	P
Bunyaviridae	spherical	+	ssRNA	3 - strands	V
Tenuivirus	rod	-	ssRNA	4 -? strands	P
Orthomyxoviridae	helical	+	ssRNA	8 - strands	V
Retroviridae	spherical	+	ssRNA	dimer 1 + strand	V

Key to Identification of Virus Families and Groups

Compiled by M.A. Mayo & C. Fauquet

1.	Host a prokaryote	2
	Host a eukaryote	13
2.	Genome of DNA	3
	Genome of RNA	12
3.	Virion DNA double-stranded	4
	Virion DNA single-stranded	11
4.	Virions with lipid-containing envelopes	5
	Virions not enveloped	7
5.	Virions rod-shaped	***Lipothrixviridae***
	Virions not rod-shaped	6
6.	Virions lemon-shaped, host an archaebacterium	**SSV1 group**
	Virions not lemon-shaped, host a mycoplasma	***Plasmaviridae***
7.	Virions isometric without tails	8
	Virions with tails	9
8.	DNA linear, >10 kbp	***Tectiviridae***
	DNA super-coiled, < 10 kbp	***Corticoviridae***
9.	Tails contractile	***Myoviridae***
	Tails not contractile	10
10.	Tails long, DNA c. 35 kbp	***Podoviridae***
	Tails short, DNA c. 50 kbp	***Siphoviridae***
11.	Virions icosahedral	***Microviridae***
	Virions rod-shaped	***Inoviridae***
12.	RNA double-stranded	***Cystoviridae***
	RNA single-stranded	***Leviviridae***
13.	Genome of DNA	14
	Genome of RNA	27
14.	Virion DNA double-stranded	15
	Virion DNA single-stranded	26
15.	DNA > 20 kbp	16
	DNA < 20 kbp	23
16.	Virions with lipid-containing envelopes	17
	Virions not enveloped or infective without an envelope	20
17.	Virions containing > 1 fusiform or cylindrical nucleocapsid and multiple DNA molecules	***Polydnaviridae***
	Virions containing a single DNA molecule	18

18.	Virions isometric, 120 - 200 nm in diameter	***Herpesviridae***
	Virions not isometric	19
19.	Virions brick-shaped, 300-450 nm x 170-260 nm	***Poxviridae***
	Virions rod-shaped, single nucleocapsids, 40-60 x 200-400 nm	***Baculoviridae***
20.	Virion diameter > 100 nm, DNA > 300kbp	21
	Virion diameter < 100 nm, DNA < 300 kbp	22
21.	Host an animal	***Iridoviridae***
	Host an alga	***Phycodnaviridae***
22.	Host a vertebrate	***Adenoviridae***
	Host a fungus	***Rhizidiovirus***
23.	DNA 5 to 8 kbp, lacking single - stranded discontinuities	***Papovaviridae***
	DNA < 5 kbp or > 7 kbp, containing single - stranded discontinuities	24
24.	DNA < 5 kbp, host a vertebrate	***Hepadnaviridae***
	DNA > 7 kbp, host a plant	25
25.	Virions isometric	***Caulimovirus***
	Virions bacilliform	**Commelina yellow mottle virus**
26.	Host an animal, DNA linear	***Parvoviridae***
	Host a plant, DNA circular	***Geminivirus***
27.	RNA double-stranded	28
	RNA single-stranded	32
28.	Virions contain > 9 RNA segments	***Reoviridae***
	Virions contain < 9 RNA segments	29
29.	RNA in 1 segment, host a fungus	***Totiviridae***
	RNA in > 1 segment	30
30.	Virions contain 2 RNAs, host an animal	***Birnaviridae***
	Host not an animal	31
31.	Virions contain 2 or more RNAs, host a plant	***Cryptovirus***
	Virions contain 3 RNAs, host a fungus	***Partitiviridae***
32.	RNA in 1 segment	33
	RNA in > 1 segment	57
33.	Virions with a lipid-containing envelope	34
	Virions lack an envelope	40
34.	RNA c. 9 kb, virions isometric, > 70 nm in diameter replication involves reverse transcription	***Retroviridae***
	RNA > 10 kb, no DNA phase during replication	35
35.	RNA negative sense	36
	RNA positive sense	38
36.	Virions c. isometric, RNA > 15 kb	***Paramyxoviridae***
	Virions not isometric	37

37.	RNA 10 to 14 kb, virions bacilliform	***Rhabdoviridae***
	RNA >15 kb, virions filamentous and/or pleomorphic	***Filoviridae***
38.	RNA > 20 kb, virions pleomorphic	***Coronaviridae***
	RNA < 15 kb	39
39.	Virion diameter 50 to 70 nm, sub-genomic RNA formed during multiplication	***Togaviridae***
	Virion diameter 40 to 50 nm, no sub-genomic RNA	***Flaviviridae***
40.	No sub-genomic RNA formed during multiplication	41
	Sub-genomic RNA formed	42
41.	Virions filamentous, host a plant	***Potyvirus***
	Virions isometric, host an animal	***Picornaviridae***
42.	Host an animal	43
	Host a plant	44
43.	Host a vertebrate	***Caliciviridae***
	Host an insect	***Tetraviridae***
44.	Virions rod-shaped, ca. 300 nm long	***Tobamovirus***
	Virions not rod-shaped	45
45.	Virions filamentous	46
	Virions isometric	49
46.	Virions > 700 nm in length	***Closterovirus***
	Virions < 700 nm in length	47
47.	Virions < 600 nm, coat protein < 30K	***Potexvirus***
	Virions > 600 nm	48
48.	Virions with prominent banding, coat protein c. 27K	***Capillovirus***
	Virions without banding, coat protein c. 32K	***Carlavirus***
49.	RNA > 9 kb, >1 coat protein	50
	RNA < 9 kb, 1 coat protein	51
50.	Virus transmitted by aphids	**Parsnip yellow fleck virus**
	Virus transmitted by leafhoppers	**Maize chlorotic dwarf virus**
51.	Virus transmitted propagatively by leafhoppers	***Marafivirus***
	Virus not leafhopper-transmitted	52
52.	Virus not mechanically transmissible, persistently aphid-transmissible	***Luteovirus***
	Virus transmitted mechanically, not by aphids	53
53.	RNA 6 kb, coat protein 20K	***Tymovirus***
	RNA < 6 kb, coat protein > 20K	54
54.	Coat protein > 35K	55
	Coat protein < 35K	56
55.	RNA 4 kb, coat protein 38K and encoded by the 3'- most ORF of the genome	***Carmovirus***
	RNA 4.7 kb, coat protein 43K not encoded by the 3'-most ORF of the genome	***Tombusvirus***

56.	RNA has a VPg, virus insect-transmitted, usually by beetles	***Sobemovirus***
	RNA does not have a VPg, virus fungus-transmitted	***Necrovirus***
57.	RNA negative sense or ambisense	58
	RNA positive sense	61
58.	Virions filamentous, host a plant	***Tenuivirus***
	Virions isometric	59
59.	RNA in > 6 segments	***Orthomyxoviridae***
	RNA in < 6 segments	60
60.	Virions contain 3 RNA segments	***Bunyaviridae***
	Virions contain 2 virus-specific RNAs + 3 host RNAs	***Arenaviridae***
61.	Virions rod-shaped	62
	Virions not rod-shaped	64
62.	Virions > 20 nm in diameter, larger virions c. 200 nm long, nematode-transmitted	***Tobravirus***
	Virions < 20 nm in diameter	63
63.	Some virions > 250 nm, largest RNA > 5 kb, virus fungus-transmitted	***Furovirus***
	Virions < 200 nm, largest RNA < 5 kb	***Hordeivirus***
64.	RNA in 2 segments	65
	RNA in > 2 segments	70
65.	RNA < 7 kb in total	66
	RNA > 7 kb in total	68
66.	Largest RNA < 4 kb, host an animal	***Nodaviridae***
	Largest RNA > 4 kb, host a plant	67
67.	Coat protein c. 40K, smaller RNA c. 1.5 kb	***Dianthovirus***
	Coat protein c. 22K, smaller RNA > 3 kb	**Pea enation mosaic virus**
68.	Virus aphid-transmitted, coat proteins 43K and 27K	***Fabavirus***
	Virus not transmitted by aphids	69
69.	Coat proteins 42K and 22K, virus beetle-transmitted	***Comovirus***
	Coat protein often 57K, sometimes > 1 species, virus usually nematode-transmitted	***Nepovirus***
70.	Virions isometric, sedimenting as 1 component	71
	Virions not isometric, sedimenting as > 1 component	72
71.	Coat protein c. 20K, virus not aphid-transmitted	***Bromovirus***
	Coat protein > 24K, virus aphid-transmitted	**Cucumovirus**
72.	Some virions bullet-shaped, virus aphid-transmitted	**Alfalfa mosaic virus**
	Virions slightly pleomorphic, virus not aphid-transmitted	***Ilarvirus***

Taxonomic status	English vernacular name	International name
	SUBFAMILIES	
	Poxviruses of vertebrates	*Chordopoxvirinae*
	Poxviruses of insects	*Entomopoxvirinae*

SUBFAMILY	POXVIRUSES OF VERTEBRATES	***CHORDOPOXVIRINAE***

PROPERTIES OF THE VIRUS PARTICLE

Pleomorphic, brick-shaped or ovoid virions chemically like other members of the family ***Poxviridae***. At least 20 major antigens in virion, one of which cross-reacts with most vertebrate poxviruses. Extensive serological cross-reactivity within each genus of vertebrate poxviruses, less obvious in *Avipoxvirus*.

GENERA

English vernacular name	International name
Vaccinia subgroup	*Orthopoxvirus*
Orf subgroup	*Parapoxvirus*
Fowlpox subgroup	*Avipoxvirus*
Sheep pox subgroup	*Capripoxvirus*
Myxoma subgroup	*Leporipoxvirus*
Swinepox subgroup	*Suipoxvirus*
Molluscum subgroup	*Molluscipoxvirus*
Yaba/Tanapox subgroup	*Yatapoxvirus*

GENUS	VACCINIA SUBGROUP	***ORTHOPOXVIRUS***
TYPE SPECIES	VACCINIA VIRUS	—

PROPERTIES OF THE VIRUS PARTICLE

Morphology Virions brick-shaped, 250-300 nm x 200 nm x 250 nm.

Physicochemical properties Infectivity of virions is ether-resistant.

Nucleic acid Single linear molecule of dsDNA, ≈ 185 kbp, G+C ≈ 36%, with complementary strands covalently linked at the ends and with sets of tandemly repeated sequences within terminal inverted repetitions.

Protein Virions released by exocytosis via the Golgi have a single membrane envelope containing several viral proteins. A glycoprotein haemagglutinin is produced in infected cells and becomes incorporated into the host cell Golgi and plasma membrane, thereby into the envelope of virions released by exocytosis and by extrusion via microvilli. Virions released by cell lysis lack the cell membrane-derived envelope and are also infectious. The envelope encloses an external coat, a lipoprotein tegument assembled from proteins and host cell lipids; at the late stage of morphogenesis tubule protein(s) are added. The external coat encases a biconcave core and lateral bodies that are in the concavities of the core. Various enzymes are located within the external coat The core encloses the genome. Proteinaceous A-type cytoplasmic inclusions made by some members may encase infectious virions, depending on strain.

Replication

Morphogenesis of immature to mature virus particles occurs in type-B cytoplasmic inclusions (viroplasms). Different species undergo genetic recombination and exhibit extensive serological cross-reactivity and nucleic acid homology. Enveloped virions contain distinct neutralization sites compared to lytically released virions and virus particles in A-type inclusions.

Biological Aspects

Host range

Monkeypox, cowpox, and vaccinia (smallpox vaccine) have wide vertebrate host range - others narrow, some limited to a single animal host in nature. All members have wide cell culture host range.

Other Members

Camelpox (camels)
Cowpox (felines, bovines, humans; rodent reservoir suspected)
Ectromelia (mousepox - isolated only from captive mice)
Monkeypox (humans, monkeys, African arboreal squirrel reservoir suspected)
Raccoonpox (North American raccoon *Procyon lotor*)
Taterapox (one isolate, African gerbil *Tatera kempi*)
Variola (humans)
Volepox (from *Microtus californicus* and *Peromyscus truei*)
Vaccinia subspecies:
- Buffalopox (milking buffalos, cattle, humans)
- Rabbitpox (isolated only from captive rabbits)

Probable members

Uasin Gishu disease (African horses)

Taxonomic status	English vernacular name	International name
GENUS	ORF SUBGROUP	***PARAPOXVIRUS***
TYPE SPECIES	ORF VIRUS	—

PROPERTIES OF THE VIRUS PARTICLE

Morphology Virions ovoid, 220-300 nm x 140-170 nm; external coat and filaments are thicker than in vaccinia virions and appear as a regular cross-hatched spiral coil of a continuous single thread.

Physicochemical properties Infectivity is ether-sensitive.

Nucleic acid One molecule dsDNA, 130-150 kbp, G+C ≈ 64%.

Antigenic properties Members show serological cross-reactivity. Hemagglutinin rare, reported for orf and contagious ecthyma isolates.

BIOLOGICAL ASPECTS

Viruses of ungulates may infect humans, sealpox might infect dog and cat.

OTHER MEMBERS

Orf virus, *synonyms* - contagious pustular dermatitis (CPD), contagious ecthyma (sheep, goats, musk oxen, humans).
Stomatitis papulosa (bovines), *synonyms* - bovine papular stomatitis (BPS).
Pseudocowpox virus (bovines), *synonyms* - milkers' nodule (humans), paravaccinia (humans, bovines).

Probable members

Parapoxvirus of New Zealand red deer (30-50% DNA homology with orf)
Ausduk disease, *synonym* -camel contagious ecthyma
Chamois contagious ecthyma
Sealpox

GENUS	FOWLPOX SUBGROUP	***AVIPOXVIRUS***
TYPE SPECIES	FOWLPOX VIRUS	—

PROPERTIES OF THE VIRUS PARTICLE

Morphology Virions brick-shaped, 330 nm x 280 nm x 200 nm.

Physicochemical properties Infectivity is usually ether-resistant.

Taxonomic status	English vernacular name	International name

Nucleic acid	One molecule dsDNA, ≈ 260 kbp.
Protein	Infected cells do not usually produce haemagglutinin.
Lipid	Certain members produce A-type inclusions with much lipid.
Antigenic properties	Members show serological cross-reactivity but cross-protection is variable.

BIOLOGICAL ASPECTS

Host range	Viruses of birds, mammalian cell infection is abortive.
Transmission	Usually mechanical by arthropods.

OTHER MEMBERS

Canarypox
Juncopox
Pigeonpox
Psittacinepox
Quailpox
Sparrowpox
Starlingpox
Turkeypox

Probable members

Peacockpox
Penguinpox
Mynahpox

GENUS	SHEEP POX SUBGROUP	***CAPRIPOXVIRUS***
TYPE SPECIES	SHEEP POX VIRUS	—

PROPERTIES OF THE VIRUS PARTICLE

Morphology	Virions brick-shaped, 300 nm x 270 nm x 200 nm.
Physicochemical properties	Infectivity is ether-sensitive.
Nucleic acid	One molecule dsDNA, 150-160 kbp.
Antigenic properties	Members show serological cross-reactivity.

Taxonomic status	English vernacular name	International name

BIOLOGICAL ASPECTS

Host range Viruses of ungulates (sheep, goats, cattle).

Transmission Usually mechanical by arthropods, also by contact, fomites and airborne.

OTHER MEMBERS

Sheeppox
Goatpox
Lumpy skin disease, *Synonym*-Neethling

GENUS	MYXOMA SUBGROUP	***LEPORIPOXVIRUS***
TYPE SPECIES	MYXOMA VIRUS	—

PROPERTIES OF THE VIRUS PARTICLE

Morphology Virions brick-shaped, 250-300 nm x 250 nm x 200 nm.

Physicochemical properties Infectivity is ether-sensitive.

Nucleic acid One molecule dsDNA, ≈ 160 kbp, G+C ≈ 40%.

Antigenic properties Members show serological cross-reactivity.

BIOLOGICAL ASPECTS

Host range Viruses of leporids and squirrels, extended range in cell cultures.

Transmission Usually mechanical by arthropods. Causes localized benign tumors in natural hosts, but myxoma viruses cause severe generalized disease in European rabbits.

OTHER MEMBERS

Hare fibroma
Rabbit (Shope) fibroma
Squirrel fibroma

Probable members

Malignant rabbit fibroma (natural history uncertain, apparently a myxoma-fibroma recombinant).

Taxonomic status	English vernacular name	International name
GENUS	SWINEPOX SUBGROUP	***SUIPOXVIRUS***
TYPE SPECIES	SWINEPOX VIRUS	—

PROPERTIES OF THE VIRUS PARTICLE

Morphology Virions brick-shaped, size like vaccinia virus.

Nucleic acid One molecular dsDNA, ≈ 170 kbp.

BIOLOGICAL ASPECTS

Host range Virus of swine, genus apparently contains one distinct member.

Taxonomic status	English vernacular name	International name
GENUS	*MOLLUSCUM CONTAGIOSUM* SUBGROUP	***MOLLUSCIPOXVIRUS***
TYPE SPECIES	*MOLLUSCUM CONTAGIOSUM* VIRUS	—

PROPERTIES OF THE VIRUS PARTICLE

Morphology Virions brick-shaped, 320 x 250 nm.

Physicochemical properties Buoyant density in CsCl ≈ 1.288 g/cm^3.

Nucleic acid One molecule of dsDNA, ≈ 188 kbp, G+C ≈ 60%, 53.02 ± 1.87 µm in length with covalently closed ends and terminal inverted repetitions.

Antigenic properties Antigenically distinct from other chordopoxviruses. Two virus types are recognised with different DNA restriction patterns.

BIOLOGICAL ASPECTS

Host range Humans.

Transmission Transmission in children by direct contact ; transmission in young adults by sexual contact. Lesions contain enlarged cells with intracytoplasmic inclusions.

Taxonomic status	English vernacular name	International name
GENUS	YABA/TANAPOX VIRUS SUBGROUP	***YATAPOXVIRUS***
TYPE SPECIES	YABA MONKEY TUMOR VIRUS	—

PROPERTIES OF THE VIRUS PARTICLE

Morphology Virions, brick-shaped like vaccinia, double-enveloped virions common (Tanapox).

Nucleic acid One molecule of dsDNA, ≈ 146 kbp, G+C ≈ 33%, with covalently closed ends.

BIOLOGICAL ASPECTS

Host range Monkeys, baboons, humans (accidental infection), rabbits (experimentally). Mature lesions in primates are epidermal histiocytomas (tumor-like masses of mononuclear cells).

OTHER MEMBERS

Tanapox virus (humans, monkeys)
Yaba-like disease virus (monkeys) - Tanapox subspecies.

SUBFAMILY	POXVIRUS OF INSECTS	***ENTOMOPOXVIRINAE***

PROPERTIES OF THE VIRUS PARTICLE

Pleomorphic, brick-shaped or ovoid virions 170-250 nm x 300-400 nm; chemically like other members of the family. Virions contain at least 4 enzymes found in vaccinia virus. Virions of several morphological types with globular surface units that give a mulberry-like appearance; some have one lateral body, others two. No serological relationships between viruses of the probable genera or with vertebrate poxviruses. Replicate in cytoplasm of cells of insects in hemocytes or adipose cells, few insect cell cultures support virus growth. Mature virions usually occluded in crystalline proteinaceous occlusion bodies. Subdivision into probable genera based on virion morphology, host range, and genome molecular weight of a few isolates.

PROBABLE GENERA

Genus A	*Entomopoxvirus* A
Genus B	*Entomopoxvirus* B
Genus C	*Entomopoxvirus* C

Taxonomic status	English vernacular name	International name
PROBABLE GENUS	POXVIRUS OF *COLEOPTERA*	***ENTOMOPOXVIRUS A***
TYPE SPECIES	POXVIRUS OF *MELOLONTHA MELOLONTHA*	—

PROPERTIES OF THE VIRUS PARTICLE

Morphology Virions ovoid, 450 nm x 250 nm, with one lateral body and unilateral concave core; globular surface units 22 nm in diameter.

Nucleic acid One molecule dsDNA, 260-370 kbp.

OTHER MEMBERS

Partial listing of members isolated from the following:

Coleoptera:
Anomala cuprea
Aphodius tasmaniae
Demodema boranensis
Dermolepida albohirtum
Figulus sublaevis
Geotrupes sylvaticus

Taxonomic status	English vernacular name	International name
PROBABLE GENUS	POXVIRUS OF *LEPIDOPTERA* AND *ORTHOPTERA*	***ENTOMOPOXVIRUS B***
TYPE SPECIES	POXVIRUS OF *AMSACTA MOOREI* (LEPIDOPTERA)	—

PROPERTIES OF THE VIRUS PARTICLE

Morphology Virions ovoid, 350 nm x 250 nm, with a sleeve-shaped lateral body and cylindrical core; globular surface units 40 nm in diameter.

Nucleic acid One molecule dsDNA, ≈ 225 kbp; G+C ≈ 18.5%.

Protein Infected cells synthesise a 116 kDa occlusion protein monomer.

Taxonomic status	English vernacular name	International name

OTHER MEMBERS

Partial listing of members isolated from the following:

Lepidoptera:
Acrobasis zelleri
Choristoneura biennis
Choristoneura conflicta
Choristoneura diversuma
Chorizagrotis auxiliaris
Operophtera brumata

Orthoptera:
Arphia conspersa
Locusta migratoria
Melanoplus sanguinipes
Oedaleus senugalensis
Schistocerca gregaria

PROBABLE GENUS	POXVIRUS OF *DIPTERA*	***ENTOMOPOXVIRUS C***
TYPE SPECIES	POXVIRUS OF *CHIRONOMUS LURIDUS* (*DIPTERA*)	—

PROPERTIES OF THE VIRUS PARTICLE

Morphology Virions brick-shaped, 320 nm x 230 nm x 110 nm, with two lateral bodies and biconcave core.

Nucleic acid One molecule dsDNA, 250-380 kbp.

OTHER MEMBERS

Partial listing of similar members isolated from the following:

Diptera:
Aedes aegypti
Camptochironomus tentans
Chironomus attenuatus
Chironomus plumosus
Goeldichironomus holoprasimus

OTHER MEMBERS OF FAMILY *POXVIRIDAE*

Not yet allocated to genera, little information available:

Albatrosspox (probably *Avipoxvirus*)
Cotia (mosquito transmitted to rodent, reservoir unknown)
Embu (mosquito transmitted to rodent, reservoir unknown)

Marmosetpox (virion morphology like *Yatapoxvirus*)
Marsupialpox (Australian 'quokkas')
Mule deer *poxvirus* (USA - *Odocoileus hemionus*, probably *Capripoxvirus*)
Volepox (USSR - *Microtus oeconomus,* Canada-*Microtus pennsylvanicus*)
Skunk poxvirus (USA - *Mephitis mephitis,* probably *Orthopoxvirus*).

Derivation of Names

pox: from old English *poc, pocc-*, plural of pock 'pustule, ulcer'
ortho: from Greek *orthos*, 'straight, correct'
avi: from Latin *avis*, 'bird'
capri: from Latin *caper, capri*, 'goat'
lepori: from Latin *lepus, leporis*, 'hare'
para: from Greek *para*, 'by side of'
entomo: from Greek *entomon*, 'insect'
sui: from Latin *sus*, 'swine'
molluscum: from Latin *molluscum*, 'clam, snail'

REFERENCES

Baxby, D.: Poxviruses. *In* Brown, F. (ed.), Principles of Bacteriology, Virology and Immunity, Vol. 4, 7th edn., (Arnold, London, 1984).

Bugert, J.; Rosen-Wolf, A.; Darai, G.: Genomic characterization of *Molluscum contagiosum* virus type I: Identification of repetitive DNA sequences in the viral genome. Virus Genes *3*:159-173 (1989).

Cooley, A.J.; Reinhard, M.K.; Gross, T.L.; Fadok, V.A.; Levy, M.: *Molluscum contagiosum* in a horse with granulomatous enteritis. J. Comp. Pathol. *97:* 29-34 (1987).

Dales, S.; Pogo, B.G.T.: Biology of Poxviruses. (Springer, Wien, New York, 1981).

Descoteaux, J-P.; Mihok, S.: Serologic study of the prevalence of murine viruses in a population of wild meadow voles (*Microtus pennsylvanicus*). J. Wildlife Dis. *22:* 314-319 (1986).

Dumbell, K.R.: *Poxviridae. In* Porterfield, J. (ed.), Andrews' Viruses of Vertebrates. 5th edn., pp. 395-427 (Bailliere Tindall, London, 1989).

Earl, P.L.; Moss, B.: Vaccinia virus. *In* "Genetic Maps 1989", pp. 1.138-1.148 (Cold Spring Harbor, New York, 1989).

Esposito, J.J.; Palmer, E.L.; Borden, E.C.; Harrison, A.K.; Obijeski, J.F.; Murphy, F.A.: Studies on the poxvirus Cotia. J. gen. Virol. *47*:37-46 (1980).

Fenner, F.; Henderson, D.A.; Arita, I.; Jezek, Z.; Ladnyi, I.D.: Smallpox and Its Eradication. (The World Health Organization, Geneva, 1988).

Fenner, F.; Nakano, J.H.: *Poxviridae*: The Poxviruses. *In* Lennette, E.H.; Halonen, P.; Murphy, F.A. (eds.), Laboratory Diagnosis of Infectious Diseases: Principles and Practice. Vol. 2, Viral, Rickettsial, and Chlamydial Diseases, pp. 177-207 (Springer, Berlin, Heidelberg, New York, Tokyo, 1988).

Fenner, F.; Wittek, R.; Dumbell, K.R.: The Orthopoxviruses (Academic Press, New York, 1989).

Gershon, P.D.; Black, D.N.: A comparison of the genomes of *Capripoxvirus* isolates of sheep, goats, and cattle. Virology *164:* 341-349 (1988).

Goebel, S.J.; Johnson, G.P.; Perkus, M.E.; Davis, S.W.; Winslow, J.P.; Paoletti, E.: The complete DNA sequence of Vaccinia virus. Virology *179*:247-266 and 517-563 (1990).

Gough, A.W.; Barsoum, N.J.; Gracon, S.T.; Mitchel, L.; Sturgess, J.M.: Poxvirus infection in a colony of common marmosets (*Callithrix jacchus*). Lab. Animal Sci. *32*:87-90 (1982).

Granados, R.R.: *Entomopoxvirus* infections in insects. *In* Davidson, I. (ed.), Pathogenesis of Invertebrate Microbial Diseases, pp. 101-129 (Allenheld Osmu, Totowa, New Jersey, 1981).

Jezek, Z.; Arita, I.; Szczeniowski, M.; Paluku, K.M.; Ruti, K.; Nakano, J.H.: Human Tanapox in Zaire: Clinical and epidemiological observations on cases confirmed by laboratory studies. Bull. W.H.O. *63:* 1027-1035 (1985).

Jezek, Z.; Fenner, F.: Human monkeypox. Monographs in Virology, Vol. 17 (Karger, Basel, 1988).

Kaminjolo, J.S.; Winqvist, G.: Histopathology of skin lesions in Uasin Gishu skin disease in horses. J. Comp. Pathol. *85:* 391-395 (1975).

Kilpatrick, D.; Rouhandeh, H.: The analysis of Yaba monkey tumor virus DNA. Virus Res. *7:* 151-157 (1987).

Knight, J.C.; Novembre, F.J.; Brown, D.R.; Goldsmith, C.S.; Esposito, J.J.: Studies on Tanapox virus. Virology *172:* 116-124 (1989).

McFadden, G.: Poxviruses of Rabbits. *In* Darai, G. (ed.), Virus Diseases in Laboratory and Captive Animals, pp. 37-61 (Martinus Nijhoff, Boston ,1988).

McKenzie, R.A.; Fay, F.R.; Prior, H. C.: Poxvirus infection of the skin of an eastern grey kangaroo. Aust. Vet. J. *55:* 188-190 (1979).

Moens, Y.; Kombe, A.H.: *Molluscum contagiosum* in a horse. Equine Vet. J. *20:* 143-145 (1988).

Moss, B.: *Poxviridae* and Their Replication. *In* Fields, B.N.; Knight, J.C. (eds.), Virology, Vol. 2, 2nd edn., pp. 2079-2111 (Raven Press, New York, 1990).

Nakano, J.H.; Esposito, J.J.: Poxviruses. *In* Schmidt, N.J.; Emmons, R.W. (eds.), Diagnostic Procedures for Viral, Rickettsial, and Chlamydial Infections, 6th edn., pp. 224-265 (American Public Health Association, Washington, D.C., 1989).

Porter, C.D.; Muhlemann, M.F.; Cream, J.J.; Archard, L.C.: *Molluscum contagiosum*: Characterization of viral DNA and clinical features. Epidemiol. Infect. *99:* 563-567 (1987).

Robinson, A.J.; Lyttle, D.J.: Parapoxviruses: Their biology and potential as recombinant vaccines. *In* Bins, M.; Smith, G., (eds.), Poxviruses as vaccine vectors (CRC Press, Boca Raton, Fl.) *In press.*

Roslaykov, A.A.: Comparative ultrastructure of viruses of camelpox, pox-like disease of camels (Ausduk) and contagious ecthyma of sheep. Voprosy Virusologii *7*:26-30 (1972).

Rouhandeh, H.; Kilpatrick, D; Vafai, A.: The molecular biology of Yaba tumor poxvirus: Analysis of lipids, proteins and DNA. J. gen. Virol. *62*:207-218 (1982).

Scholz, J.; Rosen-Wolff, A.; Bugert, J.; Reisner, H.; White, M.I.; Darai, G.; Postlethwaite, R.: Epidemiology of *molluscum contagiosum* using genetic analysis of the viral DNA. J. Med. Virol. *27:* 87-90 (1989).

Tripathy, D.N.: Pox. *In* Calnek, B.W., Barnes, H.J., Beard, C.W., Reld, W.M., Yoder, H.W. (eds.), Diseases of Poultry., 9th edn., pp. 583-596 (Iowa State University Press, Ames, Iowa, 1990).

Tripathy, D.N.; Hanson, L.E.; Crandall, R.A.: Poxviruses of Veterinary Importance: Diagnosis of Infections. *In* Kurstak, E.; Kurstak, C., (eds.), Comparative Diagnosis of Viral Diseases. Vol. 3, pp. 267-346 (Academic Press, New York, 1981).

Williams, E.S.; Becerra, V.M.; Thorne, E.T.; Graham, T.J.; Owens, M.J.; Nunmaker, C.E.: Spontaneous poxviral dermatitis and keratoconjunctivitis in free-ranging mule deer (*Odocoileus hemionus*) in Wyoming. J. Wildlife Dis. *21:* 430-432 (1985).

Wilson, T.M.; Dykes, R.W.; Tsai, K.S.: Pox in young captive harbor seals. JAVMA *161*:611-617 (1972).

Winterfield, R.W.; Reed, W.: Avian Pox: Infection and immunity with quail, psittacine, fowl, and pigeon pox viruses. Poultry Science *64*:65-70 (1985).

Wokatsch, R.: Vaccinia viruses and Variola viruses. *In* Majer, M.; Plotkin, S.A. (eds.), Strains of Human Viruses, pp. 240-268 (Karger, New York, 1972)

Taxonomic status	English vernacular name	International name
FAMILY	HERPESVIRUS GROUP	***HERPESVIRIDAE***

Reported by B. Roizman

PROPERTIES OF THE VIRUS PARTICLE

Morphology

The virion, 120-200 nm in diameter, consists of 4 structural components. The core consists of a fibrillar spool on which the DNA is wrapped. The ends of the fibers are anchored to the underside of the capsid shell. The capsid, 100-110 nm in diameter, has 162 capsomeres arranged as an icosahedron. (150 hexameric and 12 pentameric capsomeres). Capsomeres are hexagonal in cross-section with a hole running half-way down the long axis. The tegument surrounding the capsid consists of globular material which is frequently asymmetrically distributed and may be variable in amount. The envelope, a bilayer membrane surrounding the tegument, has surface projections. The intact envelope is impermeable to negative stain.

Physicochemical properties

MW > 1,000 x 10^6; buoyant density in CsCl = 1.20 - 1.29 g/cm^3.

Nucleic acid

One molecule of dsDNA, 120-220 kbp, G+C ≈ 35-75%

Protein

More than 20 structural proteins, MW = 12,000 - > 222,000.

Lipid

Probably variable; located in virion envelope.

Carbohydrate

Present, largely as glycoproteins in envelope.

Antigenic properties

The virion contains several surface glycoproteins. Neutralizing antibody reacts with major viral envelope glycoproteins. An Fc receptor may be present in the envelope.

Effect on cells

In the absence of replication, fusion and agglutination occur rarely or under very special conditions

REPLICATION

The viral envelope attaches to receptors on the plasma membrane of the host cell, fuses with the membrane, and releases the capsid which is transported to the nuclear pore. A DNA-protein complex is transported into the nuclear pore where the DNA circularizes.

Taxonomic status	English vernacular name	International name

Viral DNA is transcribed in the nucleus and the mRNA is translated in the cytoplasm. Viral DNA is replicated in the nucleus. Unit length DNA is cleaved from concatemers and spooled into preformed, immature capsids which mature by acquisition or processing of proteins that bind to the surface of the capsid.

The ability to infect cells is acquired as capsids are enveloped by budding through the inner lamella of the nuclear membrane. The virus accumulates in the perinuclear space and cisternae of the endoplasmic reticulum. Virus particles are released by transport to the cell surface through modified endoplasmic reticulum in structures bounded by cytoplasmic membranes.

BIOLOGICAL ASPECTS

Host range

Each virus has its own natural and experimental host range. Both warm and cold-blooded vertebrates and invertebrates are hosts to herpesviruses. Some herpesviruses have been reported to induce neoplasia both in their natural hosts and in experimental animals. In cell culture, some herpesviruses have been reported to convert cell strains into continuous cell lines which may cause invasive cancers in appropriate experimental hosts.

Transmission

For many herpesviruses, transmission is by contact between moist mucosal surfaces. Some herpesviruses can be transmitted transplacentally, intrapartum, via breast milk, or by transfusions. Some, are probably also transmitted by airborne and waterborne routes. Herpesviruses may remain latent in their primary hosts for the lifetime of those hosts; cells harboring latent virus may vary depending on the virus.

SUBFAMILIES

Herpes simplex virus group	*Alphaherpesvirinae*
Cytomegalovirus group	*Betaherpesvirinae*
Lymphoproliferative virus group	*Gammaherpesvirinae*

Taxonomic status	English vernacular name	International name
SUBFAMILY	HERPES SIMPLEX VIRUS	***ALPHAHERPESVIRINAE***

PROPERTIES OF THE VIRUS PARTICLE

Nucleic acid DNA = 120-180 kbp. The sequences from both or either terminus are present in an inverted form internally. The DNA packaged in virions may consist of two or four isomeric forms. Natural isolates may exhibit restriction endonuclease cleavage site polymorphism.

REPLICATION

Relatively short (< 24 h) replicative cycle.

BIOLOGICAL ASPECTS

Host range Variable, from very wide to very narrow.

Cytopathology Rapid spread of infection in cell culture results in mass destruction of susceptible cells. Establishment of carrier cultures of susceptible cells harboring nondefective genomes difficult to accomplish.

Latent infections Latent infections frequently but not exclusively demonstrated in sensory and autonomic ganglia.

GENERA

Human herpesvirus 1 group	*Simplexvirus*
Human herpesvirus 3 group	*Varicellovirus*

GENUS	HUMAN HERPESVIRUS 1 GROUP	***SIMPLEXVIRUS***
TYPE SPECIES	HUMAN (ALPHA) HERPESVIRUS 1 (HERPES SIMPLEX VIRUS 1)	—

PROPERTIES OF THE VIRUS PARTICLE

Nucleic acid DNA ≈ 152 kbp, G+C ≈ 67% . Sequences from both termini are repeated in an inverted form internally; virion DNA exists in 4 isomeric forms and shares > 50% of its sequences with human (alpha) herpesvirus 2 DNA under stringent hybridization conditions.

Protein > 30 structural proteins, including 8 glycoproteins.

Taxonomic status	English vernacular name	International name
Antigenic properties	At least 3 glycoproteins are capable of inducing neutralizing antibody.	

BIOLOGICAL ASPECTS

Host range Recovered in nature only from humans, but the virus may sustain itself and be transmitted in captive non-human primate colonies. Experimental host range, very wide.

OTHER MEMBERS

Human (alpha) herpesvirus 2 (herpes simplex virus 2)
Bovine (alpha) herpesvirus 2 (bovine mammillitis virus)

GENUS	HUMAN HERPESVIRUS 3 GROUP	***VARICELLOVIRUS***
TYPE SPECIES	HUMAN (ALPHA) HERPESVIRUS 3 (VARICELLA-ZOSTER VIRUS)	—

PROPERTIES OF THE VIRUS PARTICLE

Nucleic acid DNA ≈ 125 kbp. DNA sequences from one terminus are repeated in an inverted form internally. Virion DNA exists in 2 isomeric forms.

BIOLOGICAL ASPECTS

Host range Recovered only from humans. Experimental host range may vary from broad to highly restricted.

OTHER MEMBERS

Suid (alpha) herpesvirus 1 (pseudorabies virus)
Bovine (alpha) herpesvirus 1 (infectious bovine rhinotracheitis virus)
Equid (alpha) herpesvirus 1 (equine abortion virus)
Equid (alpha) herpesvirus 4 (respiratory infection virus)

Probable members

Cercopithecid herpesvirus 1 (B virus)
Equid herpesvirus 3 (coital exanthema)
Felid herpesvirus 1 (feline herpesvirus)
Canid herpesvirus (canine herpesvirus)

Taxonomic status	English vernacular name	International name
SUBFAMILY	CYTOMEGALOVIRUS GROUP	***BETAHERPESVIRINAE***

PROPERTIES OF THE VIRUS PARTICLE

Nucleic acid DNA = 180-250 kbp; G+C ≈ 56%. Sequences from either or both termini may be present in an inverted form internally.

REPLICATION

Relatively slow reproductive cycle (> 24 h). Slowly progressing lytic foci in cell culture. Enlargement of the infected cell *in vivo* and often *in vitro* (cytomegalia). Inclusion bodies containing DNA may be present in nuclei and cytoplasm late in infection. Carrier cultures easily established.

BIOLOGICAL ASPECTS

Host range *In vivo* - narrow, frequently restricted to the species or genus to which the host belongs. *In vitro* - replication may be restricted to a specific cell type, but exceptions exist.

Latent infections Possibly in secretory glands, lymphoreticular cells, and kidneys and other tissues.

GENERA

Human cytomegalovirus group	*Cytomegalovirus*
Murine cytomegalovirus group	*Muromegalovirus*

GENUS	HUMAN CYTOMEGALOVIRUS GROUP	***CYTOMEGALOVIRUS***
TYPE SPECIES	HUMAN (BETA) HERPESVIRUS 5 (HUMAN CYTOMEGALOVIRUS)	—

PROPERTIES OF THE VIRUS PARTICLE

Nucleic acid DNA ≈ 200 kbp.

BIOLOGICAL ASPECTS

Virus recovered only from human infections. Experimental host range narrow; grows best in human fibroblasts and less well in certain human lymphoblastoid cells.

Taxonomic status	English vernacular name	International name
GENUS	MURINE CYTOMEGALOVIRUS GROUP	***MUROMEGALOVIRUS***
TYPE SPECIES	MURID (BETA) HERPESVIRUS 1 (MOUSE CYTOMEGALOVIRUS)	—

PROPERTIES OF THE VIRUS PARTICLE

Nucleic acid DNA ≈ 200 kbp.

Possible members

Suid herpesvirus 2 (pig cytomegalovirus)
Equid herpesvirus 2
Murid herpesvirus 2 (rat cytomegalovirus)
Caviid herpesvirus 1 (guinea pig cytomegalovirus)

SUBFAMILY	LYMPHO-PROLIFERATIVE VIRUS GROUP	***GAMMAHERPESVIRINAE***

PROPERTIES OF THE VIRUS PARTICLE

Nucleic acid DNA ≈ 170 kbp; both ends of the molecule contain reiterated sequences that are not reiterated internally.

REPLICATION

Duration of the reproductive cycle is variable. All members replicate in lymphoblastoid cells, and some will also cause lytic infections in some types of epithelioid and fibroblastic cells. Viruses are specific for either B- or T-lymphocytes; in the lymphocyte, infection is frequently arrested at a prelytic stage, with persistence and minimum expression of the viral genome in the cell (latent infection), or at a lytic stage, causing death of the cell without production of complete virions. Latent infection is frequently demonstrated in lymphoid tissue.

BIOLOGICAL ASPECTS

Host range Narrow; experimental hosts usually limited to the same order as the host it naturally infects.

Cytopathology Variable.

GENERA

Human herpesvirus 4 group	*Lymphocryptovirus*
Ateline herpesvirus group	*Rhadinovirus*

Taxonomic status	English vernacular name	International name
GENUS	HUMAN HERPESVIRUS 4 GROUP	*LYMPHOCRYPTOVIRUS*
TYPE SPECIES	HUMAN (GAMMA) HERPESVIRUS 4 (EPSTEIN-BARR VIRUS)	—

PROPERTIES OF THE VIRUS PARTICLE

Nucleic acid DNA ≈ 170 kbp; some isolates lack as much as 15 kbp at specific sites.

BIOLOGICAL ASPECTS

Virus shows specificity for B-lymphocytes.

OTHER MEMBERS

Pongine herpesvirus 1 (Chimpanzee herpesvirus)
Cercopithecine herpesvirus 2 (Baboon herpesvirus)

Taxonomic status	English vernacular name	International name
GENUS	ATELINE HERPESVIRUS GROUP	*RHADINOVIRUS*
TYPE SPECIES	ATELINE HERPESVIRUS 2 (HERPESVIRUS ATELES)	—

PROPERTIES OF THE VIRUS PARTICLE

Nucleic acid DNA ≈ 105 kbp; stretch of quasi unique sequences low in GC content flanked at both ends with numerous repeat sequences of high GC content.

BIOLOGICAL ASPECTS

Host range variable but restricted to New World primates. Grows in a variety of cells in culture.

OTHER MEMBERS

Saimirine herpesvirus 1

Derivation of Name herpes: from Greek *herpes, herpetos*, 'creeping, crawling creature'; from nature of herpes febrilis lesions.

REFERENCES

Baer, R.; Bankier, A.T.; Biggin, M.D.; Deininger, P.L.; Farrell, P.J.; Gibson, T.J.; Hatfull, G.; Hudson, G.S.; Satchwell, S.C.; Seguin, C.; Tuffnell, P.S.; Barrell, B.G.: DNA sequence and expression of the B95-8 Epstein-Barr virus genome. Nature *310*:207-211 (1984).

Bornkamm, G.W.; Delius, H.; Fleckenstein, B.; Werner, F.-J.; Mulder, C.: Structure of Herpesvirus saimiri genomes: arrangement of heavy and light sequences in the M genome. J. Virol. *19*:154-161 (1976).

Buchman, T.G.; Roizman, B.: Anatomy of bovine mammillitis DNA. II. Size and arrangements of the deoxynucleotide sequences. J. Virol. *27*:239-254 (1978).

Buchman, T.G.; Roizman, B.; Nahmias, A.J.: Demonstration of exogenous genital reinfection with herpes simplex virus type 2 by restriction endonuclease fingerprinting of viral DNA. J. Infect. Dis. *140*:295-304 (1979).

Davison, A.J.; Scott, J.E.: The complete DNA sequence of Varicella-zoster virus. J. gen. Virol. *67*:1759-1816 (1986).

Fleckenstein, B.; Bornkamm, G.W.; Mulder, C.; Werner, F.-J.; Daniel, M.D.; Falk L.A.; Delius, H.: Herpesvirus ateles DNA and its homology with Herpesvirus saimiri nucleic acid. J. Virol. *25*:361-373 (1978).

Kilpatrick, B.A.; Huang, E.-S.: Human cytomegalovirus genome: partial denaturation map and organization of genome sequences. J. Virol. *24*:261-276 (1977).

McGeoch, D.J.; Dalrymple, J.M.; Davison, A.J.; Dolan, A.; Frame, M.C.; McNab, D.; Perry, L.J.; Scott, J.E. and Taylor, P.: The complete DNA sequence of the long unique region in the genome of herpes simplex virus type 1. J. gen. Virol. *69:* 1531-1574 (1988).

McGeoch, D.J.; Dolan, A.; Donald, S.; Brauer, D.H.K.: Complete DNA sequence of the short repeat region in the genome of herpes simplex virus type 1. Nuc. Acids Res. *14:* 1727-1745 (1986).

McGeoch, D.J.; Dolan, A.; Donald, S.; Rixon, F.J.: Sequence determination and genetic content of the short unique region in the genome of herpes simplex virus type 1. J. Mol. Biol. *181:* 1-13 (1985).

Mosmann, T.R.; Hudson, J.B.: Some properties of the genome of murine cytomegalovirus (MCV). Virology *54*:135-149 (1973).

Roizman, B,: *Herpesviridae*: a brief introduction. *In* Fields, B.N.; Knipe, D.M., (eds.), Virology, Vol. 2, 2nd edn., pp. 1787-1794, (Raven Press, New York, 1990).

Roizman, B.; Sears, A.E.: Herpes simplex viruses and their replication. *In* Fields, B.N.; Knipe, D.M., (eds.), Virology, Vol. 2, 2nd edn., pp. 1795-1841 (Raven Press, New York, 1990).

Roizman, B.; Batterson, W.: The replication of herpesviruses. *In* Fields, B.N.; Knipe, D.M.; Chanock, R.M.; Roizman, B.; Melnick, J.L.; Shope, R.E. (eds), Virology, pp. 497-526 (Raven Press, New York, 1985).

Roizman, B.; Carmichael, L.E.; Deinhardt, F.; de-The, G.; Nahmias, A.J.; Plowright, W.; Rapp, F.; Sheldrick, P.; Takahashi, M.; Wolf, K.: *Herpesviridae*: definition, provisional nomenclature and taxonomy. Intervirology *16*:201-217 (1981).

Ross, L.J.N.; Milne, B.; Biggs, P.M.: Restriction endonuclease analysis of Marek's disease virus DNA and homology between strains. J. gen. Virol. *64:* 2785-2790 (1983).

Taxonomic status	English vernacular name	International name

FAMILY *HEPADNAVIRIDAE*

Reported by C.R. Howard

PROPERTIES OF THE VIRUS PARTICLE

Morphology

Overall spherical particles, 40-48 nm in diameter with no surface projections. Outer 7 nm detergent-sensitive envelope surrounds an icosahedral nucleocapsid with 180 capsomeres arranged with T = 3 symmetry, made up of one major polypeptide species. The virion envelope is antigenically similar to the nucleic acid-free 22 nm lipoprotein particles (HBsAg) that occur naturally in the sera of infected patients.

Physicochemical properties

$S_{20w} \approx 280$; buoyant density in CsCl = 1.24-1.26 g/cm^3, (surface antigen particles without core = 1.18 g/cm^3). Unstable in acid pH; infectivity retained for 6 months at 30-32°C or 10 h at 60°C.

Nucleic acid

Single, circular molecule of partially ds and partially ssDNA; MW ≈ 1.6 x 10^6; $S_{20w} \approx$ 15S; G+C ≈ 48%. One strand (negative sense, complementary to mRNA) is full length (3.02-3.32 kb) and the other varies in length from 1.7 to 2.8 kb. Length of cloned DNA (fully double-stranded) ≈ 3.2 kbp.

The full length strand (negative strand) has a nick at a unique site 242 bp (or ca. 50 bp for *Avihepadnavirus*) from the 5' end of the short positive strand. Neither strand is a covalently closed circle The uniquely located 5'-ends of the two strands overlap by approximately 240 bp so that the circular configuration of the DNA is maintained by base pairing of cohesive ends. The 5' end of the full-length DNA strand has a covalently attached terminal protein. Virion core contains a DNA polymerase which uses the 3' end of the short DNA strand as a primer and repairs ss regions to make full-length (3.2 kbp) ds molecules.

Genome DNA has four ORF's, all orientated in the same direction on the long (minus) DNA strand. One ORF (the S-gene) specifies the major (MW ≈ 24 x 10^3) hepatitis B surface antigen (HBsAg) polypeptide and is preceded by a 'pre-S region' with two in-frame start codons (ATG) which are sites for initiation of the minor HBsAg polypeptides (MW ≈ 33 x 10^3, 36 x 10^3, 39 x 10^3 and 42 x 10^3). A second ORF (the C-gene) specifies the major (MW ≈ 22 x 10^3) hepatitis B core antigen (HBcAg)

Taxonomic status	English vernacular name	International name

polypeptide and is preceded by a short 'pre-C region' which can specify 29 amino acids. The longest ORF (the P-gene) covers 80% of the genome and overlaps the other three ORF's. It codes for the terminal protein, a reverse transcriptase, the viral DNA polymerase and an RNAse H. The fourth ORF, designated the X-gene, has been shown to possess transactivation properties in *in vitro* transfection experiments but its role in natural infection is unknown.

Protein

The virion coat is composed of following virus-coded proteins: S-proteins (P24, GP27), M-proteins (GP33, GP36), L-proteins (P39, GP42). The virion core is composed of one major protein, MW $\approx 22 \times 10^3$.

HBsAg particles composed of virion envelope material consist largely of S-proteins. The two major S polypeptides have MW = 24×10^3 (GP24) and 27×10^3 (GP27). They appear to have the same amino acid composition except that GP27 is glycosylated; The M-proteins GP33 and GP36 are composed of P24 with an additional 55 amino acids at the N-terminus, differ in the extent of glycosylation and bear the pre-S2 domain. The L-proteins P39 and GP42 contain a further ca. 120 amino acids, differ in glycosylation and bear the pre-S1 domain.

Enzymes: protein kinase, RNA- and DNA-dependant polymerase and RNase H. Other functional proteins: Terminal protein covalently attached to the 5'-end of the full-length DNA strand which may act as a primase.

Lipid

Demonstrated in 22 nm HBsAg particles and virions probably derived from the ER. The N-terminus of the L-proteins is myristoylated.

Carbohydrate

Demonstrated in 22 nm HBsAg particles and virions as N-linked glycans.

Antigenic properties

HBsAg, HBcAg, HBeAg antigens. HBeAg and HBcAg proteins share common epitopes but also contain epitopes which distinguish these two proteins from each other. Antigens involved in neutralization: HBsAg, HBsAg cross-reacts to a limited extent with the analogous antigens of woodchuck and ground squirrel viruses. No cross-reaction exists between HBsAg and the analogous antigen of DHBV. 'Pre-S region' may bear specific neutralization determinants. S proteins are sufficient to stimulate protective immunity.

HBcAg has been found to cross-react more strongly with the woodchuck virus core antigen than did the corresponding surface antigens.

Antigenic properties used for identification: At least 5 antigenic specificities may be found on HBsAg particles. A group determinant (a) is shared by all HBsAg preparations, and 2 pairs of subtype determinants (d, y and w, r) which are, for the most part, mutually exclusive and thus usually behave as alleles, have been demonstrated. Antigenic heterogeneity of the w determinants and additional determinants, such as q and x or g, have also been described. To date, 8 HBsAg subtypes have been identified, namely ayw, ayw_2, ayw_3, ayw_4, ayr, adw_2, adw_4 and adr. Unusual combinations of HBsAg subtype determinants such as awr, adwr, adyw, adyr and adywr, have been reported. The distribution of HBsAg subtypes occurs in uneven geographical distribution. The subtype specificity of HBsAg can be affected by mutations.

REPLICATION

Transcription: At least two major RNA transcripts are found in HBV-infected human liver. The two unspliced transcripts have different 5'-ends (both capped) and colinear 3'-ends (both polyadenylated) ending within the core protein gene. The shortest transcript (2.3 kb) is initiated in the middle of the pre-S region, and the greater than genome length longer transcript (3.4 kb) is initiated near the core gene start codon. The 2.3 kb transcript appears to be found in cells expressing HBsAg only, and both appear in cells supporting virus replication.

DNA replication: Current evidence indicates that virus replication involves the generation of a covalently closed circular DNA molecule followed by synthesis of a greater than genome length ($\approx$ 3.4 kb) plus strand RNA which is packaged in viral core particles and serves as a template for synthesis of the minus DNA strand (reverse transcription) using a protein primer. The minus DNA strand serves as template for plus DNA strand synthesis and is primed by transposition of the 5'-end of the plus strand RNA remaining after RNase H digestion from direct repeat 1 (DR1) to DR2. The plus DNA strand is incomplete in most core particles at the time of virion assembly and is released from the cell. Partially ssDNA of hepatitis B virus with properties of a replicative intermediate, has been detected in hepatocyte cytoplasm and similar material in HBV-infected liver extracts has been identified as negative sense ss DNA.

Taxonomic status	English vernacular name	International name

Site of maturation of full and empty hepatitis B virion cores appears to be in the nuclei and cytoplasm of infected hepatocytes, but no reliable information is available on the exact mechanism of hepadnavirus maturation. HBsAg has only been detected in cell cytoplasm and cytoplasmic membranes but HBcAg has been detected in both cytoplasm and nucleus (hepatitis B virus only). Integration is not required for replication.

Biological Aspects

Host range

The hepadnaviruses are exquisitely host specific. For example, the only known natural hosts of hepatitis B virus are humans, but chimpanzees and gibbons may be infected experimentally. Transmission of hepatitis B virus has also been reported in African monkeys, rhesus and wooly monkeys. Hepadnaviruses may cause acute and chronic hepatitis, cirrhosis, hepatocellular carcinoma, immune complex disease, polyarteritis, glomerulonephritis, infantile papular acrodermatitis and aplasmic anaemia. *In vitro,* hepatitis B virus, ground squirrel hepatitis B virus and woodchuck hepatitis B virus replication with production of infectious virus, has been demonstrated following transfection of tissue culture cells with cloned DNA. Replication of several hepadnaviruses has been achieved following inoculation of primary hepatocytes with serum containing virus.

Transmission

Vertical transmission has been clearly demonstrated in ducks and may occur in humans. Horizontal transmission can be by perinatal percutaneous, sexual and other routes of close contact, e.g. intravenous drug abuse, and by use of infected blood and blood products. Hepadnaviruses can survive on surfaces which may contact mucous membranes or open skin breaks, such as toothbrushes, baby bottles, toys, eating utensils, razors or hospital equipment such as respirators, endoscopes or laboratory equipment.

Although some populations of mosquitoes and bedbugs caught in Africa and the United States have been shown to contain HBsAg, there has been no direct demonstration of transmission to man by insect vectors.

Genera

Hepatitis B virus group	*Orthohepadnavirus*
Duck hepatitis B virus group	*Avihepadnavirus*

Taxonomic status	English vernacular name	International name
GENUS	HEPATITIS B VIRUS GROUP	***ORTHOHEPADNAVIRUS***
TYPE SPECIES	HEPATITIS B VIRUS (HBV)	—

PROPERTIES OF THE VIRAL PARTICLE

Spherical particles 40-42 nm diameter, internal nucleocapsid 27 nm diameter. Virion coat consists of L, M and S proteins.

BIOLOGICAL ASPECTS

Clinical manifestations include acute and chronic liver disease. Associated with the development of hepatocellular carcinoma.

OTHER MEMBERS

Woodchuck hepatitis B virus
Ground squirrel hepatitis B virus

Taxonomic status	English vernacular name	International name
GENUS	DUCK HEPATITIS GROUP	***AVIHEPADNAVIRUS***
TYPE SPECIES	DUCK HEPATITIS B VIRUS (DHBV)	—

PROPERTIES OF THE VIRUS PARTICLE

Spherical particle 46-48 nm diameter, internal nucleocapsid 35 nm diameter.

Plus strand DNA nearly full length.

Do not contain a separate X ORF. Virion coat possesses L protein, MW $\approx 36 \times 10^3$ and S protein MW $\approx 17 \times 10^3$. There is no M protein.

BIOLOGICAL ASPECTS

Clinical manifestations are rare.
Transmission is predominantly vertical.

OTHER MEMBERS

Heron hepatitis B virus (HHBV).

Derivation of Name

hepa: from hepatotropism
dna: from DNA (= the sigla for deoxyribonucleic acid)
ortho: from Greek *orthos*, 'straight, correct'
avi: from Latin *avis*, 'bird'.

REFERENCES

Galibert, F.; Mandart, E.; Fitoussi, F.; Tiollais, P.; Charnay, P.: Nucleotide sequence of the hepatitis B virus genome (subtype ayw) cloned in E. coli. Nature *281*:646-650 (1979).

Ganem, D.; Varmus, H.E.: The molecular biology of the hepatitis B viruses. Ann. Rev. Biochem. *56:* 651-693 (1987).

Howard, C.R.: The biology of hepadnaviruses. J. gen. Virol. *67:* 1215-1235 (1986).

Marion, P.L.; Robinson, W.S.: Hepadnaviruses: Hepatitis B and related viruses. Curr. Top. Microbiol. Immunol. *105*:99-121 (1983).

Schödel, F.; Sprengel, R.; Weimer, T.; Fernholz, D.; Schneider, R.; Will, H.: Animal hepatitis B viruses. Adv. Viral. Oncol. *8:* 73-102 (1989).

Summers, J.; Mason, W.S.: Replication of the genome of a hepatitis B-like virus by reverse transcription of an RNA intermediate. Cell *29:* 403-415 (1982).

Tiollais, P.; Pourcel, C.; Dejean, A.: The hepatitis B virus. Nature *317:* 489-495 (1985).

Taxonomic status	English vernacular name	International name
FAMILY	BACULOVIRUSES	***BACULOVIRIDAE***

Reported by M. Wilson

PROPERTIES OF THE VIRUS PARTICLE

Morphology

Virions consist of one or more rod-shaped electron-dense nucleocapsids enclosed within a single envelope. The nucleocapsids average 30-60 nm in diameter and 250-300 nm in length within the subfamily ***Eubaculovirinae***. The size of enveloped nucleocapsids within this subfamily is more variable. Virions of members of the subfamily ***Nudibaculovirinae*** are of one or other of two types. The virions of *Oryctes rhinoceros* virus contain a long and narrow tail-like projection (10 x 270 nm) attached to one end of the nucleocapsid which is approximately 100 x 200 nm in size. The *Heliothis zea* nonoccluded virions are morphologically similar to those of the occluded baculoviruses but are approximately twice the size, measuring 80 x 414 nm.

Physicochemical properties

Density of nucleocapsid in CsCl ≈ 1.47 g/cm^3 and of the virion, 1.18-1.25 g/cm^3. Ether and heat labile.

Nucleic acid

Single molecule of circular supercoiled dsDNA; MW 90-230 kb; 8-15 % of particle weight. G+C content is variable from 28-59%.

Protein

Virions are structurally complex and contain at least 12-30 structural polypeptides; Alkaline proteases associated with occlusions isolated from infected insects. The major protein of the viral inclusion (where present) is a single polypeptide, viral encoded, with MW = 25-33 x 10^3. This protein called polyhedrin for nuclear polyhedrosis viruses and granulin for granulosis viruses. Virions contain protein kinase activity.

Lipid

Present but not characterized.

Carbohydrate

Present but not characterized.

Antigenic properties

Antigenic determinants that cross react exist on the virion structural proteins and on the major subunit of polyhedrin and granulin polypeptides.

REPLICATION

Nuclear polyhedrosis viruses and non-occluded baculoviruses replicate exclusively in the nucleus.

Taxonomic status	English vernacular name	International name

Members of the granulosis virus genus also replicate mostly within the nucleus, but replication can occur in the cytoplasm. During infection, two forms of virions are produced. Early in infection, single nucleocapsids bud through the plasma membrane and this form of the virion is referred to as the extracellular virion (ECV). The occluded virions appear later in the infection cycle as enveloped virions embedded within a viral inclusion. The occluded form of the virus is important in the horizontal transmission of the virus. Members of the nonoccluded baculovirus subfamily do not produce inclusion bodies. For all genera, cell to cell spread is presumably by extracellular virions.

Biological Aspects

Baculoviruses have been isolated from Arthropoda: Insecta, Arachnida and Crustacea. Transmission: (i) natural - horizontal, by contamination of food, etc.; (ii) vertical on the egg; (iii) experimental - by injection of insects or by infection or transfection of cell cultures.

Subfamilies

Occluded baculoviruses	*Eubaculovirinae*
Nonoccluded baculoviruses	*Nudibaculovirinae*

Subfamily — *Eubaculovirinae*

Properties of the Virus Particle

Virions are either occluded in a crystalline protein viral occlusion which may be polyhedral in shape and contain one or many virions (genus NPV) or the inclusions are ovicylindrical and contain only one or rarely two virions (genus GV).

Genera

Nuclear polyhedrosis virus (NPV)
Granulosis virus (GV)

Genus Nuclear Polyhedrosis Virus (NPV) —

Subgenera

Multiple nucleocapsids per envelope (MNPV)
Single nucleocapsid per envelope (SNPV)

Taxonomic status	English vernacular name	International name
SUBGENUS	MULTIPLE NUCLEOCAPSID VIRUSES (MNPV)	—
TYPE SPECIES	*AUTOGRAPHA CALIFORNICA* NUCLEAR POLYHEDROSIS VIRUS (ACMNPV)	—

PROPERTIES OF THE VIRUS PARTICLE

Autographa californica nuclear polyhedrosis virus (AcMNPV) is representative of subgenus MNPV, where the virions may contain one to many or multiple (M) nucleocapsids within a single viral envelope (MNPV). All species have many virions embedded in a single viral occlusion or polyhedron. The inclusion-specific protein is referred to as polyhedrin and enveloped nucleocapsids released from polyhedra are referred to as polyhedral-derived virus (PDV). Virions that have not been occluded and released naturally from infected cells are referred to as extracellular virus (ECV).

OTHER MEMBERS

Choristoneura fumiferana MNPV (CfMNPV)
Mamestra brassicae MNPV (MbMNPV)
Orgyia pseudotsugata MNPV (OpMNPV)
and approximately 400-500 species isolated from seven insect orders and from Crustacea.

Taxonomic status	English vernacular name	International name
SUBGENUS	SINGLE NUCLEOCAPSID VIRUSES (SNPV)	—
TYPE SPECIES	*BOMBYX MORI* S NUCLEAR POLYHEDROSIS VIRUS (BMSNPV)	—

PROPERTIES OF THE VIRUS PARTICLE

Enveloped single (S) nucleocapsids with many virions embedded per viral inclusion.

OTHER MEMBERS

Heliothis zea SNPV (HzSnpv)
Trichoplusia ni SNPV (TnSnpv)
and similar viruses isolated from seven insect orders and from Crustacea.

Taxonomic status	English vernacular name	International name
GENUS	GRANULOSIS VIRUSES (GV)	—
TYPE SPECIES	*PLODIA INTERPUNCTELLA* GRANULOSIS VIRUS (PIGV)	—

PROPERTIES OF THE VIRUS PARTICLE

Enveloped single nucleocapsid with one virion per viral occlusion or granule. Granulin, the major granule or viral occlusion protein is similar in function to that of polyhedrin. Virions released from the granule are referred to as granule-derived virus (GDV). Virions that are not occluded are referred to as extracellular virions (ECV).

OTHER MEMBERS

Trichoplusia ni granulosis virus (TnGV)
Pieris brassicae granulosis virus (PbGV)
Artogeia rapae granulosis virus (ArGV)
Cydia pomonella granulosis virus (CpGV)
and similar viruses from about 50 species in the Lepidoptera

SUBFAMILY	NON-OCCLUDED BACULOVIRUSES	***NUDIBACULOVIRINAE***

PROPERTIES OF THE VIRUS PARTICLE

Enveloped single nucleocapsids. No viral occlusions are produced. Establishes persistent infections with all known host cells. Wide host range among lepidopteran cell cultures. "Standards" and defective populations can be isolated. The standard Hz-1 genome is 228 kb. The virus particle is bacilliform, measuring $414 \pm 30 \times 80 \pm 3$ nm. There are approximately 28 structural proteins ranging in molecular weight from 153,000 to 14,000. Fourteen of these are glycoproteins. The defective particles are heterogenous in length (370 ± 76 nm) and contain genomic deletions up to 100 kb. Defective virus particles contain the same structural proteins detected in standard virus particles.

Taxonomic status	English vernacular name	International name
GENUS	NON-OCCLUDED BACULOVIRUSES (NOB)	—
TYPE SPECIES	*HELIOTHIS ZEA* NOB (HZNOB)	—

PROPERTIES OF THE VIRUS PARTICLE

Enveloped single nucleocapsids. No viral occlusions produced.

OTHER MEMBERS

Oryctes rhinoceros virus

Possible members of the family Baculoviridae

A diverse group based upon morphological variation of virus structure which requires further delineation into distinct subgroups as more data become available. These are virus particles with similar general structure to baculoviruses isolated from mites, Crustacea and Coleoptera. Putative baculoviruses have been observed in a fungus (*Strongwellsea magna*), a spider, the European crab (*Carcinus maenas*), and the blue crab (*Callinectes sapidus*).

Derivation of Name

baculo from Latin *baculum*, 'stick', from morphology of virion.
eu from Greek *eu,* 'good, well, correct'.
nudi from Latin *nudus*, 'nude'.

REFERENCES

Arif, B.A.; Kuzio, J.; Faulkner, P.; Doerfler, W.: The genome of *Choristoneura fumiferans* nuclear polyhedrosis virus: Molecular cloning and mapping of the EcoRI, BamHI, SmaI, XbaI and BglII restriction sites. Virus Res. *1*:605-614 (1984).

Arnott, H.J.; Smith, K.M.: An ultrastructural study of the development of a granulosis virus in the cells of the moth *Plodia interpunctella* (Hbn.). J. Ultrastruct. Res. *21*:251-268 (1968).

Carstens, E.B.; Tjia, S.T.; Doerfler, W.: Infection of *Spodoptera frugiperda* cells with *Autographa californica* nuclear polyhedrosis virus. Virology *99*:386-398 (1979).

Carstens, E.B.; Tjia, S.T.; Doerfler, W.: Infectious DNA from *Autographa californica* nuclear polyhedrosis virus. Virology *101*:311-314 (1980).

Chakerian, R.; Rohrmann, G.F.; Nesson, M.H.; Leisy, D.J.; Beaudreau, G.S.: The nucleotide sequence of the *Pieris brassicae* granulosis virus granulin gene. J. gen. Virol. *66*:1263-1269 (1985).

Consigli, R.A.; Russell, D.L.; Wilson, M.E.: The biochemistry and molecular biology of the granulosis virus that infects *Plodia interpunctella*. Cur. Topics Microbiol. Immunol. *131*:69-101 (1986).

Couch, J.A.: An enzootic nuclear polyhedrosis virus of pink shrimp: ultrastructure, prevalence, and enhancement. J. Invertebr. Pathol. *24*:311-331 (1974).

Couch, J.A.: Viral diseases of invertebrates other than insects. *In* Davidson, D. (ed.), Pathogenesis of Invertebrate Microbial Diseases (Osmun Publishers, Allenheld, 1981).

Couch, J.A.: Inclusion body viruses. Baculoviruses of Invertebrates other than Insects. (CRC Press, Boca Raton, Fl.) (in press).

Crook, N.E.; Spencer, R.A.; Payne, C.C.; Leisy, D.J.: Variation in *Cydia pomonella* granulosis virus isolates and physical maps of the DNA from three variants. J. gen. Virol. *66*:2423-2430 (1985).

Doerfler, W.; Bohm, P.: The molecular Biology of Baculoviruses. Cur. Topics Microbiol. Immunol. *131*:1-168 (1986).

Granados, R.R.; Federici, B.A.: The Biology of Baculoviruses. (CRC Press, Boca Raton, Fl., 1986)

Granados, R.R.; Lawler, K.A.: *In vivo* pathway of *Autographa californica* baculovirus invasion and infection. Virology *108*:297-308 (1981).

Harrap, K.A.: The structure of nuclear polyhedrosis viruses. III. Virus assembly. Virology *50*:133-139 (1972).

Harrap, K.A.; Payne, C.C.: The structural properties and identification of insect viruses. Adv. Virus Res. *25*:273-355 (1979).

Hohmann, A.W.; Faulkner, P.: Monoclonal antibodies to baculovirus structural proteins: determination of specificities by Western Blot Analysis. Virology *125*:432-444 (1983).

Hull, R.; Brown, F.; Payne, C.C.: Dictionary and Directory of Animal, Plant and Bacterial Viruses. (Macmillans, London, 1989).

Krell, J.D.; Summers, M.D.: A physical map for the *Heliothis zea* SNPV genome. J. gen. Virol. *65*:445-450 (1984).

Maruniak, J.E.; Summers, M.D.: Comparative peptide mapping of baculovirus polyhedrins. J. Invertebr. Pathol. *32*:196-201 (1978).

Miller, L.K.; Adang, M.J.; Browne, D.: Protein kinase activity associated with the extracellular and occluded forms of the baculovirus *Autographa californica* nuclear polyhedrosis virus. J. Virol. *46*:275-278 (1983).

Payne, C.C.: The isolation and characterization of a virus from *Oryctes rhinoceros*. J. gen. Virol. *25*:105-116 (1974).

Payne, C.C.; Kalmakoff, J.: Alkaline protease associated with virus particles of a nuclear polyhedrosis virus: assay, purification and properties. J. gen. Virol. *26*: 84-92 (1978).

Smith, G.E.; Summers, M.D.: DNA homology among subgroup A, B, and C baculoviruses. Virology *123*:393-406 (1982).

Smith, G.E.; Summers, M.D.: Restriction maps of five *Autographa californica* MNPV variants, *Trichoplusia ni* MNPV, and *Galleria mellonella* MNPV DNAs with endonucleases *Sma*I, *Kpn*I, *Bam*HI, *Sac*I, *Xho*I, and *Eco*RI. J. Virol. *30*:828-838 (1979).

Smith, I.R.L.; Crook, N.E.: Physical maps of the genome of four variants of *Artogeia rapae* granulosis virus. J. gen. Virol. *69*: 1741-1747 (1988).

Summers, M.D.; Hoops, P.: Radioimmunoassay analysis of baculovirus granulins and polyhedrins. Virology *103*:89-98 (1980).

Summers, M.D.; Smith, G.E.; Krell, J.D.; Burand, J.P.: Physical maps of *Autographa californica* and *Rachiplusia ou* nuclear polyhedrosis virus recombinants. J. Virol. *34*: 693-703 (1980).

Tweeten, K.A.; Bulla, L.A.Jr.; Consigli, R.A.: Characterization of an extremely basic protein derived from granulosis virus nucleocapsids. J. Virol. *33*: 866-876 (1980a).

Tweeten, K.A.; Bulla, L.A.Jr.; Consigli, R.A.: Structural polypeptides of the granulosis virus of *Plodia interpunctella* . J. Virol. *33*: 877-886 (1980b).

Vlak, J.M.; Odink, K.G.: Characterization of *Autographa californica* nuclear polyhedrosis virus deoxyribonucleic acids. J. gen. Virol. *44*: 333-347 (1979).

Vlak, J.M.; Smith, G.E.: Orientation of the genome of *Autographa californica* nuclear polyhedrosis virus: a proposal. J. Virol. *41*:1118-1121 (1982).

Volkman, L.E.; Summers, M.D.; Hsieh, C.H.: Occluded and nonoccluded nuclear polyhedrosis virus grown in *Trichoplosia ni*: Comparative neutralization, comparative infectivity and *in vitro* growth studies. J. Virol. *19*:820-832 (1976).

Wiegers, F.P.; Vlak, J.M.: Physical map of the DNA of a *Mamestra brassicae* nuclear polyhedrosis virus variant isolated from *Spodoptera exigua*. J. gen. Virol. *65*: 2011-2019 (1984).

Wilson, M.E.; Consigli, R.A.: Characterization of a protein kinase activity associated with purified capsids of the granulosis virus infecting *Plodia interpunctella*. Virology *143*:516-525 (1985).

Wood, H.A.: *Autographa californica* nuclear polyhedrosis virus-induced proteins in tissue culture. Virology *102*:21-27 (1980).

Wood, H.A.; Burand, J.P.: Persistent and productive infections with the Hz-1 baculovirus. Cur. Topics Microbiol. Immunol. *131*:119-133 (1985).

Taxonomic status	English vernacular name	International name
FAMILY	PLEOMORPHIC PHAGES	***PLASMAVIRIDAE***

Revised by H.-W. Ackermann & J. Maniloff

GENUS	PLEOMORPHIC PHAGES	***PLASMAVIRUS***
TYPE SPECIES	*ACHOLEPLASMA* PHAGE L2 GROUP	—

PROPERTIES OF THE VIRUS PARTICLE

Morphology Quasi-spherical, slightly pleomorphic, with envelope, about 80 (range 50-125) nm in diameter. Size range is due to virion heterogeneity; at least three distinct virion forms are produced during infection. Sections show a small, dense core inside the envelope.

Physicochemical properties Infectivity is ether-, chloroform-, detergent-, and heat-sensitive.

Nucleic acid One molecule of circular supercoiled dsDNA; MW ≈ 7.6 x 10^6, 11970 kbp; G+C = 32% .

Protein At least 7 proteins, MW ≈ 19-68 x 10^3.

Lipid Present in envelope; similar to lipids in host cell membranes.

Carbohydrate Not known.

REPLICATION

Has both nonlytic cytocidal producing infectious cycle and lysogenic cycle. Noncytocidal infection; progeny virus released by budding from host cell membrane, with host surviving as lysogen. Lysogeny involves integration into unique site in host cell chromosome.

BIOLOGICAL ASPECTS

Host range: *Acholeplasma.*

OTHER MEMBERS

1307

Possible members

v1, v2, v4, v5, v7

Derivation of Name plasma: from Greek *plasma,* 'shaped product'

REFERENCES

Ackermann, H.-W.; DuBow, M.S.: Viruses of Prokaryotes, Vol. II, pp. 171-218 (CRC Press, Boca Raton, Fl., 1987).

Maniloff, J.: Mycoplasma viruses. CRC Crit. Rev. Microbiol. *15*: 339-389 (1988).

Taxonomic status	English vernacular name	International name
FAMILY	SSV1-TYPE PHAGES	—

Compiled by H.-W. Ackermann & W. Zillig

GENUS	SSV1 GROUP	—
TYPE SPECIES	*SULFOLOBUS* PARTICLE SSV1	—

PROPERTIES OF THE VIRUS PARTICLE

Morphology Lemon-shaped, slightly flexible particles of 60 x 100 nm; short spikes at one end.

Physicochemical properties Structure is resistant to high temperatures, acid pH, urea and ether. It is sensitive to basic pH and chloroform.

Nucleic acid One molecule of circular, positively supercoiled dsDNA of ≈ 15 kbp (15,463 bp), associated with polyamines and a virus-coded basic protein.

Protein Two hydrophobic coat proteins, MW = 7.7 and 9.7 x 10^3, one DNA-associated protein. Major coat protein is ether-soluble.

Lipid None.

Carbohydrate Not known.

REPLICATION

Genome is present in cells as a plasmid or integrated into specific sites. UV induction results in large numbers of particles which are released without lysis.

BIOLOGICAL ASPECTS

Host range *Sulfolobus shibatae* strain B12.

Possible members

Particles produced by the archaebacteria *Desulfurolobus* and *Methanococcus*.

REFERENCES

Zillig, W.; Reiter, W.-D.; Palm, P.; Gropp, F.; Neumann, H.; Rettenberger, M.: Viruses of archaebacteria. *In* Calendar. R. (ed.), The Bacteriophages, Vol. 1, pp. 517-558 (Plenum Press, New York, 1988).

Taxonomic status	English vernacular name	International name
FAMILY	TTV1 FAMILY	*LIPOTHRIXVIRIDAE*

Compiled by H.-W. Ackermann & W. Zillig

GENUS	TTV1 GROUP	*LIPOTHRIXVIRUS*
TYPE SPECIES	*THERMOPROTEUS* PHAGE TTV1	—

PROPERTIES OF THE VIRUS PARTICLE

Morphology Thick rigid rods about 400 nm long x 40 nm in diameter. Both ends have protrusions which seem to participate in adsorption. Envelope.

Physicochemical properties Ether and detergents cause disruption of particles.

Nucleic acid One molecule of linear dsDNA; MW ≈ 10 x 10^6 (16 kbp).

Protein Four proteins (MW 14-45 x 10^3). P1 and P2 are DNA-associated, P3 is the envelope protein. The location of P4 is unknown.

Lipid In addition to P3 protein, the envelope contains lipids in the same proportions as the host membranes. Bilayer structure.

Carbohydrate Glucose in glycolipid.

REPLICATION

Adsorption to pili? Infection results in virus production with lysis or the establishment of a carrier state. Pieces of TTV1 DNA may be integrated into the host genome.

BIOLOGICAL ASPECTS

Host range Host range is limited to the archaebacterium *Thermoproteus tenax*. Other rod-shaped particles of different dimensions were found associated with *Thermoproteus* cultures or were observed by electron microscopy in water from Icelandic solfataras but were not cultivated. One of these particles, TTV2, is temperate.

Derivation of Name lipo: from Greek, *lipos*, 'fat'
thrix: from Greek, *thrix*, 'hair'

REFERENCES

Zillig, W.; Reiter, W.-D.; Palm, P.; Gropp, F.; Neumann, H.; Rettenberger, M.: Viruses of archaebacteria. *In* Calendar, R. (ed.), The Bacteriophages, Vol. 1, pp. 517-558 (Plenum, Press, New York, 1988).

Taxonomic status	English vernacular name	International name
FAMILY	POLYDNAVIRUS GROUP	***POLYDNAVIRIDAE***

Compiled by D. Stoltz

PROPERTIES OF THE VIRUS PARTICLE

Morphology

Ichnovirus particles consist of nucleocapsids of usually uniform size (approximately 85 nm x 330 nm), having the form of a prolate ellipsoid, and surrounded by 2 unit membrane envelopes. The inner envelope appears to be assembled *de novo* within the nucleus of infected calyx cells and the outermost envelope is acquired by budding through the plasma membrane into the oviduct lumen. Bracovirus particles consist of enveloped cylindrical electron-dense nucleocapsids of uniform diameter but of variable length (40 nm diameter by 30-150 nm length) and may contain one or more nucleocapsids within a single viral envelope; the latter appears to be assembled *de novo* within the nucleus.

Nucleic acid

Genomes consist of multiple supercoiled double-stranded DNAs of variable MWs ranging from approximately 2.0 to more than 28 kbp. No aggregated MW for any polydnavirus genome has as yet been determined with any degree of accuracy. Estimates of genome complexity are complicated by the presence of related DNA sequences shared among the multiple DNAs.

Protein

Virions are structurally complex and contain at least 20-30 polypeptides, with MWs ranging from 10-200 x 10^3.

Lipid

Present, but uncharacterized.

Carbohydrate

Present, but uncharacterized.

Antigenic properties

Cross-reacting antigenic determinants exist on ichnoviruses within each genus; in some cases, viral nucleocapsids share at least one major conserved epitope. Antigenic relationships among the bracoviruses have not as yet been investigated.

REPLICATION

Viruses replicate in the nucleus. Replication occurs in the calyx epithelium of the ovaries of all female wasps belonging to any species. Ichnoviruses bud directly from the calyx epithelial cells into the lumen of the oviduct. The mode of release of bracovirus particles is presently

Taxonomic status	English vernacular name	International name

unclear, but may involve lysis of infected calyx epithelial cells. Viral DNAs are present in male wasps, but replication has not been demonstrated.

Biological Aspects

Host range Polydnaviruses have been isolated only from endoparasitic hymenopteran insects belonging to the families Ichneumonidae and Braconidae.

Transmission The mechanism of transmission appears to be vertical within parasitoid species. Viruses are injected into larval hosts during oviposition, but replication in parasitized host insects or in cultured cells has not been observed.

Genus	—	*Ichnovirus*
Type species	*Campoletis sonorensis* virus (CsV)	—

Properties of the Virus Particle

One nucleocapsid having the form of a prolate ellipsoid (85 x 330 nm), per virion; two envelopes; segmented DNA genome.

Other Members

A similar virus has been found in *Glypta* sp., but these particles differ in having more than one nucleocapsid per virion. Polydnaviruses having a morphology resembling that of CsV are typically but perhaps not exclusively found in ichneumonids belonging to the sub-family Campopleginae.

Genus	—	*Bracovirus*
Type species	*Cotesia melanoscela* virus (CmV)	—

Properties of the Virus Particle

Nucleocapsids are cylindrical with a uniform diameter (40 nm) but variable in length (30-150 nm). Virions have only one envelope which may surround one or more nucleocapsids; segmented DNA genome.

Derivation of Name polydna: from *poly*, 'several', *DNA virus*

REFERENCES

Blissard, G.W.; Smith, O.P.; Summers, M.D.: Two related viral genes are located on a single superhelical DNA segment of the multipartite *Campoletis sonorensis* virus genome. Virology *160:* 120-134 (1987).

Blissard, G.W.; Fleming, J.G.W.; Vinson, S.B.; Summers, M.D.: *Campoletis sonorensis* virus: expression in *Heliothis virescens* and identification of expressed sequences. J. Insect Physiol. *32:* 351-359 (1986).

Cook, D.; Stoltz, D.B.: Comparative serology of viruses isolated from ichneumonid parasitoids. Virology *130*:215-220 (1983).

Edson, K.M.; Vinson, S.B.; Stoltz, D.B.; Summers, M.D.: Virus in a parasitoid wasp: suppression of the cellular immune response in the parasitoid's host. Science *211*:582-583 (1981).

Fleming, J.G.W.; Summers, M.D.: *Campoletis sonorensis* endoparasitic wasps contain forms of *C. sonorensis* virus DNA suggestive of integrated and extrachromosomal polydnavirus DNAs. J. Virol. *57*:552-562 (1986).

Fleming, J.G.W.; Blissard, G.W.; Summers, M.D.; Vinson, S.B.: Expression of *Campoletis sonorensis* virus in the parasitized host, *Heliothis virescens.* J. Virol. *48*:74-78 (1983).

Krell, P.J.; Stoltz, D.B.: Unusual baculovirus of the parasitoid wasp *Apanteles melanoscelus*: isolation and preliminary characterization. J. Virol. *29*:1118-1130 (1979).

Krell, P.J.; Stoltz, D.B.: Virus-like particles in the ovary of an ichneumonid wasp: purification and preliminary characterization. Virology *101*:408-418 (1980).

Krell, P.J.; Summers, M.D.; Vinson, S.B.: Virus with a multipartite superhelical DNA genome from the ichneumonid parasitoid *Campoletis sonorensis.* J. Virol. *43*:859-870 (1982).

Stoltz, D.B.; Vinson, S.B.: Viruses and parasitism in insects. Adv. Virus Res. *24*:125-171 (1979).

Stoltz, D.B.: Evidence for chromosomal transmission of polydnavirus DNA. J. gen. Virol. *71:* 1051-1056 (1990).

Theilmann, D.A.; Summers, M.D.: Identification and comparison of *Campoletis sonorensis* virus transcripts in the insect hosts *Campoletis sonorensis* and *Heliothis virescens.* Virology *167:* 329-341 (1988).

Theilmann, D.A.; Summers, M.D.: Molecular analysis of *Campoletis sonorensis* virus DNA in the lepidopteran host *Heliothis virescens.* J. gen. Virol. *67:* 1961-1969 (1986).

Theilmann, D.A.; Summers, M.D.: Physical analysis of the *Campoletis sonorensis* virus multipartite genome and identification of a family of tandemly repeated elements. J. Virol. *61:* 2589-2598 (1987).

Webb, B.A.; Summers, M.D.: Venom and viral expression products of the endoparasitic wasp *Campoletis sonorensis* share epitopes and related sequences. Proc. Natl. Acad. Sci., USA *87:* 4961-4965 (1990).

Taxonomic status	English vernacular name	International name
FAMILY	ICOSAHEDRAL CYTOPLASMIC DEOXYRIBOVIRUSES	***IRIDOVIRIDAE***

Reported by A.M. Aubertin

PROPERTIES OF THE VIRUS PARTICLE

Morphology Icosahedral, 125-300 nm diameter; spherical nucleoprotein core surrounded by membrane consisting of lipid modified by morphological protein subunits; the released virions of some genera possess a plasma membrane-derived envelope that enhances, but is not required for infectivity.

Physicochemical properties MW of virions is 500-2000 x 10^6; S_{20w} = 1300-4450; density = 1.35-1.6 g/cm^3; members of *Iridovirus* and *Chloriridovirus* genera resistant to ether, all others sensitive to ether and nonionic detergents; stable at pH 3 - 10 and at 4°C for several years; inactivated by 15-30 min at 55°C.

Nucleic acid One molecule of linear dsDNA, MW = 100-250 x 10^6. Possibly two molecules in some viruses. 12-30% by weight of the virus particle. G + C ≈ 20-58%; there is no cross-hybridization between genera. The DNA of *Ranavirus*, *Lymphocystivirus*, and at least one *Iridovirus* species is circularly permuted and has direct terminal repeats; the genomic DNA of *Ranavirus* and *Lymphocystivirus* contains a high proportion of methylated cytosine.

Protein 13-35 structural polypeptides by one and two dimensional PAGE, with MWs ranging from 10-250 x 10^3. Most members that have been examined possess several virion-associated enzymes, in particular an active protein kinase.

Lipid Unenveloped particles contain 5-9% lipid (predominantly phospholipid) as an integral part of the icosahedral shell. Some members have an additional plasma-membrane-derived envelope.

Carbohydrate None has been detected.

Antigenic properties Antibodies prepared against virions are often non-neutralizing, but useful in establishing relationships between species. Neutralizing antibodies against *Tipula* iridescent virus have been produced in rabbits.

Effect on cells Generally cytocidal; most members rapidly inhibit host cell macromolecular synthesis.

Taxonomic status	English vernacular name	International name

Replication

Virus entry is by pinocytosis, with uncoating in phagocytic vacuoles. The host cell nucleus appears to be required for transcription and replication of DNA, but some DNA synthesis, and the assembly of virions into mature particles, takes place in the cytoplasm, where paracrystalline inclusion bodies are readily observed. Release is by lysis or budding. Virions that bud from the host acquire a plasma- or endoplasmic reticulum-derived envelope, but most virus remains cell-associated and unenveloped virions are infectious.

Biological Aspects

Host range Many species appear to have a restricted host range *in vivo* and *in vitro*; exceptions are the genus *Ranavirus* (frog virus 3), which grows in a wide variety of cultured cells, and *Iridovirus* (*Tipula* iridescent virus), which infects a wide range of insects. *Iridovirus* type 32 infects both terrestrial isopods and nematodes.

Transmission Both horizontal and vertical.

Genera

Small iridescent insect virus group	*Iridovirus*
Large iridescent insect virus group	*Chloriridovirus*
Frog virus group	*Ranavirus*
Lymphocystis disease virus group	*Lymphocystivirus*
Goldfish virus group (proposed)	-

GENUS	SMALL IRIDESCENT INSECT VIRUS GROUP	***IRIDOVIRUS***
TYPE SPECIES	*CHILO* IRIDESCENT VIRUS	—

Properties of the Virus Particle

Particles ≈ 120 nm diameter. Complex icosahedral shell contains lipid, but integrity is protected under protein capsid as infectivity is not sensitive to ether. Infected larvae and purified virus pellets produce a blue to purple iridescence. *Chilo* iridescent virus has circularly permuted and terminally redundant DNA.

Other Members

Insect iridescent viruses 1, 2, 6, 9, 10, 16-32.

Taxonomic status	English vernacular name	International name
GENUS	LARGE IRIDESCENT INSECT VIRUS GROUP	***CHLORIRIDOVIRUS***
TYPE SPECIES	MOSQUITO IRIDESCENT VIRUS (IRIDESCENT VIRUS - TYPE 3, REGULAR STRAIN)	—

PROPERTIES OF THE VIRUS PARTICLE

Particle diameter is ≈ 180 nm. Infected larvae and virus pellets of most members iridesce with a yellow-green color.

OTHER MEMBERS

Insect iridescent viruses 3 - 5, 7, 8, 11-15

Probable member

Chironomus plumosus iridescent

Taxonomic status	English vernacular name	International name
GENUS	FROG VIRUS GROUP	***RANAVIRUS***
TYPE SPECIES	FROG VIRUS 3 (FV3)	—

PROPERTIES OF THE VIRUS PARTICLE

Does not cause disease in natural host, adult *Rana pipiens*, but is lethal for tadpoles and Fowler toads; grows in piscine, avian, and mammalian cells from 12°C to 32°C; structural viral protein causes rapid inhibition of host macromolecular synthesis without interfering with viral replication. DNA-dependent RNA polymerase not found associated with virus particle. DNA contains a high proportion of 5-methyl cytosine and is circularly permuted and terminally redundant. DNA synthesis occurs in 2 stages: (1) synthesis of unit-length molecules in the nucleus and (2) synthesis of concatemers and virion assembly in the cytoplasm. mRNA lacks poly(A).

OTHER MEMBERS

Frog viruses 1, 2, 5 - 24, L2, L4 and L5
Tadpole edema virus from *Rana catesbriana* LT 1 - 4 and T6-T20 from newts
T21 from *Xenopus*

Taxonomic status	English vernacular name	International name
GENUS	LYMPHOCYSTIS DISEASE VIRUS GROUP	***LYMPHOCYSTIVIRUS***
TYPE SPECIES (PROPOSED)	FLOUNDER ISOLATE (LCDV-1)	—

PROPERTIES OF THE VIRUS PARTICLE

Can be transmitted by implantation or injection into centrarchid fish hosts; forms giant host connective tissue cells at 25°C. Genomic DNA is circularly permuted, terminally redundant, and highly methylated at cytosine residues.

OTHER MEMBERS

Lymphocystis disease virus dab isolate (LCDV-2)

Possible member

Octopus vulgaris disease virus

Taxonomic status	English vernacular name	International name
GENUS (PROPOSED)	GOLDFISH VIRUS GROUP	—
TYPE SPECIES (PROPOSED)	GOLDFISH VIRUS 1 (GFV-1)	—

PROPERTIES OF THE VIRUS PARTICLE

Does not cause overt disease in natural host. Has a more restricted host range *in vitro* than amphibian viruses. Produces cytoplasmic vacuolization and cell roundy in goldfish cell line (CAR) at 25°C.

DNA is highly methylated at cytosine residues, not only at CpG sequences but most likely, also at CpT.

OTHER MEMBERS

Goldfish virus 2 (GF-2)

Taxonomic status	English vernacular name	International name

Intracellular site of transcription and DNA replication is unknown.

BIOLOGICAL ASPECTS

Host range

Host range is limited to eukaryotic algae with the appropriate receptor. Three groups of viruses are delineated based on host specificity:

Paramecium bursaria Chlorella NC64A viruses (NC64A viruses)
Paramecium bursaria Chlorella Pbi viruses (Pbi viruses)
Hydra virdis Chlorella viruses (HVCV)
Chlorella strains NC64A, ATCC 30562, and N1A (originally symbionts of the protozoan *P. bursaria*), collected in the United States, are the only known hosts for NC64A viruses. *Chlorella* strain Pbi (originally a symbiont of a European strain of *P. bursaria*) collected in Germany, is the only known host for Pbi viruses. Pbi viruses do not infected *Chlorella* strains NC64A, ATCC 30562, and N1A. *Chlorella* strain Florida (originally a symbiont of *Hydra viridis*) is the only known host for HVCV. NC64A viruses are placed in 16 subgroups based on plaque size, serological reactivity, resistance to restriction endonucleases, and nature and content of methylated bases.

OTHER MEMBERS

Chlorella NC64A viruses (Thirty seven NC64A viruses including PBCV-1, the type species of the family are known: *Chlorella* virus NE-8D (CV-NE8D; synonym NE-8D), CV-NYb1, CV-CA4B, CV-AL1A, CV-NY2C, CV-NC1D, CV-NC1C, CV-CA1A, CV-CA2A, CV-IL2A, CV-IL2B, CV-IL3A, CV-IL3D, CV-SC1A, CV-SC1B, CV-NC1A, CV-NE8A, CV-AL2C, CV-MA1E, CV-NY2F, CV-CA1D, CV-NC1B, CV-NYs1, CV-IL5-2s1, CV-AL2A, CV-MA1D, CV-NY2B, CV-CA4A, CV-NY2A, CV-XZ3A, CV-SH6A, CV-BJ2C, CV-XZ6E, CV-XZ4C, CV-XZ5C, CV-XZ4A).
Chlorella Pbi viruses (CVA-1, CVB-1, CVG-1, CVM-1, and CVR-1).
Hydra viridis Chlorella viruses (HVCV-1, HVCV-2, and HVCV-3).

Derivation of Name

phyco: from Greek *phycos*, 'algae'.
dna (= sigla for deoxyribonucleic acid)

Taxonomic status	English vernacular name	International name
GENUS	AVIAN ADENOVIRUSES	*AVIADENOVIRUS*
TYPE SPECIES	FOWL ADENOVIRUS 1 (CELO)	GAL 1

Hosts: English name	Hosts: Zoological name	Serotype designation
Fowl	*Galius domesticus*	gal1-gal2
Turkey	*Meleagris gallopavo*	mel1-mel3
Goose	*Anser domesticus*	ans1-ans3
Pheasant	*Phasianus colchicus*	pha1
Duck	*Anas domestica*	ana1-ana2

Derivation of Names

adeno: from Greek *aden*, *adenos*, "gland"; adenoviruses were first isolated from human adenoid tissue
avi: from Latin *avis*, "bird"
mast: from Greek *mastos*, "breast" - a by-form is Greek and Latin *mamma*, hence mammalian.

REFERENCES

Adrian, T.; Wadell, G.; Hierholzer, J.C.; Wigand, R.: DNA restriction analysis of adenovirus prototypes 1 to 41. Arch. Virol. *91*:277-290 (1986).

Ginsberg, H.S.: The Adenoviruses (Plenum Press, New York, 1984).

Ginsberg, H.S.; Young, C.S.H.: Genetics of adenoviruses. *In* Fraenkel Conrat , H.; Wagner, R.R. (eds.) Comprehensive Virology, Vol. 9, pp. 27-88 (Plenum Press, New York, 1977).

Green, M.; Mackey, J.K.; Wold, W.S.M.; Rigden, P.: Thirty-one human adenovirus serotypes (Ad1-Ad31) form five groups (A-E) based upon DNA genome homologies. Virology *93*:481-492 (1979).

Hierholzer, J.C.; Wigand, R.; Anderson, L.J.; Adrian, T.; Gold, J.W.M.: Adenoviruses from patients with AIDS; a plethora of serotypes and a description of five new serotypes of subgenus D (types 43-47). J. Infect. Dis. *158:* 804-813 (1988).

Horner, G.W.; Hunter, K.; Bartha, A.; Benko, M.: A new subgroup of bovine adenovirus proposed as prototype strain 10. Arch. Virol. *109:* 121-124 (1989).

Horwitz, M.S.: Adenoviridae and their replication. *In* Fields, B.N.; Knipe, D.M., (eds.), Virology, Vol. 2, 2nd edit., pp. 1679-1722 (Raven Press, New York, 1990).

Kalter, S.S.: Enteric viruses of nonhuman primates. Vet. Pathol. 19 Suppl. 7:33-43 (1982).

Mautner, V.: *Adenoviridae. In* Porterfield, J.S. (ed.), Andrewes' Viruses of Vertebrates, pp. 249-284 (Bailliore Tindall, 1989).

Norrby, E.; Bartha, A.; Boulanger, P.; Dreizin, R.S.; Ginsberg, H.S.; Kalter, S.S.; Kawamura, H.; Rowe, W.P.; Russell, W.C.; Schlesinger, R.W.; Wigand, R.: *Adenoviridae.* Intervirology *7*:117-125 (1976).

Pettersson, U.; Wadell, G.: Antigenic structure of the adenoviruses. *In* van Regenmortel, M.H.V.; Neurath, A.R. (eds.), Immunochemistry of viruses. The basis for serodiagnosis and vaccines pp. 295-323 (Elsevier/North Holland, Amsterdam, 1985).

Roberts, R.J.; Akusjarvi, G.; Alestrom, P.; Gelinas, R.E.; Gingeras, T.R.; Sciaky, D.; Pettersson, U.: A consensus sequence for the adenovirus-2 genome. *In* Doerfler, W.

(ed.), Adenovirus DNA. The viral genome and its expression, pp. 1-51 (Martinus Nijhoff Publishing, Boston, 1986).

Shenk, T.; Williams, J.: Genetic analysis of adenoviruses. Cur. Topics Microbiol. Immunol. *111*:1-39 (1984).

van der Eb, A.J.; Bernards, R.: Transformation and oncogenicity by adenoviruses. Cur. Topics. Microbiol. Immunol. *110*:23-51 (1984).

Wadell, G.: Classification of human adenoviruses by SDS-polyacrylamide gel electrophoresis of structural polypeptides. Intervirology *11*:47-57 (1979).

Wadell, G.; Hammarskjold, M.-L.; Winberg, G.; Varsanyi, T.M.; Sundell, G.: Genetic variability of adenoviruses. Ann. N. Y. Acad. Sci. *354*:16-42 (1980).

Wigand, R.; Adrian, T.; Bricout, F.: A new human adenovirus of subgenus D: Candidate adenovirus type 42. Arch. Virol. *94*:283-286 (1987).

Wigand, R.; Bartha, A.; Dreizin, R.S.; Esche, H.; Ginsberg, H.S.; Green, M.; Hierholzer, J.C.; Kalter, S.S.; McFerran, J.B.; Pettersson, U.; Russell, W.C.; Wadell, G.: *Adenoviridae*; Second report. Intervirology *18*:169-176 (1982).

Zsak, L.; Kisary, J.: Grouping of fowl adenoviruses based upon the restriction patterns of DNA generated by BamH I and Hind III. Intervirology *22*:110-114 (1984).

Taxonomic status	English vernacular name	International name
GENUS	*RHIZIDIOMYCES* VIRUS GROUP (POSSIBLE AFFINITIES TO THE *ADENOVIRIDAE* FAMILY)	***RHIZIDIOVIRUS***
TYPE SPECIES	*RHIZIDIOMYCES* VIRUS (FROM *RHIZIDIOMYCES* SP. ISOLATE F)	—

Compiled by K.W Buck & S.A. Ghabrial

PROPERTIES OF THE VIRUS PARTICLE

Morphology Isometric, particles ≈ 60 nm in diameter.

Physicochemical properties S_{20w} ≈ 625 S; buoyant density in CsCl ≈ 1.314 g/cm^3.

Nucleic acid Single molecule of dsDNA, MW ≈ 16.8 x 10^6, G+C ≈ 42%.

Protein At least 14 polypeptides with MWs = 84.5-26 x 10^3 with the largest one being dominant.

Lipid None detected.

Carbohydrate None detected.

REPLICATION

Particles appear first in the nucleus.

BIOLOGICAL ASPECTS

The virus appears to be transmitted in a latent form in zoospores of the fungus host. Activation of the virus under stress conditions, such as heat, low nutrition or ageing, results in céll lysis.

Derivation of Names rhizidio from name of the host *Rhizidiomyces* sp.

REFERENCES

Dawe, V.H.; Kuhn, C.W.: Virus-like particles in the aquatic fungus, *Rhizidiomyces*. Virology *130*:10-20 (1983).

Dawe, V.H.; Kuhn, C.W.: Isolation and characterization of a double-stranded DNA mycovirus infecting the aquatic fungus, *Rhizidiomyces*. Virology *130*:21-28 (1983).

Taxonomic status	English vernacular name	International name
FAMILY	PAPOVAVIRUS GROUP	***PAPOVAVIRIDAE***

Reported by R. Frisque

PROPERTIES OF THE VIRUS PARTICLE

Morphology Nonenveloped, icosahedral particles 40-55 nm in diameter; 72 capsomers in skew arrangement; filamentous forms occur.

Physicochemical properties MW = 25-47 x 10^6; S_{20w} = 240-300; buoyant density in CsCl = 1.34 g/cm^3. Resistant to ether, acid and heat treatment.

Nucleic acid One molecule circular dsDNA, MW = 3-5 x 10^6; G+C = 40-50%, 10 - 13% of virion by weight.

Protein 6-9 polypeptides, MW = 3-82 x 10^3. Low MW components are cellular histones.

Lipid None.

Carbohydrate None.

Antigenic properties Different species antigenically distinct by neutralization and HI tests; antisera prepared against disrupted virions detect common antigens shared by other species belonging to the same genus.

Effects on cells Cytolytic in cells of host of origin; may transform cells from other species; several species of virus haemagglutinate by reacting with neuraminidase-sensitive receptors; no tissue culture systems for papillomaviruses.

REPLICATION

Virions attach to cellular receptors, are engulfed and transported to nucleus; host cell enzymes are derepressed and cellular DNA synthesis is stimulated; expression of viral genome divided into early and late events; host cell histones are incorporated into virions during maturation in nucleus; virions released by lysis of infected cells. Replication of papillomaviruses *in vivo* is dependent on the terminal differentiation of keratinocytes.

BIOLOGICAL ASPECTS

Host range Each virus has its own host range in nature and in the laboratory. Transformation tends to occur in cells which do not support replication of virus.

Taxonomic status	English vernacular name	International name
Transmission	Contact and airborne infection. Some human papillomaviruses may be sexually transmitted.	
	GENERA	
	-	*Papillomavirus*
	-	*Polyomavirus*

GENUS	—	***PAPILLOMAVIRUS***
TYPE SPECIES	RABBIT (SHOPE) PAPILLOMA VIRUS	*PAPILLOMAVIRUS SYLVILAGI*

PROPERTIES OF THE VIRUS PARTICLE

Morphology Particles 50 - 55 nm in diameter.

Physicochemical properties $S_{20w} \approx 300$.

Nucleic acid MW $\approx 5 \times 10^6$; G+C = 40-50%. ORFs located on one strand of DNA.

Antigenic properties Each virus species contains a distinct surface antigen, but all members of the genus share one common antigen revealed by disrupting the virions.

BIOLOGICAL ASPECTS

Host range Cause papillomas in natural hosts and related species. Host and tissue-specific viruses induce papillomas in skin and mucous membranes but do not grow in cell cultures. Warts may convert to malignancy.

OTHER MEMBERS

Members of this genus are known from humans (> 63 types), chimpanzee, colobus and rhesus monkeys, cow (6 types), deer, dog, horse, sheep, elephant, elk, *opossum*, multimammate and European harvest mice, turtle, chaffinch and parrot.

Taxonomic status	English vernacular name	International name
GENUS		*POLYOMAVIRUS*
TYPE SPECIES	POLYOMA VIRUS	*POLYOMAVIRUS MURIS* 1

PROPERTIES OF THE VIRUS PARTICLE

Morphology Particles 40-45 nm in diameter.

Physicochemical properties $S_{20w} \approx 240$.

Nucleic acid MW $\approx 3 \times 10^6$; G+C = 40-48%. ORFs located on both strands of DNA.

Antigenic properties Several species haemagglutinate. Whole viruses show no serological cross-reactivity between most species, but a common genus antigen can be detected in disrupted virions of all species. T antigens induced by primate viruses, cross-react.

Effects on cells Inapparent infections in most hosts. Oncogenic in hosts (chiefly immunodeficient newborn hamsters) which are often different from species of origin of virus. They have a restricted host range and replicate in cell culture. Cells which do not support replication may be transformed. Viral DNA integrates into cellular chromosomes of transformed cells.

OTHER MEMBERS

Polyomavirus muris 2 (K)
Polyomavirus hominis 1 (BK)
Polyomavirus hominis 2 (JC)
Polyomavirus sylvilagi (Rabbit kidney vacuolating)
Polyomavirus maccacae 1 (SV40)
Polyomavirus papionis 1 (SA12)
Polyomavirus papionis 2
Polyomavirus cercopitheci (lymphotropic)
Polyomavirus bovis (WRSV)
Other possible members have been found in pigs and hamsters.

Derivation of Names	papova: sigla, from *pa*pilloma, *po*lyoma, and *va*cuolating agent (early name for SV40). papilloma: from Latin *papilla*, 'nipple, pustule', and Greek suffix *-oma*, used to form nouns denoting 'tumors' polyoma: from Greek *poly*, 'many', and *-oma*, denoting 'tumors'.

References

Melnick, J.L.; Allison, A.C.; Butel, J.S.; Eckhart, W.; Eddy, E.E.; Kit, S.; Levine, A.J.; Miles, J.A.R.; Pagano, J.S.; Sachs, L.; Vonka, V.: *Papovaviridae*. Intervirology *3*:106-120 (1974).

Salzman, N.P.: The *Papovaviridae*: The Polyomaviruses, Vol. 1 (Plenum Press, New York, 1986).

Salzman, N.P.; Howley, P.M.: The *Papovaviridae*: The Papilloma-viruses, Vol. 2 (Plenum Press, New York, 1987).

Taxonomic status	English vernacular name	International name
GROUP	CAULIFLOWER MOSAIC VIRUS (295)	*CAULIMOVIRUS*

Revised by R. Hull

TYPE MEMBER	CAULIFLOWER MOSAIC VIRUS (CAMV) (24; 243) (CABBAGE B, DAVIS ISOLATE)	—

PROPERTIES OF THE VIRUS PARTICLE

Morphology Isometric particles ≈ 50 nm in diameter.

Physicochemical properties MW ≈ 20 x 10^6; S_{20w} ≈ 208; D ≈ 0.75 x 10^{-7} cm^2/s; apparent partial specific volume ≈ 0.704; buoyant density in CsCl ≈ 1.37 g/cm^3; particles very stable.

Nucleic acid One molecule of dsDNA; open circular molecule with single-strand discontinuities at specific sites, the transcribed (α) strand with one and the non-transcribed (β) strand with two discontinuities; DNAs of four CaMVs (isolates Cabb S with 8,024 bp, CM1841 with 8,031 bp, D/H with 8,016 bp and Xinjiang with 8,060 bp) have been sequenced. Six or possibly 8 ORFs (putative genes) are present on the α strand. The β strand is noncoding.

Protein Capsid protein is translated from ORF IV, and assembled into capsids as 57 x 10^3 phosphorylated polypeptide. Rapid degradation occurs *in vivo* (and perhaps also during purification) to give several polypeptide forms, MW predominantly ≈ 42 x 10^3 and 37 x 10^3.

Lipid None.

Carbohydrate Coat protein has some glycosylation.

Antigenic properties Efficient immunogens; serological relationships among some members.

REPLICATION

Transcription occurs in the nucleus from a DNA template with properties of a minichromosome. Two major transcripts (19S and 35S) are found. The 19S transcript is from ORF VI, and translates to a MW = 62 x 10^3 protein found in the cytoplasmic viral inclusion body in which mature virus particles accumulate; these electron-dense inclusion bodies are characteristic of the group. The 35S transcript has not been translated *in vitro* but is though to be the mRNA of several of the ORFs. The 35S transcript is 180 nucleotides longer than the full length viral DNA (i.e., it contains a 180 nucleotide terminal repeat), and is thought to be a template for replication of the viral genome by reverse transcription. ORF V may code for the replication enzyme.

BIOLOGICAL ASPECTS

Host range Narrow.

Transmission Transmissible experimentally by mechanical inoculation; transmitted by aphids in a semipersistent manner. Transmission of CaMV requires a virus-coded protein (the product of ORF II) also located within inclusion bodies.

OTHER MEMBERS

Blueberry red ringspot (327)
Carnation etched ring (182)
Dahlia mosaic (51)
Figwort mosaic
Horseradish latent
Mirabilis mosaic
Peanut chlorotic streak
Soybean chlorotic mottle (331)
Strawberry vein banding (219)
Thistle mottle

Possible members

Aquilegia necrotic mosaic
Cassava vein mosaic
Cestrum virus
Petunia vein clearing
Plantago virus 4
Sonchus mottle

Derivation of Name caulimo: sigla from *cauli*flower *mo*saic

References

Covey, S.N.; Hull, R.: Advances in cauliflower mosaic virus research. Oxford Surveys of Plant Mol. Cell. Biol. *2*:339-346 (1985).

Covey, S.N.: Organization and expression of the cauliflower mosaic virus genome, *In* Davies, J.W. (ed.), Molecular Plant Virology, Replication and Gene Expression, Vol. II, pp. 121-160 (CRC Press, Boca Raton, Fl, 1985).

Francki, R.I.B.; Milne, R.G.; Hatta, T.: Caulimovirus group, *In* Atlas of plant viruses, Vol. I, pp. 17-32, (CRC Press, Boca Raton, Fl, 1985).

Frank, A.; Guilley, H.; Jonard, G.; Richards, K.; Hirth, L.: Nucleotide sequence of cauliflower mosaic virus DNA. Cell *21*:285-294 (1980).

Gardner, R.C.; Howarth, A.J.; Hahn, P.; Brown-Luedi, M.; Shepherd, R.J.; Messing, J.: The complete nucleotide sequence of an infectious clone of cauliflower mosaic virus by M13mp7 shotgun sequencing. Nuc. Acids Res. *9*:2871-2888 (1981).

Kruse, J.; Timmins, P.; Witz, J.: The spherically averaged structure of a DNA isometric plant virus: cauliflower mosaic virus. Virology *159*: 166-168 (1987).

Maule, A.J.: Replication of caulimoviruses in plants and protoplasts, *In* Davies, J.W. (ed.), Molecular Plant Virology, Replication and Gene Expression, Vol. II, pp. 161-190 (CRC Press, Boca Raton, Fl., 1985).

Pfeiffer, P.; Hohn, T.: Involvement of reverse transcription in the replication of cauliflower mosaic virus: A detailed model and test of some aspects. Cell *33*:781-789 (1983).

Richins, R.D.; Shepherd, R.J.: Physical maps of the genomes of dahlia mosaic virus and mirabilis mosaic virus - two members of the caulimovirus group. Virology *124*:208-214 (1983).

Taxonomic status	English vernacular name	International name
GROUP	COMMELINA YELLOW MOTTLE VIRUS GROUP	—

Compiled by B.E.L. Lockhart & R. Hull

TYPE MEMBER	COMMELINA YELLOW MOTTLE VIRUS (COYMV)	—

PROPERTIES OF THE VIRUS PARTICLE

Morphology Bacilliform particles ≈ 130 x 30 nm.

Physicochemical properties CoYMV has a density in CsCl of 1.37 g/cm^3, cacao swollen shoot virus has a S_{20w} of 218.

Nucleic acid One molecule of dsDNA: open circular molecules with single-strand discontinuities at specific sites, one in each strand. Mealybug transmitted viruses have genomes ≈ 7.5 kbp (7489 bp in CoYMV), and rice tungro bacilliform virus has a genome of ≈ 8.0 kbp.

Protein Two protein species ≈ 40 x 10^3 and 35 x 10^3.

Lipid None determined.

Carbohydrate None detected.

Antigenic properties Moderately efficient immunogens, serological relationships among some members.

REPLICATION

Mechanism not determined but, as the genome has various properties in common with caulimoviruses, it is thought to involve reverse transcription.

BIOLOGICAL ASPECTS

Host range Narrow.

Transmission Most members and possible members not transmissible mechanically; those that are, are only transmitted with difficulty. Members and possible members for which a vector is known are all transmitted by mealybugs in a semi-persistent manner except for rice tungro bacilliform virus which is leafhopper transmitted in association with rice tungro spherical virus, and rubus yellow net which is aphid transmitted.

Taxonomic status	English vernacular name	International name
	OTHER MEMBERS	
	Banana streak	
	Sugarcane bacilliform	
	Possible members	
	Aucuba ringspot	
	Cacao swollen shoot (10)	
	Canna yellow mottle	
	Colocasia bacilliform	
	Dioscorea bacilliform	
	Kalanchoe top-spotting	
	Mimosa bacilliform	
	Rubus yellow net (188)	
	Rice tungro bacilliform	
	Schefflera ringspot	
	Yucca bacilliform	

REFERENCES

Lockhart, B.E.L.: Evidence for a double-stranded circular DNA genome in a second group of plant viruses. Phytopathology *80:* 127-131 (1990).

Taxonomic status	English vernacular name	International name
FAMILY	PHAGES WITH DOUBLE CAPSIDS	*TECTIVIRIDAE*

Revised by H.-W. Ackermann

GENUS	PHAGES WITH DOUBLE CAPSIDS	*TECTIVIRUS*
TYPE SPECIES	PHAGE PRD1 GROUP	—

PROPERTIES OF THE VIRUS PARTICLE

Morphology Icosahedral, 63 nm diameter. Some show single, 20 nm long spikes on vertices. Double capsid consisting of a rigid outer shell 3 nm thick and a flexible inner coat 5-6 nm thick. The latter is destroyed by lipid solvents. Upon nucleic acid ejection, a tail-like structure of about 60 nm in length appears. No envelope.

Physicochemical properties Particle weight ≈ 70 x 10^6 (ϕNS11), S_{20w} ≈ 390; buoyant density in CsCl ≈ 1.28 g/cm^3. Infectivity is ether- and chloroform-sensitive.

Nucleic acid One molecule of linear dsDNA; MW ≈ 10 x 10^6, about 14 % of particle, G+C ≈ 50%.

Protein At least 6 proteins; MW ≈ 11-70 x 10^3.

Lipid 10-20 % by weight of particle; seems to be located in the inner coat and differs partly from that of the host; 5-6 species.

Carbohydrate Not known.

REPLICATION

Virions adsorb to tips of plasmid-dependent pili of gram-negative bacteria . Assembly in nucleoplasm; capsid is assembled first and later filled with DNA. Virulent, lysis.

BIOLOGICAL ASPECTS

Host range Gram-negative bacteria carrying certain drug-resistance plasmids (enterobacteria, *Acinetobacter, Pseudomonas, Vibrio*) and *Bacillus*.

OTHER MEMBERS

L17, PR3, PR4, PR5, PR772 (gram-negatives); AP50 series (6 isolates), Bam35, ϕNS11 (*Bacillus*).

Derivation of Name	tecti: from Latin *tectus*, 'covered'

References

Ackermann, H.-W.; DuBow, M.S.: Viruses of Prokaryotes, Vol. II, pp 171-218 (CRC Press, Boca Raton, Fl., 1987).

Bamford, D.H.; Rouhiainen, L.; Takkinen, K.; Söderlund, H.: Comparison of the lipid-containing bacteriophages PRD1, PR3, PR4, PR5, and L17. J. gen. Virol. *57*:365-373 (1981).

Taxonomic status	English vernacular name	International name
FAMILY	PM2 PHAGE GROUP	***CORTICOVIRIDAE***

Revised by H.-W. Ackermann

GENUS	PM2 PHAGE GROUP	***CORTICOVIRUS***
TYPE SPECIES	*ALTEROMONAS* PHAGE PM2	—

PROPERTIES OF THE VIRUS PARTICLE

Morphology Icosahedral, ≈ 60 nm in diameter, with brush-like spikes on vertices. Multilayered capsid. No envelope.

Physicochemical properties MW ≈ 49 x 10^6; S_{20w} = 230; buoyant density in CsCl = 1.28 g/cm^3. Infectivity is ether-, chloroform-, and detergent-sensitive.

Nucleic acid One molecule of circular supercoiled dsDNA, MW ≈ 6 x 10^6, 13% by weight of particle; G+C = 43% .

Protein Four proteins with MWs = 4.7-44 x 10^3. Protein I forms spikes, II forms outer shell; inner shell of virion contains a transcriptase (protein IV?). Proteins III and IV behave as lipoproteins.

Lipid About 13% of particles; forms a bilayer between outer and inner shell and differs from that of the host; over 90% is phospholipid: 5 species.

Carbohydrate Protein IV is a glycoprotein.

REPLICATION

Adsorption to cell wall. Assembly near plasma membrane, no inclusion bodies. Virulent, lysis.

BIOLOGICAL ASPECTS

Host range *Alteromonas*

Possible member

06N-58P (*Vibrio*)

Derivation of Name cortico: from Latin *cortex, corticis*, 'bark, crust'

REFERENCES

Ackermann, H.-W.; DuBow, M.S.: Viruses of Prokaryotes, vol. II, pp 171-218 (CRC Press, Boca Raton, Fl., 1987).

Franklin, R.M.; Marcoli, R.; Satake, H.; Schäfer, R.; Schneider, D.: Recent studies on the structure of bacteriophage PM2. Med. Microbiol. Immunol. *164*:87-95 (1977).

Mindich, L.: Bacteriophages that contain lipid. *In* Fraenkel-Conrat, H.; Wagner, R.R. (eds.), Comprehensive Virology, Vol. 12, pp. 271-335 (Plenum Press, New York 1978).

Taxonomic status	English vernacular name	International name
ORDER (POSSIBLE)	TAILED PHAGES	—

Compiled by H.-W. Ackermann

GENERAL

Tailed phages are extremely variable in dimensions and physico-chemical properties. Over 3,000 descriptions have been published. Three families are distinguished by tail structure, but no basis for genus definition is apparent. Each family includes large numbers of species.

PROPERTIES OF THE VIRUS PARTICLE

Morphology

Virions consist of head (capsid), tail, and fixation organelles. No envelope. Heads are isometric or elongated and are icosahedra or derivatives thereof (proposed triangulation numbers T = 1, T = 9, T = 13, T = 81). Capsomers are seldom visible and heads usually appear smooth. Isometric heads are 45-170 nm in diameter. Elongated heads are up to 230 nm long. Tails are helical and contractile, long and noncontractile, or short. They may have base plates, spikes, or fibers, and undergo functional changes. Some phages have collars and head or collar appendages. Aberrant structures are frequent.

Physicochemical properties

MW = 29-470 x 10^6, may be higher; S_{20w} = 226-1,230, may be higher; buoyant density in CsCl = 1.41-1.55 g/cm^3. Infectivity is generally ether- and chloroform-resistant. Detergent sensitivity is variable.

Nucleic acid

One molecule of linear dsDNA; MW = 11-490 x 10^6; 25-62% by weight of particle. G+C = 27-72% and usually close to that of the host. DNA may contain unusual bases, which replace normal bases partially or completely, and unusual sugars. It may be circularly permuted, terminally redundant, or nicked and may have cohesive ends, strands of different weight, or terminal proteins. Genes with related functions frequently cluster together.

Protein

Virions contain many different polypeptides (5-50?) (MW = 4-200 x 10^3). Lysozyme is located at the tail tip; other enzymes may be present.

Lipid

Reported in a few phages, mostly of *Mycobacterium*. Presence in others is controversial.

Taxonomic status	English vernacular name	International name

Carbohydrate Glycoproteins, glycolipids, hexosamine, and a polysaccharide have been found in a few cases.

Antigenic properties Virions are antigenically complex and efficient immunogens.

Replication

Tailed phages are virulent or temperate. Temperate phages have a vegetative and a prophage state. Prophages are integrated in, and replicate synchronously with the host genome, or are in the cytoplasm and behave as plasmids. Some phages have transduction or conversion ability. Virions adsorb tail first to cell wall, capsule, flagella, or pili. The cell wall is digested by phage lysozyme. Infecting DNA replicates in a semiconservative way. Replicative intermediates are concatemers or circles. Replication depends on host polymerases (exceptions). Assembly is complex and includes prohead formation and several pathways for separate phage components. DNA is cut to size and packed into preformed capsids. Maturing phages are usually dispersed through the cell; some form regular arrays. Lysis.

Biological Aspects

Host range Over 100 genera of eubacteria and archaebacteria.

Families

Phages with contractile tails	*Myoviridae*
Phages with long, non-contractile tails	*Siphoviridae*
Phages with short tails	*Podoviridae*

Taxonomic status	English vernacular name	International name
FAMILY	PHAGES WITH CONTRACTILE TAILS	***MYOVIRIDAE***

Compiled by H.-W. Ackermann

MAIN CHARACTERISTICS

Tail is contractile, long (80-455 nm) and complex, consisting of a central tube and a contractile sheath separated from the head by a neck. Contraction seems to require ATP. Relatively large capsids.

GENUS	—	—
TYPE SPECIES	COLIPHAGE T4 GROUP	—

PROPERTIES OF THE VIRUS PARTICLE

Morphology Elongated head of about 111 x 78 nm; 152 capsomers (T = 13). Tail of 113 x 16 nm; provided with a collar, base plate, 6 short spikes and 6 long fibers.

Physicochemical properties MW ≈ 210 x 10^6; S_{20w} ≈ 1,030; buoyant density in CsCl = 1.51 g/cm^3. Infectivity is ether- and chloroform-resistant.

Nucleic acid One molecule of linear dsDNA; MW ≈ 120 x 10^6; 48% by weight of particle; contains hydroxymethyl-cytosine instead of thymine; G+C = 35% ; contains glucose. DNA is circularly permuted and terminally redundant. 150-160 genes.

Protein At least 42 polypeptides with MW = 8-155 x 10^3; 1,600-2,000 copies of major capsid protein (MW ≈ 43 x 10^3); 3 proteins are located inside the head. Various enzymes are present, e.g. dehydrofolate reductase, thymidylate synthetase.

Other constituents ATP, folate and polyamines.

REPLICATION

Adsorption site is cell wall; virulent infection. Host chromosome breaks down and viral DNA replicates as concatemer, giving rise to forked replicative intermediates. Heads, tails, and tail fibers are assembled by 3 different pathways. Morphologically aberrant particles are frequent.

Taxonomic status	English vernacular name	International name

BIOLOGICAL ASPECTS

Host range — Enterobacteria.

OTHER MEMBERS

T2, T4, T6, C16, DdVI, PST, SMB, SMP2, α1, 3, 3T+, 9/0, 11F, 50, 66F, 5845, 8893 and about 70 others.

Other members of the family include the following phages listed by host genus or group:

Actinomycetes: SK1, 108/106
Aeromonas: Aeh2, 29, 37, 43, 44RR2.8t, 51, 59.1
Agrobacterium: PIIBNV6
Alcaligenes: A6
Bacillus: G, MP13, PBS1, SP3, SP8, SP10, SP15, SP50, SPy-2, SST
Clostridium: HM3, CEβ
Coryneforms: A19
Cyanobacteria: AS-1, N1, S-6(L)
Enterobacteria: Beccles, FC3-9, K19, Mu, O1, P1, P2, Vil, ϕ92, 121, 16-19, 9266
Lactobacillus: fri, hv, hw, 222a
Listeria: 4211
Mollicutes: Br1
Mycobacterium: I3
Pasteurella: AU
Pseudomonas: PB-1, PP8, PS17, ϕKZ, ϕW-14, ϕ1, 12S
Rhizobium: CM_1, CT4, m, WT1, ϕgal-1-R
Staphylococcus: Twort
Xanthomonas: XP5
Vibrio: kappa, nt-1, X29, VP1, 06N-22P, II

Derivation of Name — myo: from Greek *mys, myos*, 'muscle', relating to contractile tail

---

REFERENCES

Ackermann, H.-W.; DuBow, M.S.: Viruses of Prokaryotes, Vol. II, pp. 1-161 (CRC Press, Boca Raton, Fl., 1978).

Taxonomic status	English vernacular name	International name
FAMILY	PHAGES WITH LONG, NON-CONTRACTILE TAILS	*SIPHOVIRIDAE*

Compiled by H.-W. Ackermann

MAIN CHARACTERISTICS

Tail is noncontractile, long (64?-570 nm).

GENUS	—	—
TYPE SPECIES	COLIPHAGE λ GROUP	—

PROPERTIES OF THE VIRUS PARTICLE

Morphology Isometric head of about 60 nm in diameter; 72 capsomers (T = 7). Flexible tail of 150 x 8 nm with short terminal and subterminal tail fibres.

Physicochemical properties MW ≈ 60 x 10^6; S_{20w} = 388; buoyant density in CsCl = 1.50 g/cm^3. Infectivity is ether-resistant.

Nucleic acid One molecule of linear dsDNA; MW ≈ 33 x 10^6; 54% by weight of particle; G+C = 52%; cohesive ends. About 50 genes.

Protein Nine structural proteins; MWs = 17-130 x 10^3; about 420 copies of major capsid protein (MW = 38 x 10^3).

REPLICATION

Adsorption site is cell wall. Temperate infection; infecting DNA circularizes and integrates into host genome. Bidirectional replication as θ ring is followed by unidirectional replication via rolling-circle mechanism. No breakdown of host DNA. Heads and tails assemble by 2 pathways.

BIOLOGICAL ASPECTS

Host range Enterobacteria.

OTHER MEMBERS

HK97, HK022, PA-2, φD328, φ80

Taxonomic status	English vernacular name	International name

Possible members

T1

Other members of family include the following phages, listed by host genus or group, which probably represent as much species:

Actinomycetes: A1-Dat, Bir, M_1, MSP8, P-a-1, R_1, R_2, SV2, VP5, φC, φ31C, φUW21, φ115-A, φ150-A, 119
Agrobacterium PS8, PT11, ψ
Alcaligenes: A5/A6, 8764
Bacillus: BLE, IPy-1, MP15, mor1, PBP1, SPP1, SPβ, type F, α, φ105, 1A, II
Clostridium: F1, HM7
Coryneforms: A, Arp, BL3, CONX, MT, β, φA8010
Cyanobacteria: S-2L, S-4L
Enterobacteria: H-19J, Jersey, ZG/3A, T5, ViII, β4, χ
Lactobacillus: 1b6, PL-1, y5, φFSW, 223
Lactococcus: BJ5-T, c2, P087, P107, P335, 936, 949, 1358, 1483
Leuconostoc: pro2
Listeria: H387, 2389, 2671, 2685
Micrococcus: N1, N5
Mycobacterium: lacticola, Leo, R1-Myb
Pasteurella: C-2, 32
Pseudomonas: D3, Kf1, M6, PS4, SD1
Rhizobium: NM1, NT2, φ2037/1, 5, 7-7-7, 16-2-12, 317
Staphylococcus: 3A, B11-M15, 77, 107, 187, 2848A
Streptococcus: A25, PE1, VD13, ω3, 24
Vibrio: VP3, VP5, VP11, α3a, OXN-52P, IV

Derivation of Name sipho: from Greek *siphon*, 'tube'

References

Ackermann, H.-W.; DuBow, M.S.: Viruses of Prokaryotes, Vol. II, pp. 1-161 (CRC Press, Boca Raton, Fl., 1987).

Jarvis, A.W.; Fitzgerald, G.F.; Mata, M.; Mercenier, A.; Neve, H.; Powell, I.B.; Ronda, C.; Saxelin, M.; Teuber, M.: Type species and type phages of lactococcal bacteriophages. Intervirology (in press).

Taxonomic status	English vernacular name	International name
FAMILY	PHAGES WITH SHORT TAILS	***PODOVIRIDAE***

Compiled by H.-W. Ackermann

MAIN CHARACTERISTICS

Tail is short (about 20 nm) and noncontractile.

GENUS	—	—
TYPE SPECIES	COLIPHAGE T7 GROUP	—

PROPERTIES OF THE VIRUS PARTICLE

Morphology Isometric head of about 60 nm diameter; 72 capsomers (T = 7). Short tail of 17 x 8 nm with 6 short fibers.

Physicochemical properties MW ≈ 48 x 10^6; S_{20w} = 507; buoyant density in CsCl = 1.50 g/cm^3. Infectivity is ether- and chloroform-resistant.

Nucleic acid One molecule of linear dsDNA; MW ≈ 25 x 10^6; 51% by weight of particle; G+C = 50% and is non-permuted and terminally redundant. 40-50 genes.

Protein About 12 proteins, MW ≈ 14-150 x 10^3; about 450 copies of major capsid protein (MW = 38 x 10^3); 1 or 2 proteins are located inside the head.

REPLICATION

Adsorption site is cell wall. Virulent infection. Host chromosome breaks down and viral DNA replicates as concatemer.

BIOLOGICAL ASPECTS

Host range Enterobacteria.

OTHER MEMBERS

H, PTB, R, T3, W31, ϕI, ϕII.

Other members of the family include the following phages, listed by host genus or group, which probably represent a number of different species:

Actinomycetes: AV-1, Ta_1, 114
Aeromonas: AA-1
Agrobacterium: PIIBNV6-C

Taxonomic status	English vernacular name	International name
	Bacillus: GA-1, ϕ29 *Brucella:*. Tb *Clostridium:* HM2 Coryneforms: AN25S-1, 7/26 Cyanobacteria: AC-1, A-4(L), SM-1, LPP-1 Enterobacteria: Esc-7-11, N4, P22, sd, Ω8, 7-11, 7480b *Lactococcus:* KSY1, P034 Mollicutes: C3, L3 *Mycobacterium:* ϕ17 *Pasteurella:* 22 *Pseudomonas:* F116, gh-1 *Rhizobium:* ϕ2042, 2 *Staphylococcus:* 44AHJD *Streptococcus:* Cp-1, Cvir, H39, 2BV, 182 *Xanthomonas:* RR66 *Vibrio:* OXN-100P, 4996, I, III	

Derivation of Name podo: from Greek *pous, podos*, 'foot', for short tail

REFERENCES

Ackermann, H.-W.; DuBow, M.S.: Viruses of Prokaryotes, Vol. II, pp. 1-161 (CRC Press, Boca Raton, Fl., 1987).

Jarvis, A.W.; Fitzgerald, G.F.; Mata, M.; Mercenier, A.; Neve, H.; Powell, I.B.; Ronda, C.; Saxelin, M.; Teuber, M.: Type species and type phages in lactococcal bacteriophages. Intervirology (in press).

Taxonomic status	English vernacular name	International name
FAMILY	—	***PARVOVIRIDAE***

Reported by G. Siegl

PROPERTIES OF THE VIRUS PARTICLE

Morphology Nonenveloped isometric particles 18-22 nm in diameter, with icosahedral symmetry.

Physicochemical properties MW = 5.5-6.2 x 10^6; S_{20w} = 110-122; buoyant density = 1.39-1.42 g/cm^3 in CsCl. Infectious particles with buoyant densities near 1.45 g/cm^3 may represent conformational variants or precursors to the mature particles. The mature particle is stable in the presence of lipid solvents, at pH 3-9, and in most species at 56 °C for at least 60 min.

Nucleic acid Single molecule of ssDNA of MW = 1.5-2.0 x 10^6; G+C = 41-53%. Members of the genus *Parvovirus* preferentially encapsidate single-stranded DNA of negative polarity. However, under as yet unknown conditions, the percentage of particles encapsidating the positive strand can vary from 1 to 50%. In the genera *Dependovirus* and *Densovirus* complementary plus and minus strands are encapsidated with about equal frequency. After extraction the complementary strands may hybridize *in vitro* to form a double strand.

Protein Three polypeptides, MW = 60-90 x 10^3, can usually be demonstrated in conventionally purified mature virions of the genera *Parvovirus* and *Dependovirus*. Probably all are derived from a common sequence. Densoviruses were shown to have four structural polypeptides. 60-72 protein molecules account for 63-81 % of the weight of the virions. Enzymes are probably lacking.

Lipid None reported.

Carbohydrate None reported.

Polyamines Spermidine, spermine, and putrescine have been demonstrated in mature *Densovirus* particles.

Antigenic properties The polypeptides of the virion are immunologically distinguishable; they are, however, antigenically related. In general, antisera to polypeptides do not show neutralization or react with whole virion using HI, complement-fixation or immune electrophoresis. For the genus *Parvovirus* hemagglutinating, complement-fixing,

Taxonomic status	English vernacular name	International name

and neutralizing antigens are type specific without cross-reaction. Cross-reactions, however, have been observed in the fluorescent antibody test for several parvoviruses. This may be due to the existence of at least one nonstructural, highly conserved antigen. Dependoviruses share a similar common antigen and common antigens were also demonstrated for Densoviruses by fluorescent antibody staining and by immunodiffusion.

Replication

Viral replication takes place in the nucleus where viral proteins in the form of empty capsid structures and progeny infectious virions accumulate. For multiplication, members of the genus *Parvovirus* require one or more cellular functions generated during the S phase of the cellular division cycle. Members of the genus *Dependovirus* require helper virus coinfection (adenoviruses, herpesviruses) for efficient replication, but recent data suggest that cells may become at least partially competent in independent replication of the viruses during a narrow window (presumably within the S phase) of the cell cycle.

Genera

Parvovirus group	*Parvovirus*
Adeno-associated virus group	*Dependovirus*
Insect parvovirus group	*Densovirus*

GENUS	PARVOVIRUS GROUP	***PARVOVIRUS***
TYPE SPECIES	MINUTE VIRUS OF MICE (MVM)	—

Properties of the Virus Particle

Nucleic acid The linear molecule of ssDNA has hairpin structures at both the 5'- and 3'-ends (3'-terminal hairpin: 115-116 nucleotides, 5'-palindromic structure: 200-242 nucleotides). In most members of the genus all mature virions contain minus-strand DNA. In other members, plus-strand DNA is also encapsidated in variable (1-50 %) proportions.

Effect on cells Characteristic CPE induced by many viruses during replication in cell culture. Many species contain a hemagglutinin which has different activities for a variety of red blood cells.

REPLICATION

The virus multiplies in the nucleus, and replication is dependent upon certain helper functions provided by the host cell. In consequence, viruses multiply preferentially in actively dividing cells.

BIOLOGICAL ASPECTS

Host range

In nature: cat, cattle, chicken, dog, goose, man, mink, mouse, pig, rabbit, raccoon, rat. Under experimental conditions the host range may be extended to homologous or closely related hosts. Rodent viruses and LuIII also replicate in Syrian hamsters.

Transmission

Transplacental transmission has been detected for a number of species. Vertical passage by ova is indicated for goose parvovirus. Transmission by mechanical vectors is also possible.

OTHER MEMBERS

Aleutian mink disease parvovirus
B19
Bovine parvovirus
Feline parvovirus
    Species host range variants
    FPLV (feline panleukopenia virus)
    MEV (mink enteritis virus)
    CPV (canine parvovirus)
    RPV (raccoon parvovirus)
Goose parvovirus
H-1
Lapine parvovirus
LuIII
Porcine parvovirus
Rat virus
RT
TVX

Probable members

Chicken parvovirus
HB
Minute of canines (MVC)
RA-1

Derivation of Name

parvo: from Latin *parvus*, "small"
adeno: from Greek *aden, adenos*, "gland"
dependo: from Latin *dependere*, "depending"
denso: from Latin *densus*, "thick, compact"

REFERENCES

Berns, K.I.: The Parvoviruses (Plenum Press, New York, London, 1984).

Kurstak, E.: Small DNA densonucleosis virus (DNV). Adv. Virus Res. *17*:207-241 (1972).

Lightner, D.V.; Redman, R.M.: A parvo-like virus disease of Penaeid shrimp. J. Invertebr. Pathol. *45*:47-53 (1985).

Mari, J.; Bonami, J.-R.: PC84, a parvo-like virus from the crab *Carcinus mediterraneus*: pathological aspects, ultrastructure of the agent, and first biochemical characterization. J. Invertebr. Pathol. *51*:145-156 (1988).

Pattison, J.R.: Parvoviruses and human disease (CRC Press, Boca Raton, Fl., 1988).

Siegl, G: The parvoviruses. Virology Monographs, Vol. 15 (Springer, Wien, New York, 1976).

Siegl, G.; Bates, R.C.; Berns, K.I.; Carter, B.J.; Kelly, D.C.; Kurstak, E.; Tattersall, P.: Characteristics and Taxonomy of *Parvoviridae*. Intervirology *23*:61-73 (1985).

Tijssen, P.: Handbook of Parvoviruses, Vols. I and II. (CRC Press, Boca Raton, Fl., 1990).

Tijssen, P.; Kurstak, E.; Su, T.M.; Garzon, S.: Proc. Colloqu. Invertebr. Pathol., Brighton, England, p. 148 (1982).

Ward, D.; Tattersall, P.: Replication of mammalian parvoviruses. (Cold Spring Harbor Press, Cold Spring Harbor, 1978).

Taxonomic status	English vernacular name	International name
GROUP	—	***GEMINIVIRUS***

Revised by R. Hull, J. Stanley & R.W. Briddon

PROPERTIES OF THE VIRUS PARTICLE

Morphology
Geminate particles, ≈ 18 x 30 nm, consisting of two incomplete icosahedra with T = 1 surface lattice with a total of 22 capsomers.

Physicochemical properties
$S_{20w} \approx 70$ (for particle pairs).

Nucleic acid
One (subgroup I and II) or two (subgroup III) molecules of single-stranded DNA, MW = 7-8 x 10^5 (2.5-3.0 kb). Open reading frames occur on both the viral strand and its complement.

Protein
Single coat polypeptide, MW = 27-34 x 10^3.

Lipid
None reported.

Carbohydrate
None reported.

Antigenic properties
Efficient immunogens. Single precipitin line in gel-diffusion. Some serological relationship between members of subgroup III.

REPLICATION

Genome is replicated via dsDNA which can be isolated from infected tissues. Virus particles accumulate in nucleus, producing large aggregates.

BIOLOGICAL ASPECTS

Host range
Members of the subgroup I are confined primarily to the Graminae. Members of subgroup II and III infect dicotyledonous plants. Individual members tend to have narrow host-ranges except BCTV which has a wide host range.

Transmission
Transmitted naturally by leafhoppers (subgroup I and II) or the whitefly *Bemisia tabaci* (subgroup III) in a persistent manner. Some members are also mechanically transmissible, usually with difficulty.

Taxonomic status	English vernacular name	International name

HOMOLOGIES BETWEEN SUBGROUPS

DNA A of the viruses in subgroup III bear some sequence similarities and possible structural and functional similarities to the DNA genome of viruses in subgroups I and II, suggesting a distant evolutionary relationship between the subgroups.

SUBGROUP I	—	—
TYPE MEMBER	MAIZE STREAK VIRUS (MSV) (133)	—

PROPERTIES OF THE VIRUS PARTICLE

Nucleic acid One molecule of circular single stranded DNA, MW = 7-8 x 10^5 (2.7-3.0 kb). Open reading frames (putative genes) occur on both the viral strand and its complement. DNAs of five members (MSV, WDV, DSV, MiSV and CSMV) have been sequenced .

Protein Single coat polypeptide, MW = 28-34 x 10^3.

Antigenic properties Serological tests show lack of interrelationship among subgroup members.

BIOLOGICAL ASPECTS

Host range Subgroup members have narrow host ranges limited to the Graminae.

Transmission Transmitted in nature by leafhoppers in a persistent manner. Not transmitted by mechanical inoculation. Some members have been transmitted by *Agrobacterium*-mediated transfer using recombinant DNA methods (MSV, DSV and WDV).

OTHER MEMBERS

Chloris striate mosaic (221)
Digitaria streak
Miscanthus streak
Wheat dwarf

Probable members

Bajra streak
Bromus striate mosaic
Digitaria striate mosaic
Oat chlorotic stripe
Paspalum striate mosaic

Taxonomic status	English vernacular name	International name
SUBGROUP II	—	—
TYPE MEMBER	BEET CURLY TOP VIRUS (BCTV)(210)	—

PROPERTIES OF THE VIRUS PARTICLE

Nucleic acid One molecule of circular single stranded DNA, MW = 8 x 10^5 (2993 b). Open reading frames (putative genes) occur on both the viral strand and its complement. DNA of BCTV has been sequenced.

Protein Single coat polypeptide, MW = 30 x 10^3.

Antigenic properties Serological tests show distant relationship with most subgroup III members (BCTV and TPCTV and TLRV; BSDV and TYDV are closely related).

BIOLOGICAL ASPECTS

Host range Subgroup member, BCTV, has a wide host range.

Transmission Transmitted in nature by leafhoppers in a persistent manner, except for TPCTV which is transmitted by treehopper. Transmitted with difficulty by mechanical inoculation. BCTV has been transmitted by *Agrobacterium*-mediated transfer using recombinant DNA methods.

OTHER MEMBERS

Tomato pseudo-curly top virus
Bean summer death virus
Tobacco yellow dwarf virus
Tomato leafroll virus

SUBGROUP III	—	—
TYPE MEMBER	BEAN GOLDEN MOSAIC VIRUS (BGMV) (192)	—

PROPERTIES OF THE VIRUS PARTICLE

Nucleic acid Two molecules of circular single-stranded DNA, each MW = 7-8 x 10^5 (2.4-2.8 kb). Open reading frames (putative genes) occur on both the viral strand and its complement. DNAs of six members (ACMV, BGMV, TGMV, TYLCV, AbMV and MYMV) have been sequenced.

Protein Single coat polypeptide, MW = 27-30 x 10^3.

Taxonomic status	English vernacular name	International name
Antigenic properties	Serological tests show interrelationships among some subgroup members.	

BIOLOGICAL ASPECTS

Host range Individual subgroup members generally have narrow host ranges, among dicotyledonous plants, but as a subgroup the viruses have hosts in a wide spectrum of higher plant families.

Transmission Transmitted in nature by the whitefly *Bemisia tabaci* in a persistent manner and experimentally, by mechanical inoculation.

OTHER MEMBERS

Abutilon mosaic virus
African cassava mosaic virus
Cotton leaf crumple virus
Euphorbia mosaic virus
Honeysuckle yellow vein mosaic virus
Horsegram yellow mosaic virus
Indian cassava mosaic virus
Jatropha mosaic virus
Limabean golden mosaic virus
Malvaceous chlorosis virus
Melon leaf curl virus
Mungbean yellow mosaic virus
Potato yellow mosaic virus
Rhynochosia mosaic virus
Squash leaf curl virus
Tigre disease virus
Tobacco leaf curl virus
Tomato golden mosaic virus
Tomato leaf curl virus
Tomato yellow dwarf virus
Tomato yellow leaf curl virus
Tomato yellow mosaic virus
Watermelon chlorotic stunt virus
Watermelon curly mottle virus

Probable members

Cotton leaf curl virus
Cowpea golden mosaic virus
Eggplant yellow mosaic virus
Eupatorium yellow vein virus
Lupin leaf curl virus
Solanum apical leaf curl virus
Soybean crinkle leaf virus
Wissadula mosaic virus

Derivation of Name	gemini: from Latin *gemini*, "twins", from characteristic double particle.

REFERENCES

Davies, J.W.; Townsend, R.; Stanley, J.: The structure, expression, functions and possible exploitation of geminivirus genomes; *In* Hohn, T.; Schell, J. (eds.), Plant DNA Infectious Agents pp. 31-52 (Springer, Wien, New York, 1987).

Francki, R.I.B.; Milne, R.G.; Hatta, T.: Geminivirus group, *In* Atlas of Plant Viruses, Vol. I, pp. 33-46 (CRC Press, Boca Raton, Fl., 1985).

Goodman, R.M.: Geminiviruses. J. gen. Virol. *54:*9-21 (1981).

Harrison, B.D.: Advances in geminivirus research. Annu. Rev. Phytopathol. *23:*55-82 (1985).

Howarth, A.J.; Goodman, R.M.: Divergence and evolution of geminivirus genomes. J. Molec. Evol. *23:*313-319 (1986).

Lazarowitz, S.G.: The molecular characterization of geminiviruses Plant Mol. Biol. Reptr. *4:* 177-192 (1987).

Roberts, I.M.; Robinson, D.J.; Harrison, B.D.: Serological relationships and genome homologies among geminiviruses. J. gen. Virol. *65:*1723-1730 (1984).

Stanley, J.; Davies, J.W.: Structure and function of the DNA genome of Geminiviruses; *In* Davies, J.W. (ed.), Molecular Plant Virology, Replication and gene expression, Vol. II, pp 191-218 (CRC Press, Boca Raton, Fl., 1984).

Thomas, J.E.; Massalski, P.R.; Harrison, B.D.: Production of monoclonal antibodies to African cassava mosaic virus and differences in their reactivities with other whitefly-transmitted geminiviruses. J. gen. Virol. *67:*2739-2748 (1986).

Taxonomic status	English vernacular name	International name
FAMILY	ISOMETRIC PHAGES WITH SSDNA	*MICROVIRIDAE*

Revised by H.-W. Ackermann

PROPERTIES OF THE VIRUS PARTICLE

Morphology Isometric, 25-27 nm in diameter. No envelope.

Nucleic acid One molecule of circular ssDNA, MW = 1.6-1.7 x 10^6.

Lipid None or not reported.

Carbohydrate None or not reported.

REPLICATION

Phage DNA is converted to circular RF. Virulent, lysis.

GENERA

ϕX-type phages	*Microvirus*
SpV4-type phages	*Spiromicrovirus*
MAC-1-type phages (proposed)	-

GENUS	ϕX PHAGE GROUP	*MICROVIRUS*
TYPE SPECIES	COLI PHAGE ϕX174 GROUP	—

PROPERTIES OF THE VIRUS PARTICLE

Morphology Icosahedral, about 27 nm in diameter; 12 conspicuous capsomers (T = 1) and knob-like spikes on vertices. No envelope.

Physicochemical properties MW = 6.7 x 10^6; S_{20w} = 114; buoyant density in CsCl = 1.41 g/cm^3. Infectivity is chloroform-resistant.

Nucleic acid Positive-sense ssDNA; MW ≈ 1.7 x 10^6, about 26% by weight of particle; G+C = 44% ; 11 partly overlapping genes.

Protein 60 molecules of capsid protein (MW = 50 x 10^3); 3 other proteins.

Lipid None.

Carbohydrate None.

Taxonomic status	English vernacular name	International name

REPLICATION

Adsorption to cell wall. Progeny ssDNA is generated by displacement from RF DNA. Three sections of the genome code for three different proteins in different reading frames. Capsid is assembled in nucleoplasm around scaffolding protein. No inclusion bodies.

BIOLOGICAL ASPECTS

Host range Enterobacteria.

OTHER MEMBERS

BE/1, dϕ3, dϕ4, dϕ5, G4, G6, G13, G14, 1ϕ1, 1ϕ3, 1ϕ7, 1ϕ9, M20, St-1, S13, WA/1, WF/1, WW/1, U3, 2D/13, α3, α10, δ1, ζ3, η8, o6, ϕA, ϕR (several species)

GENUS	SPV4-TYPE PHAGES	***SPIROMICROVIRUS***
TYPE SPECIES	*SPIROPLASMA* PHAGE SPV4	—

PROPERTIES OF THE VIRUS PARTICLE

Morphology Isometric, about 27 nm in diameter, no envelope.

Physiochemical properties Buoyant density in CsCl = 1.40 g/cm^3. Infectivity is ether-, chloroform-, and detergent-resistant.

Nucleic Acid One molecule of circular ssDNA; MW $\approx$ 1.7 x 10^6, G+C = 32%, at least 9 partly overlapping genes.

Protein Single capsid protein of 64 x 10^3.

Lipid Not known.

Carbohydrate Not known.

BIOLOGICAL ASPECTS

Host range *Spiroplasma.*

Taxonomic status	English vernacular name	International name
POSSIBLE GENUS	MAC-1 TYPE PHAGES	—
TYPE SPECIES	*BDELLOVIBRIO* PHAGE MAC-1 GROUP	—

PROPERTIES OF THE VIRUS PARTICLE

Morphology Isometric, about 25 nm diameter, no envelope.

Physiochemical properties S_{20w} = 94, buoyant density in CsCl = 1.36 g/cm^3.

Nucleic Acid One molecule of circular ssDNA; MW ≈ 1.6 x 10^6.

Protein Single capsid protein of 64 x 10^3.

Lipid Not known.

Carbohydrate Not known.

BIOLOGICAL ASPECTS

Host range *Bdellovibrio.*

MEMBERS

MAC-1', MAC-2, MAC-4, MAC-4', MAC-5, MAC-7.

Derivation of Name micro: from Greek *mikros*, "small"

REFERENCES

Ackermann, H.-W.; DuBow, M.S.: Viruses of Prokaryotes, Vol. II, pp. 171-218 (CRC Press, Boca Raton, Fl., 1987).

Denhardt, D.T.: The isometric single-stranded DNA phages. *In* Fraenkel-Conrat, H.; Wagner, R.R. (eds.), Comprehensive Virology, Vol. 7, pp. 1-104 (Plenum Press, New York, 1977).

Maniloff, J.: Mycoplasma Viruses. CRC Crit. Rev. Microbiol. *15:* 339-389 (1988).

Roberts, R.C.; Keefer, M.A.; Ranu, R.S.: Characterization of *Bdellovibrio bacteriovorus* bacteriophage MAC-1. J. Gen. Microbiol. *133:* 3065-3070 (1987).

Taxonomic status	English vernacular name	International name
FAMILY	ROD-SHAPED PHAGES	*INOVIRIDAE*

Revised by H.-W. Ackermann & J. Van Duin

PROPERTIES OF THE VIRUS PARTICLE

Morphology Long or short helical rods, which are or seem to be tubules. DNA is located in particle centre. Particles of abnormal length are frequent. No envelope.

Physicochemical properties Infectivity is chloroform-sensitive and heat-resistant.

Nucleic acid One molecule of circular positive-sense ssDNA.

REPLICATION

Infecting viral DNA is converted into a dsRF which replicates in a semiconservative way. No inclusion bodies. Phages extruded through host membranes; no lysis, host survives and enters a carrier state.

GENERA

Filamentous phages	*Inovirus*
Rod-shaped phages	*Plectrovirus*

GENUS	FILAMENTOUS PHAGES	*INOVIRUS*
TYPE SPECIES	COLIPHAGE fd GROUP	—

PROPERTIES OF THE VIRUS PARTICLE

Morphology Usually flexible rods, 760-1,950 nm long x 6-8 nm in diameter.

Physicochemical properties MW = 12-23 x 10^6; S_{20w} = 41-45; buoyant density in CsCl = 1.3 g/cm^3. Infectivity is sensitive to sonication; ether sensitivity is variable.

Nucleic acid MW = 1.9-3.0 x 10^6; 6-21% by weight of particle; G+C = 40-60%. So far as known, 9 partly overlapping genes.

Protein One major coat protein (MW ≈ 5 x 10^3) and 3 or 4 copies of adsorption protein (MW ≈ 65-70 x 10^3). Coat proteins appear to lack cysteine and histidine.

Lipid None.

Carbohydrate None.

Taxonomic status	English vernacular name	International name

REPLICATION

Particles adsorb slowly to pili or poles (?) of bacteria and enter the cells; many plasmid specificities. Progeny viral ssDNA is produced by displacement from RF DNA. Overlapping genes code for different proteins in different reading frames. Mature phages are assembled at the plasma membrane as the particles leave the cell. Some phages are temperate.

BIOLOGICAL ASPECTS

Host range Enterobacteria, *Pseudomonas, Vibrio, Xanthomonas*.

OTHER MEMBERS

The genus probably includes 10-11 species differentiated by particle length, host range, antigenic properties and chemical composition.

a. Phages of enterbacteria:

- fd group: AE2, Ec9, f1, HR, M13, ZG/2, ZJ/2, δA
- other: C-2, If1, If2, Ike, I_2-2, PR64FS, SF, tf-1, X

b. Pf1, Pf2, Pf3 (*Pseudomonas*).

c. Cf, Cf1t, Xf, Xf2 (*Xanthomonas*).

d. v6, Vf12, Vf33 (*Vibrio*).

GENUS	ROD-SHAPED PHAGES	***PLECTROVIRUS***
TYPE SPECIES	*ACHOLEPLASMA* PHAGE	—
	L51 GROUP	

PROPERTIES OF THE VIRUS PARTICLE

Morphology Short, straight rods, ≈ 85 to 280 nm long x 14 nm in diameter; one round end; may be derived from icosahedra (T = 1).

Physicochemical properties Buoyant density in CsCl ≈ 1.38 g/cm^3. Infectivity is ether- and detergent-resistant.

Nucleic acid MW = 2.5-5.2 x 10^6 (4.4-8.5 kb).

Protein Four proteins (MW = 19-70 x 10^3, shown in L51 only).

Lipid None reported.

Taxonomic status	English vernacular name	International name

Carbohydrate None reported.

Replication

Infecting viral DNA is converted into a dsRF which replicates in a semiconservative way. No inclusion bodies. Phages extruded through host membranes, no lysis, host survives and enters a carrier state.

Biological Aspects

Host range *Acholeplasma* and *Spiroplasma*

Other Members

a. MV-L1, MVG51, 0c1r, 10tur, others (over 50 isolates, *Acholeplasma*, 85 x 14 nm).
b. SV1 (*Spiroplasma*, 230-280 x 10-15 nm).

Derivation of Name ino: from Greek *is, inos*, 'muscle'

References

Ackermann, H.-W.; DuBow, M.S.: Viruses of Prokaryotes, Vol. II, pp. 171-218 (CRC Press, Boca Raton, Fl., 1987).

Kuo, T.-T.; Lin, Y.-H.; Huang, C.-M.; Chang, S.-F.; Dai, H.; Feng, T.-Y.: The lysogenic cycle of the filamentous phage Cflt from *Xanthomonas campestris* pv. *citri*. Virology *156*: 305-312 (1987).

Maniloff, J.: Mycoplasma viruses. CRC Crit. Rev. Microbiol. *15*: 339-389 (1988).

Ray, D.S.: Replication of filamentous bacteriophages; *In* Fraenkel-Conrat, H.; Wagner, R.R. (eds.), Comprehensive Virology, Vol. 7, pp. 105-178 (Plenum Press, New York, 1977).

Zinder, N.D.; Horiuchi, K.: Multiregulatory element of filamentous bacteriophages. Microbiol. Rev. *49*: 101-106 (1985).

Taxonomic status	English vernacular name	International name
GENUS	—	***ORBIVIRUS***
TYPE SPECIES	BLUETONGUE VIRUS	—

PROPERTIES OF THE VIRUS PARTICLE

Morphology

Particles 65-80 nm in diameter. Outer capsid shells and cores have no projections, but exhibit 32 large ring-shaped capsomers which are visible when the outer shells are still present.

Physicochemical properties

MW ≈ 80 x 10^6; S_{20w} = 550. Infectivity is lost at pH 3.0. Lipid solvents reduce infectivity ≈ 10-fold.

Nucleic acid

10 segments with MW = 0.5-2.8 x 10^6; total MW ≈ 15 x 10^6; 20% by weight of virus; G+C = 42-44%.

Protein

Seven proteins with MW = 35-150 x 10^3; 80% by weight. Removal of outer shell required for activation of the RNA-dependent RNA polymerase. Major core proteins are VP3 and VP7 (MW ≈ 103 and 38 x 10^3, respectively); the latter is the major component of capsomeres on the core surface. Cores also contain VP1, VP4 and VP6. Outer capsid layer contains VP2 (MW ≈ 111 x 10^3) and VP5 (MW ≈ 59 x 10^3). Three non-structural proteins NS1-3 have MW ≈ 64.4, 41 and 25.6 x 10^3; NS2 is a phosphoprotein.

Antigenic properties

Protein VP2 is main antigenic determinant for neutralization, although VP5 is also involved in type specificity. VP7 is the main group-specific antigen.

REPLICATION

Removal of outer capsid is required for activation of the RNA-dependent RNA polymerase. Replication takes place in cytoplasmic viroplasms. There are 3 nonstructural proteins, NS1-3 with MWs ≈ 64.4, 41 and 25.6 x 10^3. Morphogenesis is accompanied by formation of regularly structured filaments and tubules. The latter, at least, consist of NS1.

BIOLOGICAL ASPECTS

Host range

Insects and other arthropods; vertebrate hosts include man, horses, monkeys, rabbits, cattle, deer, suckling mice.

Transmission

Vectors; *Culicoides*, mosquitoes, phlebotomines and ticks.

Taxonomic status	English vernacular name	International name

OTHER MEMBERS

There are 12 serological groups in the genus *Orbivirus* (number of serotypes and vectors are indicated where known):

African horse sickness	9	
Bluetongue	24	*Culicoides*
Changuinola	7	Phlebotomines
Corriparta	3	Mosquitoes
Epizootic hemorrhagic disease	7	*Culicoides*
Equine encephalosis	5	
Eubenangee	3	Mosquitoes
Kemerovo	20	Ticks
Palyam	6	*Culicoides*
Wallal	2	
Warrego	2	*Culicoides*

GENUS	—	***COLTIVIRUS***
TYPE SPECIES	COLORADO TICK FEVER VIRUS (FLORIO STRAIN)	—

PROPERTIES OF THE VIRUS PARTICLE

Morphology Sperical particles 80 nm in diameter with two outer capsid shells and a core which possesses no projections. Surface capsomeric structure of core particles differs from orbiviruses.

Physicochemical properties Infectivity is lost at pH 3.0. Lipids solvents reduce infectivity.

Nucleic Acid 12 segments with MW = $0.24\text{-}2.5 \times 10^6$; total MW $\approx 18 \times 10^6$.

Protein Unknown.

Carbohydrate Unknown.

Antigenic properties Only two known serotypes exist, and these are represented by North American isolates and the European isolate, Eyach. Antigenic variants of the North American serotype have been reported.

REPLICATION

Replication takes place in cytoplasmic viroplasms. Morphogenesis is accompanied by formation of regularly structured filaments and tubules.

Taxonomic status	English vernacular name	International name

BIOLOGICAL ASPECTS

Host range — Primarily Ixodidae ticks, but 12 segmented dsRNA viruses have been isolated from mosquitoes. Vertebrate species in which the virus replicates and has been isolated include man, deer, and small animals. Man represents an accidental host. Isolations have been made repeatedly in North America from *Dermacentor andersoni*. Indonesian isolates have been made from mosquitoes, and Chinese isolates, from ticks, cattle, pigs and man.

Transmission — Vectored by ticks, and possibly mosquitoes (Indonesian isolates).

OTHER MEMBERS

Eyach
Ar 577 (Eyach antigenic variant)
Ar 578 (Eyach antigenic variant)

Probable members

Indonesian isolates	JKT-6423 JKT-6969 JKT-7041 JKT-7075
Chinese isolates	M14 HN59 HN131 HN199 HN295

GENUS	—	*ROTAVIRUS*
TYPE SPECIES	HUMAN ROTAVIRUS	—

PROPERTIES OF THE VIRUS PARTICLE

Morphology — Particles 65-75 nm in diameter with two outer capsid shells and a core without spikes. The capsomers are composed of shared subunits; symmetry is T = 13 (laevo).

Physicochemical properties — $S_{20w} \approx 525$; buoyant densities of particles and cores are 1.36-1.38 and 1.44 g/cm^3, respectively. Infectivity is stable at pH 3.0 and relatively heat-stable. Resistant to ether.

Nucleic acid — 11 segments with MW = 0.4-2.1 x 10^6; total MW $\approx$ 12 x 10^6; 12-15% by weight of virus for group A. Short conserved sequences at all 5' ends, and (distinct) at all 3' ends.

Protein

6 structural proteins (MW = 34-125 x 10^3) for group A. Removal of outer capsid (at low Ca^{++} concentration) required for transcriptase activity. Capping enzymes present.

Antigenic properties

6 serogroups described; the major group antigen is the major inner capsid protein VP6. Within group A, two VP6 subgroups and 11 distinct serotypes based on outer capsid glycoprotein VP7 (designated G1-11) are recognized. There are about 9 VP4 (P) "serotypes" based on sequences as there is partial antigenic overlap. Characterization of groups B-F is limited as most strains grow only in their original hosts.

REPLICATION

Unlike orthoreoviruses, rotaviruses enter the cytoplasm directly through the plasma membrane, not via endocytotic vesicles. Penetration depends on VP4 after specific cleavage by trypsin. Transcription like that of *Orthoreovirus*. 5 nonstructural proteins, one mediates budding of single capsid particles into endoplasmic reticulum. Final assembly occurs within cisternae of ER, after separate secretion of the glycoprotein VP7. *In vivo*, replication restricted to intestinal ephithelial cells.

BIOLOGICAL ASPECTS

Host range

Mammals and birds. Diarrhoeal disease is caused by homologous virus in humans, mice, calves, piglets, turkeys etc. Group A serotypes G3, G4, G8 and G9 found in both humans and other mammals, host specificity mainly dependent on VP4.

Transmission

Horizontal. No vectors. Environmental contamination.

OTHER MEMBERS

Group A rotaviruses have been identified in most mammalian and avian species studied. Group B rotaviruses occur in humans, pigs, cattle, sheep and rats. Group C viruses are found in pigs and rarely in humans; groups D and F in poultry, and group E in pigs.

Taxonomic status	English vernacular name	International name
GENUS	—	***AQUAREOVIRUS***
TYPE SPECIES	GOLDEN SHINER VIRUS (GSV)	—

PROPERTIES OF THE VIRUS PARTICLE

Morphology External appearance similar to *Orthoreovirus*, ≈ 75 nm in diameter; core ≈ 50 nm.

Physicochemical properties Buoyant density 1.36 g/cm^3. Infectivity resistant to ether and proteolytic enzymes.

Nucleic acid 11 segments with MW = 0.3-2.5 x 10^6, total MW = ≈ 15 x 10^6.

Protein 5 major structural proteins with MW = 34-135 x 10^3. At least 2 other minor virion proteins present.

REPLICATION

In cytoplasm, probably like orthoreoviruses.

BIOLOGICAL ASPECTS

Host range Poikilotherm vertebrates and invertebrates (fish and shellfish). Efficient replication in fish cell lines.

Transmission Horizontal; no vectors identified.

OTHER MEMBERS

13p2 reovirus (13p2)
Chum salmon virus (CSV)
Channel catfish reovirus (CRV)

Probable members

Tench reovirus
Chub reovirus
Coho salmon reovirus
Hard clam reovirus

Possible members

Grass carp reovirus
Turbot reovirus

Taxonomic status	English vernacular name	International name
GENUS	CYTOPLASMIC POLYHEDROSIS VIRUS GROUP	*CYPOVIRUS*
TYPE SPECIES	CYTOPLASMIC POLYHEDROSIS VIRUS (CPV) FROM *BOMBYX MORI*	—

PROPERTIES OF THE VIRUS PARTICLE

Morphology

Spherical particles, 50-65 nm diameter with 12 apparently hollow spikes located at icosahedral vertices. Dense core surrounded by an outer shell, but no clearly defined outer capsid structure like that of orthoreoviruses.

Physicochemical properties

MW ≈ 50 x 10^6; S_{20w} = 370-440. Stable at pH 3.0; infectivity lost after 10 min at 80-85 °C; resistant to ether. Capsid resistant to proteolytic enzymes such as chymotrypsin.

Nucleic acid

10 segments of dsRNA with MW = 0.3-2.7 x 10^6; 25-30% by weight of virus; G+C = 36-42%. Segments have no homology with members of other genera. Positive strands of virion RNA are methylated and capped at the 5' terminus.

Protein

3-5 proteins, MW = 30-151 x 10^3. 70-75% by weight of virus. Transcriptase in particle does not require treatment with proteolytic enzymes for activation. Also present: nucleotide phosphohydrolase; capping enzymes; exonuclease; hemagglutinin for chick, sheep, and mouse erythrocytes.

REPLICATION

Probably like orthoreoviruses. Many virus particles are occluded with 'polyhedra' composed of one major protein, MW = 25-30 x 10^3, which is probably a glycoprotein.

BIOLOGICAL ASPECTS

Pronounced cellular tropism for midgut epithelial cells.

Host range

Insects: Lepidoptera, Diptera, Hymenoptera.
Crustacea: Simocephalus.

Transmission

Horizontal.

Taxonomic status	English vernacular name	International name

Other Members

Eleven 'types' defined by the distinctive electrophoretic profiles of their RNA genome segments (in addition to type 1, the type species):

Type 2 from *Inachis io*
Type 3 from *Spodoptera exempta*
Type 4 from *Actias selene*
Type 5 from *Trichoplusia ni*
Type 6 from *Biston betularia*
Type 7 from *Triphena pronuba*
Type 8 from *Abraxas grossulariata*
Type 9 from *Agrotis segetum*
Type 10 from *Aporophylla lutulenta*
Type 11 from *Spodoptera exigua*
Type 12 from *Spodoptera exempta*

Probable members

Viruses from ≈ 150 different insect species.

GENUS	PLANT REOVIRUS SUB-GROUP 1	***PHYTOREOVIRUS***
TYPE SPECIES	WOUND TUMOR VIRUS (WTV) (34)	—

Properties of the Virus Particle

Morphology Particles 65-70 nm in diameter. WTV possesses an outer amorphous layer (2 proteins), an outer layer of distinct capsomers, and a smooth core (3 proteins, MWs ≈ 58, 118, and 160 x 10^3). The core is 45-60 nm in diameter and lacks spikes.

Physicochemical properties MW ≈ 75 x 10^6; S_{20w} ≈ 510. Optimal stability at pH 6.6. Resistant to freon and CCl_4.

Nucleic acid 12 segments with MW = 0.3-3.0 x 10^6, with total MW ≈ 16 x 10^6; 22% by weight of virus; G+C = 38-44%. Each of the 12 WTV genomic segments contains the conserved oligonucleotides (+) 5'GGUAUU...UGAU3'. The genomic segments of all three phytoreoviruses contain conserved terminal oligonucleotides with the consensus sequence (+) 5'GGU/CA...U/CGAU3'.

Protein Seven proteins with MWs ranging from 35-160 x 10^3; 78% by weight of virus. Removal of outer shell not required for activation of transcriptase.

Antigenic Properties

All three members of the genus are antigenically unrelated to each other.

REPLICATION

In cytoplasm, probably like that of orthoreoviruses. Continuous propagation in plants without access to vectors can lead to the selection of mutants that lack some genome segments and which may no longer replicate in the insect.

BIOLOGICAL ASPECTS

Host range

In nature, WTV was originally found once, in the leafhopper *Agalliopsis novella*. A second isolate (New Jersey strain) was recently detected in a single *Vinca major* plant set out as a bait plant in a blueberry field. Experimental host range of WTV is wide among dicotyledonous plants. Rice dwarf and rice gall dwarf viruses have narrow host ranges among the *Gramineae*. WTV grows in cell lines derived from embryonic tissues of vectors.

Transmission

Phytoreoviruses are transmitted only by cicadellid leafhoppers (*Agallia*, *Agalliopsis*, *Nephotettix*, etc.). Transmission is propagative; acquisition after 1 min or more; latent period ≈ 2 weeks, then lifelong transmission by insects to plants. Transovarial in insect vectors.

OTHER MEMBERS

Rice dwarf virus (102)
Rice gall dwarf virus (296)

GENUS	PLANT REOVIRUS SUBGROUP 2	***FIJIVIRUS***
TYPE SPECIES	FIJI DISEASE VIRUS (FDV) (119)	—

PROPERTIES OF THE VIRUS PARTICLE

Morphology

Particles 65-70 nm in diameter (in uranyl acetate). 12 external knobs ≈ 11 nm in diameter and 9-11 nm long (A spikes); particles break down spontaneously *in vitro* to give spiked cores 54 nm in diameter, which have 12 icosahedrally located spikes (B spikes, ≈ 8 nm high, 11-13.5 nm wide). Treatment of maize rough dwarf virus (MRDV) with various reagents produces smooth (spikeless) cores, 50-57 nm in diameter containing 2 proteins, MW ≈ 126 and 139 x 10^3.

Taxonomic status	English vernacular name	International name

Physicochemical properties Not established.

Nucleic acid 10 segments, MW = 1.0-2.9 x 10^6 with total MW = 18-20 x 10^6; G+C.= 45%. Three RNA segments of maize rough dwarf virus and one of rice black streaked virus have been shown to have the same conserved oligonucleotides (+) 5'AAGUUUUUU...UGUC3' which differ from the phytoreoviruses.

Protein For MRDV, 6 proteins MW = 64-139 x 10^3.

Antigenic properties Serological studies complicated by presence of antibodies to dsRNA in many antisera. Viruses fall into three serologically unrelated clusters (FDV, MRDV and CSDV) based on protein antigens associated with B-spiked subviral particles.

REPLICATION

In cytoplasmic viroplasms, probably like that of orthoreoviruses. Morphogenesis is accompanied by formation of regularly structured filaments and tubules in at least some hosts.

BIOLOGICAL ASPECTS

Host range Flowering plants; *Gramineae*. Insects: plant hoppers (*Delphacidae, Auchenorrhyncha, Hemiptera*).

Transmission In nature only by Delphacid plant hoppers, e.g., *Laodelphax*, *Javesella*, *Delphacodes*, *Sogatella*, *Perkinsiella*, *Unkanodes*, etc. Transmission is propagative; acquisition after some hours feeding; latent period ≈ 2 weeks; then lifelong transmission by the insect to plants. FDV can be transmitted inefficiently through the egg; probably no transovarial transmission in other members.

OTHER MEMBERS

Maize rough dwarf
(Pangola stunt (175), Rice black streaked dwarf (135), cereal tillering disease and mal de Rio Cuarto disease are considered geographical races of maize rough dwarf virus).

Oat sterile dwarf (= *Arrhenatherum* blue dwarf and *lolium* enation viruses).

Taxonomic status	English vernacular name	International name
POSSIBLE GENUS	PLANT REOVIRUS SUBGROUP 3	—
TYPE SPECIES	RICE RAGGED STUNT (RRSV) (248)	

PROPERTIES OF THE VIRUS PARTICLE

Morphology Particle lacks a complete outer capsid; core ≈ 50 nm in diameter; there are 12 spikes 15-20 nm wide and 8 nm long that represent a partially formed outer capsid.

Physicochemical properties Not determined.

Nucleic acid 10 segments, MW = 0.5-3.0 x 10^6, with total MW ≈ 18 x 10^6. RNA polymerase present in virions.

Protein Not determined.

REPLICATION

In cytoplasmic viroplasms, similar to other ***Reoviridae***.

BIOLOGICAL ASPECTS

Host range Flowering plants; *Gramineae*. Insects: plant hoppers (*Delphacidae*, *Auchenorrhyncha*, *Hemiptera*).

Transmission Propagative, in the plant hopper *Nilaparvata*; acquisition after some 3 h feeding, latent period ≈ 10 days; then intermittent, lifelong transmission. No transovarial transmission.

OTHER MEMBERS

Possible member

Echinochloa ragged stunt virus.

NOTE

RRSV is suggested as a possible new genus because it has a distinct morphology unlike any other reo-like virus (though it somewhat resembles *Cypovirus* without the matrix protein). The size distribution of the 10 dsRNA segments is unlike that of other ***Reoviridae***. No serological relationships with any other plant reo-like virus. On the other hand, the symptoms, cytopathology, vector type and number of RNA segments, are similar to those of *Fijivirus*.

Derivation of Names

reo: sigla from *r*espiratory *e*nteric *o*rphan
orbi: from Latin *orbis*, 'ring'
colti: from *Co*lorado *ti*ck fever
rota: from Latin *rota*, 'wheel'
aqua: from Latin *aqua*, 'water'
cypo: *cy*toplasmic *po*lyhedrosis
phyto: from Greek *phyton*, 'plant'
fiji: from name of country from which virus was first described

REFERENCES

Barber, T.L.; Jochim, M.M.: Bluetongue and related orbiviruses (Liss, New York 1985).

Belloncik, S.: Cytoplasmic polyhedrosis viruses - *Reoviridae*. Adv. Virus Res. *37:* 173-209 (1989).

Boccardo, G.; Milne, R.G.: Plant reovirus group. CMI/AAB Descriptions of Plant Viruses No. 294 (1984).

Bridger, J.C.: Novel rotaviruses in animals and man. *In* Novel Diarrhoea Viruses, Ciba Found. Sympos. *128:* 5-23 (Wiley, Chichester 1987).

Estes, M.K.; Cohen, J.: Rotavirus gene structure and function. Microbiol. Revs. *53:* 410-449 (1989).

Francki, R.I.B.; Boccardo, G.: The plant *reoviridae. In* Joklik, W.K. (ed.), The *Reoviridae*, pp. 505-563 (Plenum Press, New York, 1983).

Francki, R.I.B.; Milne, R.G.; Hatta, T.: Plant *reoviridae, In* Atlas of Plant Viruses, Vol. I, pp. 47-72 (CRC Press, Boca Raton, Fl., 1985).

Gorman, B.M.; Taylor, J.; Walker, P.J.: Orbiviruses. *In* Joklik, W.K. (ed.) The *Reoviridae*, pp. 287-358 (Plenum Press, New York, 1983).

Hull, R.; Brown, F.; Payne, C.: Virology - Directory and dictionary of animal, bacterial and plant viruses, pp. 266-271 (Macmillan, London 1989).

Joklik, W.K.: The *Reoviridae*, (Plenum Press, New York, 1983).

Joklik, W.K.: Recent progress in reovirus research. Ann. Rev. Genet. *19:*537-575 (1985).

Kapikian, A.Z.; Chanock, R.M.: Rotaviruses. *In* Fields, B.N.; Knight, J.C. (eds.), Virology, 2nd edn., pp. 1353-1404 (Raven Press, New York, 1990).

Knudson, D.L.: Genome of Colorado tick fever virus. Virology *112:* 361-364 (1981).

Knudson, D.L.; Monath, T.P.: Orbiviruses. *In* Fields, B.N.; Knight, J.C. (eds.), Virology, 2nd edn., pp. 1405-1433 (Raven Press, New York, 1990).

Mertens, P.P.C.; Crook, N.E.; Rubinstein, R.; Pedley, S.; Payne, C.C.: Cytoplasmic polyhedrosis virus classification by electropherotype: validation by serological analyses and agarose gel electrophoresis. J. gen. Virol. *70:* 173-185 (1989).

Mertens, P.P.C.; Pedley, S.; Cowley, J.; Burroughs, J.N.; Corteyn, A.H.; Jeggo, M.H.; Jennings, D.M.; Gorman, B.M.: Analysis of the roles of bluetongue virus outer capsid proteins VP2 and VP5 in determination of virus serotype. Virology *170:* 561-565 (1989).

Nuss, D.L.; Dall, D.J.: Structural and functional properties of plant reovirus genomes. Adv. Virus Res. *38:* 249-306 (1990).

Payne, C.C.; Mertens, P.P.C.: Cytoplasmic polyhedrosis viruses. *In* Joklik, W.K. (ed.), The *Reoviridae*, pp. 425-504 (Plenum Press, New York, 1983).

Pedley, S.; Bridger, J.C.; Brown, J.F.; McCrae, M.A.: Molecular characterization of rotaviruses with distinct group antigens. J. gen. Virol. *64:*2093-2101 (1983).

Ray, P.; Gorman, B.M.: Bluetongue viruses. Cur. Topics. Microbiol. Immunol. *162* (Springer, Berlin, Heidelberg, New York, Tokyo, 1990).

Schiff, L.A.; Fields, B.N.: Reoviruses and their replication. *In* Fields, B.N.; Knight, J.C. (eds.), Virology, 2nd edn., pp. 1275-1306 (Raven Press, New York, 1990).

Tyler, K.L.; Fields, B.N.: *Reoviridae*: A brief introduction. *In* Fields, B.N.; Knight, J.C. (eds.), Virology, 2nd edn., pp. 1271-1274 (Raven Press, New York, 1990).

Tyler, K.L.; Fields, B.N.: Reoviruses. *In* Fields, B.N.; Knight, J.C (eds.), Virology, 2nd edn., pp. 1307-1328 (Raven Press, New York, 1990).

Winton, J.R.; Lannan, C.N.; Fryer, J.L.; Hedrick, R.P.; Meyers, T.R.; Plumb, J.A.; Yamamoto, T.: Morphological and biochemical properties of four members of a novel group of reoviruses isolated from aquatic animals. J. gen. Virol. *68:* 353-364 (1987).

Taxonomic status	English vernacular name	International name
FAMILY	BISEGMENTED DSRNA VIRUS GROUP	***BIRNAVIRIDAE***

Reported by P. Dobos

GENUS	—	***BIRNAVIRUS***
TYPE SPECIES	INFECTIOUS PANCREATIC NECROSIS VIRUS (IPNV)	—

PROPERTIES OF THE VIRUS PARTICLE

Morphology Icosahedral particles ≈ 60 nm in diameter with 92 morphological subunits with no envelope or surface projections. Cores are 45 nm in diameter as seen in thin sections of infected cells. Cores cannot be generated by treating purified virus with EDTA, trypsin or chymotrypsin.

Physicochemical properties MW ≈ 55 x 10^6; S_{20w} ≈ 435; buoyant density in CsCl = 1.33 g/cm^3. Stable at pH 3-9, resistant to 1% SDS at 20°C at pH 7.5 for 30 min.

Nucleic acid Two segments of linear dsRNA, 9.7% by weight of virus particle, not infectious. Both segments contain a 94 kDa genome linked protein. Genome segment A (3092 bp) contains a large ORF that encodes a 104 kDa polyprotein in the order of 5'-pre VP2-NS-VP3-3'. A small 17 kDa ORF which overlaps the large ORF at its 5'-end has been identified in IPNV and IBDV sequences; gene product of this small ORF has not been identified in either systems. Segment B (2784 bp) contains a single large ORF that encodes a 94 kDa polypeptide, the putative RNA polymerase.

Protein Four structural polypeptides: VP1 (94 kDa) a minor protein present both as free and as genome linked protein. VP2 (54 kDa), VP3 (31 kDa) and VP4 (29 kDa). The latter represents a truncated VP3 in IPNV whereas it is a unique polypeptide in IBDV and DXV. All viruses contain a dsRNA dependent RNA polymerase activity.

Lipid No lipids in virion.

Carbohydrate VP2 is glycosylated.

Antigenic properties The major capsid protein VP2 contains the virus neutralizing epitopes. Monoclonal Ab's to VP3 do not

Taxonomic status	English vernacular name	International name

neutralize virus infectivity. Monoclonals against VP1 and VP4 have not been reported. IPNV haemagglutinates mouse (Balb c) erythrocytes at pH 5.

Effect on cells

Tissue tropism of pancreas, gonad, kidney for IPNV, bursa Fabricius for IBDV.

REPLICATION

A sinlge cycle of replication takes 16-20 h. Replication is cytoplasmic. No inhibition of host cell macromolecular synthesis. Transcription of viral RNA involves synthesis of two genome length mRNA species (one from each genome segment) that lack 3'-poly A tails. It is not known if genome replication follows a conservative or semi conservative mechanism. Peak rates of viral RNA and protein synthesis are reached approximately 6-8 hours post infection. Four virus-specific polypeptides are found in infected cells: VP1 the product of genome segment B, and preVP2 (62 kDa), NS (27 kDa) and VP3 the product of genome segment A. These three polypeptides are generated by cotranslational cleavage by the virus coded endoprotease (NS in IPNV; NP4 in IBDV and DXV) which cleaves the polyprotein at two places. The exact cleavage sites have not been mapped but the carboxy end of the NS polypeptide comprises the active site of the viral protease. PreVP2 is later trimmed to VP2 during virus maturation. Cells lyse, but about half the progeny remains cell associated. Genome segment reassortment between IPNV serotypes has been demonstrated in laboratory experiments.

BIOLOGICAL ASPECTS

Host range

Different viruses infecting fish, mostly salmonids (IPNV); molluscs (OV and TV); chickens, ducks and turkeys (IBDV); and *Drosophila* (DXV).

Transmission

Both horizontal and vertical for all viruses. No vectors known.

OTHER MEMBERS

Oyster virus (OV)
Tellina virus (TV)
Infectious bursal disease virus (IBDV)
Drosophila X Virus (DXV)

Derivation of Name	bi: signifies double strandedness, as well as the bisegmented nature of the virus genome. rna: indicates the nature of the viral nucleic acid.

REFERENCES

Becht, H.: Infectious Bursal Disease Virus. Cur. Topics Microbiol. Immunol., *90*:107-121 (1980).

Dobos, P.; Hill, B.J.; Hallett, R.; Kells, D.T.C.; Becht, H.; Teninges, D.: Biophysical and biochemical characterization of five animal viruses with bisegmented double-stranded RNA genomes. J. Virol. *32*:593-605 (1979).

Dobos, P.; Roberts, T.E.: The molecular biology of infectious pancreatic necrosis virus: a review. Can. J. Microbiol. *29*:377-384 (1983).

Duncan, R.; Nagy, E.; Krell, P.J.; Dobos, P.: Synthesis of the infectious pancreatic necrosis virus polyprotein, detection of a virus-coded protease, and fine structure mapping of genome segment A coding regions. J. Virol. *61:* 3655-3664 (1987).

Pilcher, K.S.; Fryer, J.L.: Viral diseases of fish: a review through 1978. CRC Critical Reviews in Microbiology, Vol. 1, p.287. (CRC Press, West Palm Beach, 1980).

Taxonomic status	English vernacular name	International name
FAMILY	MONOPARTITIE DSRNA MYCOVIRUS GROUP	***TOTIVIRIDAE***

Revised by K.W Buck & S.A. Ghabrial

GENUS	MONOPARTITIE DSRNA MYCOVIRUS GROUP	***TOTIVIRUS***
TYPE SPECIES	*SACCHAROMYCES CEREVISIAE* VIRUS L-A (SCV-L-A) (SYNONYM SCV-L1)	—

PROPERTIES OF THE VIRUS PARTICLE

Morphology Isometric, 40-43 nm in diameter .

Physicochemical properties S_{20w} = 160-190S. Buoyant density in CsCl = 1.40-1.43 g/cm^3. Additional components with different sedimentation coefficients and buoyant densities are present in virus isolates with satellite or defective RNAs. Particles lacking nucleic acid have S_{20w} = 98-113S.

Nucleic acid Single molecule of dsRNA, MW = 3.3-4.2 x 10^6 (4.7-6.7 kbp). Some virus isolates contain additional satellite dsRNAs which encode "killer" proteins; these satellites are encapsidated separately in capsids encoded by the helper virus genome. Some virus isolates may contain additionally or alternatively to the satellites, subgenomic or defective dsRNAs which arise from the satellite dsRNAs; these additional dsRNAs are also encapsidated separately in capsids encoded by the helper virus genome.

The complete nucleotide sequence of ScV-L-A (ScV-L1) dsRNA is deposited as EMBL accession number J04692 (X13426). The plus strand (4579 bases) has two large ORFs that overlap by 130 bases. The first ORF encodes the viral major capsid polypeptide with a predicted size of 76 x 10^3. The two ORF together encode via translational frame shift, the putative RNA-dependent RNA polymerase as a fusion protein (analogous to gag-pol fusion proteins of the retroviruses) with a predicted size of 171 x 10^3.

Protein Single major capsid polypeptide species, MW = 73-88 x 10^3. RNA polymerase (transcriptase) present.

Lipid None detected.

Carbohydrate None detected.

Taxonomic status	English vernacular name	International name

Antigenic properties Efficient immunogens.

REPLICATION

The virion-associated RNA polymerase catalyses *in vitro* end-to-end transcription of dsRNA by a conservative mechanism to produce mRNA for capsid polypeptide. The (+) ssRNA transcript of ScV-L-A is the species encapsidated to form new virus particles having a replicase activity. These particles synthesize (-) strand on the (+) template to produce dsRNA, thus completing the replication cycle. Virions accumulate in the cytoplasm.

BIOLOGICAL ASPECTS

Transmission Intracellular during cell division, sporogenesis and cell fusion. In some ascomycetes, e.g. *Gaeumannomyces graminis*, virus is usually eliminated during ascospore formation.

OTHER MEMBERS

Helminthosporium victoriae 190S virus
Ustilago maydis viruses (P1, P4 and P6)

Probable members

Gaeumannomyces graminis virus 87-1-H (Ggv-87-1-H)
Mycogone perniciosa virus (MpV)
Yarrowia lipolytica virus (YlV)

Possible members

Aspergillus foetidus virus S (AfV-S)
Aspergillus niger virus S (AnV-S)
Saccharomyces cerevisiae virus La (ScV-La; synonym ScV-LB/C)
Thielaviopsis basicola viruses

Derivation of Name totus: from Latin 'whole' or 'undivided'

REFERENCES

Adler, J.J.; Wood, H.A.; Bozarth, R.F.: Virus-like particles from killer, neutral and sensitive strains of *Saccharomyces cerevisiae*. J. Virol. *17*:472-476 (1976).

Barton, R.J.: *Mycogone perniciosa* virus. Report of the Glasshouse Crops Research Institute for 1977, p. 133 (1978).

Bozarth, R.F.; Goenaga, A.: Complex of virus-like particles containing double-stranded RNA from *Thielaviopsis basicola*. J. Virol. *24*:846-849 (1977).

Bozarth, R.F.; Koltin, Y.; Weissman, M.B.; Parker, R.L.; Dalton, R.E.; Steinlauf, R.: The molecular weight and packaging of dsRNAs in the mycovirus from *Ustilago maydis* killer strains. Virology *113*:492-502 (1981).

Brennan, V.E.; Field, L.J.; Cizdziel, P.; Bruenn, J.A.: Sequences at the 3' ends of yeast viral dsRNAs: proposed transcriptase and replicase initiation sites. Nuc. Acids Res. *9*:4007-4021 (1981).

Buck, K.W.; Girvan, R.F.; Ratti, G.: Two serologically distinct double-stranded RNA viruses isolated from *Aspergillus niger*. Biochem. Soc. Trans. *1*:1138-1140 (1973).

Buck, K.W.; Lhoas, P.; Border, D.J.; Street, B.K.: Virus particles in yeast. Biochem. Soc. Trans. *1*:1141-1142 (1973).

Buck, K.W.; Ratti, G.: Biophysical and biochemical properties of two viruses isolated from *Aspergillus foetidus*. J. gen. Virol. *27*:211-224 (1975).

Bussey, H.: Proteases and the processing of precursors to secreted proteins in yeast. Yeast *4*:17-26 (1988).

Diamond, M.E.; Dowhanick, J.J.; Nemeroff, M.E.; Pietras, D.F.; Tu, C.C.; Bruenn, J.A.: Overlapping genes in a yeast double-stranded RNA virus. J. Virol. *63:* 3983-3990 (1989).

Fujimura, T.; Wickner, R.B.: L-A double-stranded RNA virus-like particle replication cycle in *Saccharomyces cerevisiae*: Particle maturation *in vitro* and effects of *mak*10 and *pet*18 mutations. Mol. Cell. Biol. *7:* 420-426 (1987).

Fujimura, T.; Wickner, R.B.: Gene overlap results in a viral protein having an RNA binding domain and a major coat protein domain. Cell *55:* 663-671 (1988).

Ghabrial, S.A.; Bibb, J.A.; Price, K.H.; Havens, W.M.; Lesnaw, J.A.: The capsid polypeptides of the 190S virus of *Helminthosporium victoriae*. J. gen. Virol. *68*:1791-1800 (1987).

Ghabrial, S.A.; Havens, W.M.: Conservative transcription of *Helminthosporium victoriae* 190S virus double-stranded RNA *in vitro*. J. gen. Virol. *70:* 1025-1035 (1989).

Groves, D.P.; Clare, J.J.; Oliver, S.G.: Isolation and characterisation of a double-stranded RNA virus-like particle from the yeast *Yarrowia lipolytica*. Curr. Genet. *7:* 185-190 (1983).

Hopper, J.E.; Bostian, K.A.; Rowe, L.B.; Tipper, D.J.: Translation of the L-species dsRNA genome of the killer-associated virus-like particles of *Saccharomyces cerevisiae*. J. Biol. Chem. *252*:9010-9017 (1977).

Icho, T.; Wickner, R.B.: The double-stranded RNA genome of yeast virus L-A encodes its own putative RNA polymerase by fusing two open reading frames. J. Biol. Chem. *264:* 6716-6723 (1989).

Jamil, N.; Buck, K.W.: Apparently identical viruses from *Gaeumannomyces graminis* var. *tritici* and *Phialophora* sp. (lobed hyphopodia). Trans. Brit. Mycol. Soc. *83*:519-522 (1984).

Jamil, N.; Buck, K.W.: Capsid polypeptides in a group III virus from *Gaeumannomyces graminis* var. *tritici* are related. J. gen. Virol. *67*:1717-1720 (1986).

Koltin, Y.: Virus-like particles in *Ustilago maydis*: mutants with partial genomes. Genetics *86*:527-534 (1977).

Nemeroff, M.E.; Bruenn, J.A.: Conservative replication and transcription of *Saccharomyces cerevisiae* viral double-stranded RNA *in vitro*. J. Virol. *57*:754-758 (1986).

Shelbourn, S.L.; Day, P.R.; Buck, K.W.: Relationships and functions of virus double-stranded RNA in a P4 killer strain of *Ustilago maydis*. J. gen. Virol. *69:* 975-982 (1988).

Wickner, R.B.; Yeast virology. FASEB J. *3*:2257-2265 (1989).

GENUS	GIARDIA VIRUS GROUP (POSSIBLE AFFINITIES TO THE *TOTIVIRIDAE* FAMILY)	***GIARDIAVIRUS***
TYPE SPECIES	GIARDIA LAMBLIA VIRUS (GLV)	—

Compiled by S.A. Ghabrial & C.C. Wang

PROPERTIES OF THE VIRUS PARTICLE

Morphology Isometric, 33 nm in diameter.

Physicochemical properties Buoyant density in CsCl ≈ 1.368 g/cm^3.

Nucleic acid Single molecule of dsRNA ≈ 7.0 kbp in length.

Protein Single major capsid species, MW ≈ 100 x 10^3.

Lipid None reported.

Carbohydrate None reported.

REPLICATION

The virus is present in the nuclei of infected *G. lamblia.* It replicates without inhibiting the growth of *G. lamblia* trophozoites. It is also extruded into the culture medium and the extruded virus can infect many virus-free isolates of the protozoan host. There are isolates of the protozoan parasite, however, that are resistant to infection by GLV. A single-stranded copy of the viral dsRNA genome is present in infected cells. The concentration of the ssRNA observed during the time course of GLV infection is consistent with a role as a viral replicative intermediate or mRNA. The ssRNA does not appear to be polyadenylated.

BIOLOGICAL ASPECTS

The virus infects many isolates of *G. lamblia*, a flagellated protozoan human parasite. The virus does not seem to be associated with the virulence of the parasite. It is not observed in the cyst form of the parasite and it is not known whether it can be carried through the transformation between cyst and trophozoite. The virus is infectious as purified particles and can infect uninfected *G. lamblia.*

Possible member

Trichomonas vaginalis virus (TVV)

REFERENCES

Furfine, E.S.; White, T.C.; Wang, A.L.; Wang, C.C.: A single-stranded RNA copy of the *Giardia lamblia* virus double-stranded RNA genome is present in the infected *Giardia lamblia*. Nuc. Acids Res. *17:* 7453-7467 (1989).

Miller, R.L.; Wang, A.L.; Wang, C.C.: Purification and characterization of the *Giardia lamblia* double-stranded RNA virus. Mol. Biochem. Parasitol. *28:* 189-196 (1988).

Wang, A.L.; Wang, C.C.: Discovery of a specific double-stranded RNA virus in *Giardia lamblia*. Mol. Biochem. Parasitol. *21:* 269-276 (1986).

Wang, A.L.; Wang, C.C.: The double-stranded RNA in *Trichomonas vaginalis* may originate from virus-like particles. Proc. Natl. Acad. Sci. USA *83:* 7956-7960 (1986).

White, T.C.; Wang, C.C.: RNA dependent RNA polymerase activity associated with the double-stranded RNA virus of *Giardia lamblia*. Nuc. Acids Res. *18:* 553-559 (1990).

Taxonomic status	English vernacular name	International name
FAMILY	DSRNA MYCOVIRUSES WITH DIVIDED GENOMES	***PARTITIVIRIDAE***

Revised by K.W. Buck & S.A. Ghabrial

GENUS	BIPARTITE DSRNA MYCOVIRUS GROUP	***PARTITIVIRUS***
TYPE SPECIES	*GAEUMANNOMYCES GRAMINIS* VIRUS 019/6-A (GGV-019/6A)	—

PROPERTIES OF THE VIRUS PARTICLE

Morphology Isometric, diameter 30-35 nm.

Physicochemical properties S_{20w} = 101-145S. Buoyant density in CsCl = 1.35-1.36 g/cm^3. In preparations of some viruses, e.g. PsV-S, additional sedimenting and density components are found. These consist of particles with ssRNA (mRNA) and particles with both ss and dsRNA and are probably replicative intermediates. Particles lacking nucleic acid have S_{20w} = 66-100S; buoyant density in CsCl = 1.29-1.30 g/cm^3.

Nucleic acid Two unrelated segments of dsRNA, MW = 0.9-1.6 x 10^6 (for individual segments), one encoding the capsid polypeptide and the other an unrelated polypeptide, probably the virion RNA polymerase. Approximately the whole of the coding capacity of each dsRNA is required for each polypeptide i.e. each dsRNA is probably monocistronic. Each dsRNA is encapsidated in a separate particle. Additional segments of dsRNA (satellite or defective) may be present in some virus isolates.

Protein Single major capsid polypeptide species, MW = 42-73 x 10^3. RNA polymerase present.

Lipid None detected.

Carbohydrate None detected.

Antigenic properties Efficient immunogens. Single precipitin line in gel diffusion tests. Members and probable members which are serologically related, e.g. PsV-S, DrV and Aov, may be strains of a virus species.

REPLICATION

The virion-associated RNA polymerase catalyses *in vitro*, replication, and/or end-to-end transcription of each dsRNA to produce mRNA, by a semi-conservative mechanism. Particles accumulate in the cytoplasm.

BIOLOGICAL ASPECTS

Transmission Intracellular during cell division, sporogenesis and cell fusion. In some ascomycetes, e.g. *Gaeumannomyces graminis*, virus is usually eliminated during ascospore formation.

OTHER MEMBERS

Agaricus bisporus virus 4 (AbV-4, mushroom virus 4)
Aspergillus ochraceous virus (AoV)
Gaeumannomyces graminis virus T1-A (GgV-T1-A)
Penicillium stoloniferum virus S (PsV-S)
Rhizoctonia solani virus (RsV)

Probable members

Diplocarpon rosae virus (DrV)
Phialophora sp. (lobed hyphopodia) virus 2-2-A
(*Phialophora radicicola* virus 2-2-A, PrV-2-2-A)

Possible member

Penicillium stoloniferum virus F (PsV-F)

Taxonomic status	English vernacular name	International name
POSSIBLE GENUS	*PENICILLIUM CHRYSOGENUM* VIRUS GROUP	—
POSSIBLE TYPE SPECIES	*PENICILLIUM CHRYSOGENUM* VIRUS (PCV)	—

PROPERTIES OF THE VIRUS PARTICLE

Morphology Isometric, 35-40 nm in diameter

Physicochemical properties S_{20w} = 145-150.

Nucleic acid Three unrelated dsRNA components with MW in the range 1.9 x 10^6 - 2.4 x 10^6, each probably monocistronic, each separately encapsidated. The number of dsRNAs required for replication is not known. Some virus isolates contain additional dsRNAs, probably satellite or defective RNAs.

Taxonomic status	English vernacular name	International name

Protein Single major capsid polypeptide species, MW $\approx$ 125 x 10^3. RNA polymerase present.

Lipid None detected.

Antigenic properties Efficient immunogens. Single precipitin line in gel diffusion tests. All members are serologically related and may be strains of a single virus species.

REPLICATION

Particles accumulate in the cytoplasm.

BIOLOGICAL ASPECTS

Transmission Intracellular during cell division, sporogenesis and cell fusion.

OTHER MEMBERS

Penicillium brevi compactum virus (PbV)
Penicillium cyaneo-fulvum virus (Pc-fV)

Possible member

Helminthosporium victoriae 145S virus

Derivation of Name partitus: from Latin 'divided'

REFERENCES

Barton, R.J.; Hollings, M.: Purification and some properties of two viruses infecting the cultivated mushroom *Agaricus bisporus*. J. gen. Virol. *42*:231-240 (1979).

Bozarth, R.F.: The physicochemical properties of mycoviruses. *In* Lemke, L. (ed.): Viruses and plasmids in fungi, pp. 43-91 (Dekker, New York 1979).

Bozarth, R.F.; Wood, H.A.; Goenaga, A.: Virus-like particles from a culture of *Diplocarpon rosae*. Phytopathology *62*:493 (1972).

Bozarth, R.F.; Wood, H.A.; Mandelbrot, A.: The *Penicillium stoloniferum* virus complex: two similar double-stranded RNA virus-like particles in a single cell. Virology *45*:516-523 (1971).

Buck, K.W.: Current problems in fungal virus taxonomy. *In* Matthews, R.E.F. (ed.): A critical appraisal of virus taxonomy, pp. 139-176 (CRC Press, Boca Raton, Fl., 1983).

Buck, K.W.; Almond, M.R.; McFadden, J.J.P.; Romanos, M.A.; Rawlinson, C.J.: Properties of thirteen viruses and virus variants obtained from eight isolates of the wheat take-all fungus, *Gaeumannomyces graminis* var. *tritici*.. J. gen. Virol. *53*:235-245 (1981).

Buck, K.W.; Girvan, R.F.: Comparison of the biophysical and biochemical properties of *Penicillium cyaneo-fulvum* virus and *Penicillium chrysogenum* virus. J. gen. Virol. *34*:145-154 (1977).

Buck, K.W.; Kempson-Jones, G.F.: Biophysical properties of *Penicillium stoloniferum* virus S. J. gen. Virol. *18*:223-235 (1973).

Buck, K.W.; Kempson-Jones, G.F.: Capsid polypeptides of two viruses isolated from *Penicillium stoloniferum.* J. gen. Virol. *22*:441-445 (1974).

Buck, K.W.; McGinty, R.M.; Rawlinson, C.J.: Two serologically unrelated viruses isolated from a *Phialophora* sp. J. gen. Virol. *55*:235-239 (1981).

Edmondson, S.P.; Lang, D.; Gray, D.M.: Evidence for sequence heterogeneity among the double-stranded RNA segments of *Penicillium chrysogenum* mycovirus. J. gen. Virol. *65*:1591-1599 (1984).

Finkler, A.; Ben-Zvi, B.; Koltin, Y.; Barash, I.: Transcription and *in vitro* translation of the dsRNA virus isolated from *Rhizoctonia solani.* Virus Genes *1:* 205-219 (1988).

Finkler, A.; Koltin, Y.; Barash, I.; Sneh, B.; Pozniak, D.: Isolation of a virus from virulent strains of *Rhizoctonia solani.* J. gen. Virol. *66:* 1221-1232 (1985).

Kim, J.W.; Bozarth, R.F.: Intergeneric occurrence of related fungal viruses: the *Aspergillus ochraceous* virus complex and its relationship to the *Penicillium stoloniferum* virus S. J. gen. Virol. *66:* 1991-2002 (1985).

McFadden, J.J.P.; Buck, K.W.; Rawlinson, C.J.: Infrequent transmission of double-stranded RNA virus particles but absence of DNA proviruses in single ascospore cultures of *Gaeumannomyces graminis.* J. gen. Virol. *64*:927-937 (1983).

Sanderlin, R.S.; Ghabrial, S.A.: Physicochemical properties of two distinct types of virus-like particles from *Helminthosporium victoriae.* Virology *87*:142-151 (1978).

Tavantzis, S.M.; Bandy, B.P.: Properties of a mycovirus from *Rhizoctonia solani* and its virion-associated RNA polymerase. J. gen. Virol. *69:* 1465-1477 (1988).

Wood, H.A.; Bozarth, R.F.: Properties of virus-like particles of *Penicillium chrysogenum*: One double-stranded RNA molecule per particle. Virology *47*:604-609 (1972).

Wood, H.A.; Bozarth, R.F.; Mislivec, P.B.: Virus-like particles associated with an isolate of *Penicillium brevi-compactum.* Virology *44:* 592-598 (1971).

Taxonomic status	English vernacular name	International name
GROUP	CRYPTIC VIRUS GROUP	***CRYPTOVIRUS***

Revised by R. Milne

PROPERTIES OF THE VIRUS PARTICLE

Morphology Isometric particles, 30-38 nm in diameter.

Physicochemical properties One nucleoprotein component.

Nucleic acid Two molecules of linear dsRNA.

Protein Single polypeptide.

Lipid None reported.

Carbohydrate None reported.

Antigenic properties No serological relationship between members of different subgroups; some members in each subgroup are related.

REPLICATION

Virus particles of some members have been shown to contain transcriptase activity. Two proteins, MW $\approx$ 52 and 67 x 10^3 have been translated *in vitro* from the genomic dsRNAs of beet cryptic virus 1, and MW $\approx$ 54 and 68 x 10^3 from white clover cryptic virus 1. The larger protein, derived from dsRNA 1, may be involved in dsRNA replication. The smaller protein was precipitated by antiserum to virus particles, suggesting that it is the capsid protein.

BIOLOGICAL ASPECTS

Host range Narrow for individual viruses but different viruses occur in a wide range of plant families. Usually occur in very low concentration in cytoplasm and/or nucleus of host; induce no symptoms of infection, except in a few cases.

Transmission Only through seed or pollen. Viruses are possibly unable to move from cell to cell, propagating probably only via cell multiplication.

Taxonomic status	English vernacular name	International name
SUBGROUP I	WHITE CLOVER CRYPTIC VIRUS 1 GROUP	—
TYPE MEMBER	WHITE CLOVER CRYPTIC VIRUS 1 (WCCV-1)	—

PROPERTIES OF THE VIRUS PARTICLE

Morphology Isometric particles, 30 nm in diameter.

Physicochemical properties One nucleoprotein component of density ≈ 1.392 g/cm^3 in CsCl

Nucleic acid Two molecules of linear dsRNA of MW ≈ 1.20 and 0.97 x 10^6; ≈ 25% by weight of the virus.

Protein Single polypeptide of MW ≈ 55 x 10^3.

Lipid None reported.

Carbohydrate None reported.

Antigenic properties Some members are serologically related.

OTHER MEMBERS

Alfalfa cryptic 1
Beet cryptic 1
Beet cryptic 2
Beet cryptic 3
Carnation cryptic 1 (315)
Carrot temperate 1
Carrot temperate 3
Carrot temperate 4
Hop trefoil cryptic 1
Hop trefoil cryptic 3
Radish yellow edge (298)
Ryegrass cryptic
Spinach temperate
Vicia cryptic
White clover cryptic 3

Possible members

Carnation cryptic 2
Fescue cryptic
Garland chrysanthemum temperate
Mibuna temperate
Poinsettia cryptic

Taxonomic status	English vernacular name	International name
	Red pepper cryptic 1 Red pepper cryptic 2 Rhubarb temperate Santosai temperate	
SUBGROUP II	WHITE CLOVER CRYPTIC VIRUS 2 GROUP	—
TYPE MEMBER	WHITE CLOVER CRYPTIC VIRUS 2 (WCCV-2) (332)	—

PROPERTIES OF THE VIRUS PARTICLE

Morphology Isometric particles, ≈ 38 nm in diameter with prominent morphological subunits.

Physicochemical properties One nucleoprotein component of density ≈ 1.375 g/cm^3 in CsCl

Nucleic acid Two molecules of linear dsRNA of MW ≈ 1.49 and 1.38 x 10^6; ≈ 24% by weight of the virus.

Protein Not characterized.

Lipid None reported.

Carbohydrate None reported.

Antigenic properties Present members are serologically related.

OTHER MEMBERS

Carrot temperate 2
Hop trefoil cryptic 2
Red clover cryptic 2

Possible member

Alfalfa cryptic 2

Derivation of Name crypto: from Greek *kryptos*, 'hidden, covered or secret'.

REFERENCES

Abou-Elnasr, M.A.; Jones, A.T.; Mayo, M.A.: Detection of dsRNA in particles of *vicia* cryptic virus and in *Vicia faba* tissues and protoplasts. J. gen. Virol. *66:*2453-2460 (1985).

Accotto, G.P.; Boccardo, G.: The coat proteins and nucleic acids of two beet cryptic viruses. J. gen. Virol. *67:*363-366 (1986).

Accotto, G.P.; Brisco, M.J.; Hull, R.: *In vitro* translation of the double-stranded RNA genome from beet cryptic virus 1. J. gen. Virol. *68:*1417-1422 (1987).

Antoniw, J.F.; Linthorst, H.J.M.; White, R.F.; Bol, J.F.: Molecular cloning of the double-stranded RNA of beet cryptic viruses. J. gen. Virol. *67:*2047-2051 (1986).

Boccardo, G.; Accotto, G.P.: RNA-dependent RNA polymerase activity in two morphologically different white clover cryptic viruses. Virology *163:* 413-419 (1988).

Boccardo, G.; Lisa, V.; Luisoni, E.; Milne, R.G.: Cryptic plant viruses. Adv. Virus Res. *32:*171-214 (1987).

Boccardo, G.; Milne, R.G.; Luisoni, E.; Lisa, V.; Accotto, G.P : Three seedborne cryptic viruses containing double-stranded RNA isolated from white clover. Virology *147:*29-40 (1985).

Kassanis, B.; White, R.F.; Woods, R.D.: Beet cryptic virus. Phytopath. Z. *90:*350-360 (1977).

Luisoni, E.; Milne, R.G.; Accotto, G.P.; Boccardo, G.: Cryptic viruses in hop trefoil (*Medicago lupulina*) and their relationships to other cryptic viruses in legumes. Intervirology *28:* 144-156 (1987).

Marzachi, C.; Milne, R.G.; Boccardo, G.: *In vitro* synthesis of double-stranded RNA by carnation cryptic virus-associated RNA-dependent RNA polymerase. Virology *165:* 115-121 (1988).

Natsuaki, T.; Natsuaki, K.T.; Okuda, S.; Teranaka, M.; Milne, R.G.; Boccardo, G.; Luisoni, E.: Relationships between the cryptic and temperate viruses of alfalfa, beet and white clover. Intervirology *25:*69-75 (1986).

Natsuaki, T.; Yamashita, S.; Doi, Y.; Okuda, S.; Teranaka, M. Two seed-borne double-stranded RNA viruses, beet temperate virus and spinach temperate virus. Ann. Phytopathol. Soc. Jpn. *49:*709-712 (1983).

Xie, W.; Antoniw, J.F.; White, R.F.: Detection of beet cryptic viruses 1 and 2 in a wide range of beet plants using cDNA probes. Plant Pathology *38:* 527-533 (1989).

Taxonomic status	English vernacular name	International name
FAMILY	—	*TOGAVIRIDAE*

Reported by J.H. Strauss

PROPERTIES OF THE VIRUS PARTICLE

Morphology Spherical. 60-70 nm in diameter, with an envelope tightly applied to a proven or presumed icosahedral nucleocapsid 35-40 nm in diameter. Surface projections are demonstrable in most togaviruses.

Physicochemical properties $S_{20w} = 280$; buoyant density in sucrose, 1.2 g/cm^3.

Nucleic acid Single molecule of positive-sense ssRNA, MW = 4×10^6; 8-9% by weight of virus. Where characterized, genes for nonstructural proteins are located at the 5' end. The 5' terminus is capped, and the 3' end is polyadenylated.

Protein Two or three envelope proteins, one or more of which are glycosylated, and a smaller core protein.

Lipid and Carbohydrate The virus-specific glycoproteins are inserted in the lipoprotein envelope, whose lipids are cell-derived.

Antigenic properties Members of a genus are antigenically related to each other but not to members of other genera of the family.

Effect on cells Members of the genera *Alphavirus* and *Rubivirus* show ion-dependent hemagglutinating activity.

REPLICATION

Multiply in cytoplasm and have been shown or are presumed to mature by budding. Structural proteins are translated from subgenomic mRNAs.

BIOLOGICAL ASPECTS

Host range All species of the genus *Alphavirus* replicate in arthropod vectors as well as in a wide range of vertebrates.

Transmission Members of the genera *Rubivirus* and *Arterivirus*, and other possible members of the family, are not arthropod-borne.

GENERA

Arbovirus group A	*Alphavirus*
Rubella virus	*Rubivirus*
Equine arteritis virus	*Arterivirus*

Taxonomic status	English vernacular name	International name
GENUS	ARBOVIRUS GROUP A	*ALPHAVIRUS*
TYPE SPECIES	SINDBIS VIRUS	—

PROPERTIES OF THE VIRUS PARTICLE

Morphology

Overall diameter of 70 nm; glycoprotein subunits are arranged in trimer clusterings of E1-E2 heterodimers to form icosahedral particles with T = 4. In most preparations, only the surface fringe of spikes is visible by negative staining. The lipid bilayer of the virus envelope is polyhedral and surrounds a smooth T = 3 nucleocapsid.

Physicochemical properties

MW $\approx$ 46 x 10^6; $S_{20w} \approx 280$; buoyant density in sucrose 1.2 g/cm^3.

Nucleic acid

MW $\approx$ 4 x 10^6 (12 kb) capped (type 0 cap) and polyadenylated; 8.7% by weight of particle. The gene order is 5'-nsPl-nsP2-nsP3-nsP4-C-E3-E2-E1-3', established by nucleotide sequencing. RNAs of 5 viruses have been sequenced completely.

Protein

The capsid protein C (MW $\approx$ 30 x 10^3), and two envelope glycoproteins E1 and E2 (MW = 50-59 x 10^3), plus glycoprotein E3 (MW $\approx$ 10 x 10^3) in some members. These comprise 60-64% by weight of particle.

Lipid

27-31% by weight located in the viral membrane, derived from the host cell.

Carbohydrate

7% by weight located in the viral membrane. Both high mannose and complex glycans are N-linked to the envelope glycoproteins.

Antigenic properties

E1 and E2 function as a heterodimer, but most neutralizing monoclonal antibodies are directed against E2. Members may be assigned to one of at least seven antigenic complexes, each comprising one or more species.

REPLICATION

Virions mature by budding of preassembled nucleocapsids through the plasma membrane. A full length minus strand RNA is synthesized by the virus-coded polymerase (nonstructural proteins); this RNA provides the template for synthesis of progeny genome and the subgenomic 26S messenger RNA ($\approx$ 4.1 kb) representing the 3' terminal one-third of the genomic RNA. These RNA species are synthesized under independent regulation in membranous cytoplasmic membranes. Mapping of temperature sensitive

Taxonomic status	English vernacular name	International name

mutants has shown that all four nonstructural proteins nsP1 - nsP4 (MW = 60-90 x 10^3) are required for RNA replication. These are generated from the 5' end of genomic RNA (a minor messenger) as a polyprotein which is post-translationally cleaved by a proteinase in nsP2 that acts primarily in trans.

Three of them, nsP1, nsP2, and nsP4, share sequence homology with nonstructural proteins of several groups of plant viruses, including tobamoviruses (tobacco mosaic virus), suggesting a common origin for the replicase of these viruses. The structural proteins are translated from the amplified and capped subgenomic messenger, commencing with the C protein which is cleaved first, autocatalytically, from the nascent polyprotein. The translation of C is followed by that of PE2 (subsequently cleaved to E3 and E2), and E1; PE2 and E1 are inserted into the endoplasmic reticulum via signal sequences and are glycosylated, and fatty acid acylated, during passage to the Golgi apparatus and the plasma membrane. Translation of host cell messengers is inhibited during infection of permissive vertebrate cell cultures, but not during infection of mosquito cells.

OTHER MEMBERS

Aura
Barmah Forest
Babanki
Bebaru
Buggy Creek
Chikungunya
Eastern equine encephalitis
Everglades
Fort Morgan
Getah
Highlands J
Kyzylagach
Mayaro
Middelburg
Mucambo
Ndumu
Ockelbo
O'nyong-nyong
Pixuna
Ross River
Sagiyama
Semliki Forest
Una

Taxonomic status	English vernacular name	International name
	Venezuelan equine encephalitis Western equine encephalitis Whataroa	
GENUS	RUBELLA VIRUS	***RUBIVIRUS***
TYPE SPECIES	RUBELLA VIRUS	—

PROPERTIES OF THE VIRUS PARTICLE

Morphology Virions 60 nm in diameter.

Physicochemical properties Similar to the *Alphavirus* genus but serologically unrelated. Only the type species recognized so far.

Nucleic acid MW = 3.4 x 10^6, comprising 9755 nucleotides, capped and polyadenylated. The gene order is 5' nonstructural (replicase) proteins/C-E2-E1 3'.

Protein Structural proteins comprise two glycoproteins, E1 (MW = 58-59 x 10^3) and E2 (MW = 42-48 x 10^3) (size range represents heterogeneous glycosylation) and a capsid protein C (MW = 33-34 x 10^3). High mannose and complex glycans are N-linked to E1 and E2. The three proteins are cleaved from a polyprotein translated from a subgenomic 24S mRNA; the order of translation is NH_2-C-E2-E1-COOH.

REPLICATION

Virus matures by budding through intracytoplasmic membranes or the plasma membrane. A 24S mRNA (≈ 3.3 kb) is synthesized during infection, probably using a full-length minus strand RNA as template. The structural proteins are synthesized from this amplified, capped and polyadenylated subgenomic mRNA, commencing with the C protein, followed by E2 and E1. Cleavage occurs cotranslationally. E2 and E1 are inserted via independent signal sequences in the endoplasmic reticulum, where they are N-glycosylated. Host cell protein synthesis is not inhibited during infection.

BIOLOGICAL ASPECTS

Host range No invertebrate host; man is the only known vertebrate host.

Transmission Spread principally by aerosolization, but congenital transmission can occur.

Taxonomic status	English vernacular name	International name
GENUS	EQUINE ARTERITIS VIRUS	***ARTERIVIRUS***
TYPE SPECIES	EQUINE ARTERITIS VIRUS	—

PROPERTIES OF THE VIRUS PARTICLE

Morphology Virions are 60 nm diameter with 12-15 nm ring-like subunits on the surface.

Physicochemical properties Similar to those of *Alphavirus*.

Nucleic acid MW ≈ 4 x 10^6 (≈ 12.7 kb); polyadenylated. At least six open reading frames have been identified by nucleotide sequencing. The capsid gene is located at the 3' end of the genome.

Protein Structural proteins comprise a glycosylated envelope protein E1 (MW ≈ 21 x 10^3), a nonglycosylated E2 (MW ≈ 14 x 10^3) and a core protein C (MW ≈ 12 x 10^3).

REPLICATION

Maturation occurs by budding through cytoplasmic membranes into cisternae. In addition to the genome, five polyadenylated RNA species (MW = 0.2-1.0 x 10^6) are synthesized in infected cells. These subgenomic RNAs, which may be derived from a larger precursor RNA, form a 3'-terminal nested set and contain a common leader sequence of about 208 nucleotides. The leader sequences on the subgenomic RNAs are identical to the sequence at the 5' end of the genome. The smallest subgenomic RNA encodes the capsid protein and the other RNAs probably all function as mRNAs. Translation of the 5' unique region of the genome (i.e. not present in any of the subgenomic RNAs) involves ribosomal frameshifting.

BIOLOGICAL ASPECTS

Host range Equines are the only hosts, and the single virus species is distributed world wide, producing symptoms associated with characteristic necrosis in muscle cells of small arteries, and abortion in pregnant mares.

Transmission Vertical and horizontal.

NOTE ON CLASSIFICATION

The presence of a nested set of subgenomic mRNAs with a common leader sequence, and a 3'-terminal capsid protein

Taxonomic status	English vernacular name	International name

gene, suggests that the arteriviruses are more closely related to the coronaviruses than to the togaviruses. In the future these viruses will almost certainly be reclassified either as a genus in the ***Coronaviridae*** or in a new family ***Arteriviridae***.

OTHER MEMBERS

Possible members

Carrot mottle virus
Lactic dehydrogenase

Derivation of Names

toga: from Latin *toga*, 'gown, cloak'
alpha: from Greek letter 'A'.
rubi: from Latin *rubeus*, 'reddish'.
arteri: from equine *arter*itis, the disease caused by type member virus

---

REFERENCES

Calisher, C.H.; Karabatsos, N.: Arbovirus serogroups: definition and geographic distribution. *In* Monath, T.P. (ed.), The arboviruses: epidemiology and ecology, Vol. 1, pp. 19-57 (CRC Press, Boca Raton, Fl., 1988).

de Vries, A.A.F.; Chirnside, E.D.; Bredenbeek, P.J.; Gravestein, L.A.; Horzinek, M.C.; Spaan, W.J.M.: A common leader sequence is spliced to all subgenomic mRNAs of equine arteritis virus. Nuc. Acids Res. *18*:3241-3247 (1990).

Dominguez, G.; Wang, C.-Y.; Frey, T.K.: Sequence of the genomic RNA of rubella virus: Evidence for genetic rearrangement during togavirus evolution. Virology *177*:225-238 (1990).

Frey, T.K.; Marr, L.D.: Sequence of the region coding for virion proteins C and E2 and the carboxy terminus of the nonstructural proteins of rubella virus: Comparison with alphaviruses. Gene *62*: 85-99 (1988).

Fuller, S.D.: The T=4 envelope of Sindbis virus is organized by interactions with a complementary T=3 capsid. Cell *48*:923-934 (1987).

Grakoui, A.; Levis, R.; Raju, R.; Huang, H.V.; Rice, C.M.: A *cis*-acting mutation in the Sindbis virus junction region which affects subgenomic RNA synthesis. J. Virol. *63*: 5216-5227 (1989).

Hahn, Y.S.; Strauss, E.G.; Strauss, J.H.: Mapping of RNA-temperature-sensitive mutants of Sindbis virus: Assignment of complementation groups A, B and G to nonstructural proteins. J. Virol. *63*: 3142-3150 (1989).

Hardy, W.R.; Strauss, J.H.: Processing the nonstructural polyproteins of Sindbis virus: Nonstructural proteinase is in the C-terminal half of nsP2 and functions both in *cis* and in *trans*. J. Virol. *63*: 4653-4664 (1989).

Karabatsos, N.: International catalogue of arboviruses. 3rd edn. (American Society of Tropical Medicine and Hygiene, San Antonio, 1985).

Pettersson, R.F.; Oker-Blom, C.; Kalkkinen, N.; Kallio, A.; Ulmanen, I.; Kaariainen, L.; Partanen, P.; Vaheri, A.: Molecular and antigenic characteristics and synthesis of rubella virus structural proteins. Rev. Infect. Dis. 7:S140-S149 (1985).

Schlesinger, S.; Schlesinger, M.J.: The *Togaviridae* and *Flaviviridae* (Plenum Press, New York, 1985).

Spaan, W.J.M.; den Boon, J.A.; Bredenbeek, P.J.; et al.: Comparative and evolutionary aspects of coronaviral, arteriviral, and togaviral genome structure and expression. *In* Brinton, M.A.; Heinz, F.X. (eds.), Positive stranded RNA viruses (American Society for Microbiology, Washington, D.C., 1990).

Strauss, E.G.; Rice, C.M.; Strauss, J.H.: Complete nucleotide sequence of the genomic RNA of Sindbis virus. Virology *133*: 92-110 (1984).

Strauss, J.H.; Strauss, E.G.: Evolution of RNA viruses. Annu. Rev. Microbiol. *42*: 657-683 (1988).

Takkinen, K.; Vidgren, G.; Ekstrand, J.; Hellman, U.; Kalkkinen, N.; Wernstedt, C.; Pettersson, R.F.: Nucleotide sequence of the rubella virus capsid protein gene reveals an unusually high G/C content. J. gen. Virol. *69*: 603-612 (1988).

Theiler, M.; Downs, W.G.: The arthropod-borne viruses of vertebrates (Yale University Press, New Haven, 1973).

van Berlo, M.F.; Rottier, P.J.M.; Spaan, W.J.M.; Horzinek, M.C.: Equine arteritis virus-induced polypeptide synthesis. J. gen. Virol. *67*:1543-1549 (1986).

Westaway, E.G.; Brinton, M.A.; Gaidamovich, S.Ya.; Horzinek, M.C.; Igarashi, A.; Kaariainen, L.; Lvov, D.K.; Porterfield, J.S.; Russell, P.K.; Trent, D.W.: *Togaviridae*. Intervirology *24*:125-139 (1985).

Taxonomic status	English vernacular name	International name
FAMILY	—	***FLAVIVIRIDAE***

Reported by G. Wengler

PROPERTIES OF THE VIRUS PARTICLE

Morphology Spherical 40-60 nm in diameter; enveloped.

Physicochemical properties S_{20w} = 140-200.

Nucleic acid A single molecule of infectious ssRNA. No poly(A) tract at the 3'-end. A single long ORF codes for a polyprotein which is processed into all the virus-coded proteins. Structural and nonstructural proteins derived from the 5'- and 3'-terminal sequences, respectively.

Protein Two or three membrane-associated proteins and a core protein.

Lipid and carbohydrate The membrane-associated proteins are inserted in the lipoprotein envelope, whose lipids are cell derived.

Antigenic properties Members of each genus are serologically related to each other but not to members of the other genera.

REPLICATION

Multiply in cytoplasm in association with membranes and mature into cytoplasmic vesicles. Replication commonly accompanied by a characteristic proliferation of intracellular membranes. The only viral messenger is the genome.

Genera

Arbovirus group B	*Flavivirus*
Mucosal disease virus group	*Pestivirus*
Hepatitis C virus group	-

GENUS	ARBOVIRUS GROUP B	***FLAVIVIRUS***
TYPE SPECIES	YELLOW FEVER VIRUS	—

PROPERTIES OF THE VIRUS PARTICLE

Morphology Spherical, 40-50 nm in diameter with an envelope tightly applied to a spherical core 25-30 nm in diameter. Surface projections are demonstrable.

Taxonomic status	English vernacular name	International name

Physiological properties

S_{20w} = 170-210; buoyant density in CsCl = 1.22-1.24 g/cm^3 and 1.15-1.20 g/cm^3 in sucrose.

Nucleic acid

MW ≈ 4 x 10^6 (≈ 11 kb). Capped at the 5'-end; no poly(A) tract at 3' end. The gene order is 5'-C-preM-E-NS1-ns2a-ns2b-NS3-ns4a-ns4b-NS5-3', established by nucleotide and partial amino acid sequence determination.

Protein

Since flaviviruses mature into cytoplasmic vesicles two types of virus particles can be defined: cell-associated virus and extracellular virus. Extracellular virus contains two envelope proteins E and M and an internal RNA-associated protein C. Instead of the M protein cell-associated virus particles contain a larger precursor protein preM which is cleaved during or shortly after release of virus from infected cells; only the carboxy-terminal part of preM remains associated to the extra-cellular virus particle as M protein. The E membrane protein (MW = 51-59 x 10^3) is usually glycosylated. It contains twelve conserved cysteine residues which form six disulfide bridges. The M membrane protein (MW = 7-9 x 10^3) is an unglycosylated protein containing no disulfide bridges. The preM protein (MW = 20-24 x 10^3) is glycosylated containing six disulfide bridges. The C core protein (MW = 14-16 x 10^3) is rich in arginine and lysine throughout its complete primary sequence.

Lipid and carbohydrate

About 17% and 9% by weight, respectively. Located in the viral membrane. The carbohydrate moieties of E comprise both high mannose and complex glycans.

Antigenic properties

A structural model of protein E assigns monoclonal antibody-defined antigenic domains and epitopes to distinct sequence elements and protein domains; these induce antibodies with type or subtype, complex, or group reactivity, measurable by ELISA, RIA, immunofluorescence, virus neutralization, passive protection, inhibition of haemagglutination, or enchancement of infectivity.

REPLICATION

In cytoplasm, associated with proliferation of rough and smooth endoplasmic reticulum forming organelles; no nucleocapsids identified in cells and virus particles accumulate within lamellae and vesicles. RNA replication occurs in foci in the perinuclear region through a minus strand intermediate. The only messenger is the genomic RNA, which is translated into a polyprotein from a single open reading frame on membrane-bound polysomes.

Polyprotein processing has been difficult to observe in infected cells but has been studied in cell-free translation systems. The structural proteins are N-terminal in the order C, pre M and E. Signal peptidase is believed to make the three cleavages that separate the structural proteins. The nonstructural proteins NS1 (a glycoprotein), NS2A, NS2B, NS3, NS4A, NS4B, and NS5 follow. At least three, and probably four, of the cleavages to separate these proteins are made by a trypsin-like proteinase present in the N-terminus of NS3; signal paptidase probably makes the two other cleavages required to separate the nonstructural proteins. NS3 and NS5 are believed to be components of the RNA replicase. In vertebrate cells, the latent period is 12-16 h and virus production continues over 3-4 days. Host cell RNA and protein synthesis continue throughout infection.

BIOLOGICAL ASPECTS

Natural host range

Most members are arboviruses, maintained in nature by bidirectional transfer between haematophagous arthropod vectors (either mosquitoes <u>or</u> ticks, not both) and vertebrate hosts (mammals or birds). Replicate in susceptible species of both phyla. Some viruses have limited vertebrate host range (e.g. only human and simian), for others it can cover a wide variety. The non-arbovirus members of the genus have been isolated either from arthropods or from vertebrates, not both.

Transmission

The majority are transmitted by arthropod bite; transovarial transmission in arthropods has been demonstrated for some members, as has transplacental and horizontal transmission in vertebrates.

Pathogenicity

For <u>arthropods</u> essentially none. In <u>vertebrates</u> highly variable: about 30 viruses cause disease in man, varying from febrile illnesses, rashes, to life-threatening, such as hemmorrhagic fevers, encephalitis, hepatitis. Some 8 to 10 cause severe and economically important disorders in domestic animals.

Experimental hosts

Initial isolation in mice (preferably newborn) by inter-cranial inoculation; after "adaptation", many other hosts may be susceptible. In certain inbred mouse strains, a single dominant gene determines resistance specific for flaviviruses. Genetic resistance associated with generation of DI RNAs and particles. Arthropods can be infected with some by feeding or inoculation.

Taxonomic status	English vernacular name	International name

Cell structures Many vertebrate and arthropod cells support replication, some with, others without CPE or plaque formation or syncytium formation in arthropod cell cultures. Persistent infection is common.

Haema-gglutination Red blood cells from adult geese or 1-2 day-old chicks are agglutinated optimally at slightly acid pH.

Other Members

Based on cross neutralization tests with single polyclonal hyperimmune mouse ascitic fluids prepared against each of the viruses listed, except where indicated otherwise. "Unassigned" denotes viruses which gave no significant cross neutralization in these experiments but are designated as flaviviruses on basis of some serological cross-reaction with at least one accepted member of the genus.

SUBGROUP	NAME OF VIRUS
Tick-borne encephalitis	Tick-borne encephalitis (European subtype and Far Eastern subtype)
	Omsk hemorrhagic fever
	Louping ill
	Kyasanur forest disease
	Langat
	Negishi
	Powassan
	Karshi
	Royal farm
	Carey Island
	Phnom Penh bat (no known vector).
Rio Bravo	Rio Bravo
	Entebbe bat
	Dakar bat
	Bukalasa bat
	Saboya
	Apoi (no known vector).
Japanese encephalitis	Japanese encephalitis
	Murray Valley encephalitis
	Kokobera
	Alfuy
	Stratford
	St. Louis encephalitis
	Usutu
	West Nile
	Kunjin
	Koutango (all mosquito-borne).

Taxonomic status	English vernacular name	International name
	Tyuleniy	Tyuleniy Saunarez Reef Meaban (all tick-borne) (based on CF tests).
	Ntaya	Ntaya Tembusu Yokase Israel turkey meningoencephalitis Bagaza (all mosquito-borne).
	Uganda S	Uganda S Banzi Bouboui Edge Hill (all mosquito-borne).
	Dengue	Dengue types 1, 2, 3, 4 (all mosquito-borne).
	Modoc	Modoc Cowbone Ridge Jutiapa San Vieja San Perlita (no known vectors).
	Unassigned	Gadget's Gully Kadam (tick-borne) Bussuquara Ilheus Jugra Naranjal Rocio Sepik Spondweni Yellow fever Zika Wesselsbron (all mosquito-borne) Aroa Cacipacore Montana myotis leukoencephalitis Sokoluk Tamana bat (no known vectors).

Taxonomic status	English vernacular name	International name
GENUS	MUCOSAL DISEASE VIRUS GROUP	***PESTIVIRUS***
TYPE SPECIES	BOVINE VIRAL DIARRHEA VIRUS (BVDV) (MUCOSAL DISEASE VIRUS)	—

PROPERTIES OF THE VIRUS PARTICLE

Morphology Spherical, 40-60 nm in diameter with an envelope containing 10-12 nm ring-like subunits on the surface.

Physicochemical properties $S_{20w} \approx 140$; buoyant density in sucrose = 1.12-1.13 g/cm^3.

Nucleic acid MW $\approx 4.3 \times 10^6$, ($\approx$ 12.5 kb). The 5'-end has not yet been characterized; no poly(A) tract at 3'-end. Sequencing reveals a single large ORF encoding a poly-protein of about 4,000 amino acids. The tentative gene order is 5'-p20-gp48-gp25-gp55-p125-(p54/p80)-p10-X-(unidentified)-p133(p58/p75)-3', established by sequence-specific antibody reactivities. For cytopathic biotypes of BVDV, a small and variable segment of host cell nucleic acid may be integrated into one particular region (p54) of the viral genome. This insertion maintains the ORF.

Protein Establishment of "structural" proteins is not yet conclusive. There are three viral glycoproteins (MW = 53-57 x 10^3, 44-48 x 10^3, and 23-31 x 10^3) probably in the virus envelope. The core protein is likely to be the first (amino terminal) polypeptide of the polyprotein (MW = 20-31 x 10^3). The hydrophobicity plot of BVDV exhibits a pattern very similar to that seen in most flaviviruses.

Lipid and carbohydrate No reports have described the lipid composition. Virus glycoproteins contain N-linked glycans.

Antigenic properties Monoclonal antibodies reactive with at least one virus glycoprotein (MW = 55-57 x 10^3) neutralize virus infectivity. A conserved, immunodominant nonstructural protein (MW $\approx$ 80 x 10^3) probably represents the virus "soluble antigen".

REPLICATION

Replication occurs in association with membranes. Replication is uniquely sensitive to proflavine and acriflavine. No subgenomic mRNA is found in infected cells. The genomic RNA is believed to be translated into a polyprotein that is rapidly processed cotranslationally and post-translationally, although translation initiation at sites other than the first methionine of the open reading frame has not been ruled out. Differences exist in polyprotein processing by noncytopathic and cytopathic biotypes of BVDV. Both cellular and virus-encoded proteinases are probably involved in polyprotein processing. Candidate virus proteins possessing proteolytic activity for cytopathic BVDV are p20 (MW ≈ 20 x 10^3) and p80 (MW ≈ 80 x 10^3). Based on sequence comparisons, proteins p125 (p54/p80) and p133 (p58/p75) are believed to be components of the RNA replicase. Host cell RNA and protein synthesis continue throughout infection.

BIOLOGICAL ASPECTS

Host range All members have a limited host range (mammals). No invertebrate hosts.

Transmission No known vectors. Field spread occurs by direct and indirect contact (e.g. faecal contaminated feed, urine, nasal secretions) and by transplacental and congenital transmission.

Pathogenicity Highly variable; including inapparent infection, acute or persistent subclinical infection, acute fatal disease (mucosal disease), fetal death or congenital abnormalities, and chronic wasting disease. In mucosal disease, two natural virus biotypes (cytopathic and noncytopathic) must collaborate to induce fatal disease. *Pestivirus* infections of domestic animals represent economically important disease situations worldwide.

Experimental hosts No experimental infection models have been established outside the natural mammalian hosts.

Cell structures Only cells derived from host species (bovine, porcine, ovine) support virus replication. Most virus isolates do not cause CPE. Many cause persistent infections of cell cultures. For BVDV, cytopathic viruses are routinely identified capable of plaque formation and extensive CPE.

Haemagglutination No hemagglutinating activity has been found associated with pestiviruses.

OTHER MEMBERS

Border disease (of sheep)
Hog cholera (European swine fever)

Taxonomic status	English vernacular name	International name
GENUS	HEPATITIS C VIRUS GROUP	—
TYPE SPECIES	HEPATITIS C VIRUS (HCV)	—

PROPERTIES OF THE VIRUS PARTICLE

Morphology

Virus particles have not been visualized by electron microscopy. Virus is enveloped or lipid-containing (inferred from chloroform-sensitivity). Virus diameter estimated to be 40-60 nm extrapolated from filtration and chimpanzee titration studies.

Physicochemical properties

$S_{20w} \geq 150$; buoyant density in sucrose = 1.09-1.11 g/cm^3. Stable in TEN buffer (0.05 M Tris, 0.001 M EDTA, 0.1 M NaCl) at pH 8.0-8.7.

Nucleic acid

MW ≈ 3.5 x 10^6 (≈ 10 kb). The entire genome has been sequenced; a sequence containing 7310 bases located near the 3'-end of the genome has been published (European patent EPO No. 318,216). Single ORF encodes a polyprotein of about 3000 amino acids. No poly(A) tract at terminal 3'-end but several poly(A) rich regions located near 3'-end. The tentative gene order of HCV (inferred from comparative analysis of published sequence and unreported characterization of putative structural gene sequence) (M. Houghton, personal communication) is: 5'-C-preM/E-NS1-NS2A-NS2B-NS3-NS4A-NS4B-NS5-3', where preM/E may represent a fusion of preM and E or a more conventional preM and a truncated E protein peculiar to HCV. Several flavivirus-like concensus sequences are found in HCV, including GXGGXP (amino terminus of HCV-"NS3"), and GDD, a sequence found in the HCV-"NS5" protein that (by comparison to other single-stranded RNA viruses) probably represents the viral RNA-dependent RNA polymerase.

Protein

The existence of "structural" proteins has not been established by conventional gene mapping and Western blot techniques. Putative NS2AB, NS3, NS4AB and NS5 proteins have MW ≈ 41, 62, 42 and 101 x 10^3, respectively, based on hydrophobicity plots and location of known cleavage sites in flavivirus polyproteins. The hydrophobicity plot of HCV exhibits a pattern very similar to that seen in most flaviviruses.

Lipid and carbodydrates

None reported.

Antigenic properties

A highly conserved nonstructural protein (derived from the putative NS4 region and expressed as a fusion protein) has been shown to identify virus-specific antibodies in a wide variety of individuals infected with HCV. No other epitopes or expressed antigens have been described to date.

REPLICATION

Replication appears to occur within hepatocyte cytoplasm of experimentally infected chimpanzees and involves a conspicuous proliferation of endoplasmic reticulum and formation of characteristic ultrastructural alterations. Some of these structures, including convoluted membranes and dense reticular inclusion bodies, mimic those found in cells infected by known flaviviruses. No subgenomic RNA has been detected in infected liver tissues by Northern blot analysis.

BIOLOGICAL ASPECTS

Host range

Man is the natural host and apparent reservoir of HCV. No other natural host has been identified.

Transmission

Approximately 5-10% of all disease caused by HCV occurs as a result of blood transfusion. About 40% of cases of acute sporadic HCV infection have a history of i.v. drug abuse. One half of all other cases do not have any apparent risk of parenteral exposure. Serologic studies of blood donors for virus-specific antibody suggest that about 0.5-1.0% are infected with HCV. About one third of all acute hepatitis in the US is caused by HCV.

Pathogenicity

Highly variable, ranging from inapparent subclinical infection to fulminant disease resulting in hepatic failure and death. Persistent infection occurs in approximately 60% of HCV infected individuals and approximately 20% develop chronic active hepatitis and/or cirrhosis. Persistent HCV infection has been serologically linked to primary liver cancer, cryptogenic cirrhosis, and some forms of autoimmune disease.

Experimental hosts

The chimpanzee remains the only proven model of experimental HCV infection.

Cell culture

None reported.

POSSIBLE MEMBERS

Aedes albopictus cell fusing agent
Simiam hemorrhagic fever virus.

Derivation of Name

flavi: from Latin *flavus*, 'yellow'
pesti: from Latin *pestis,* 'plague'

REFERENCES

Bradley, D.W.; McCaustland, K.A.; Cook, E.H.; Schable, C.A.; Ebert, J.W.; Maynard, J.E.: Post-transfusion non-A, non-B hepatitis in chimpanzees; physicochemical evidence that the tubule-forming agent is a small, enveloped virus. Gastroenterology *88:* 773-779 (1985).

Calisher, C.H.; Karabatsos, N.; Dalrymple, J.M.; Shope, R.E.; Porterfield, J.S.; Westaway, E.G.; Brandt, W.E.: Antigenic relationships between flaviviruses as determined by cross-neutralization tests with polyclonal antisera. J. gen. Virol. *70:* 37-43 (1989).

Chamberlain, R.W.: Epidemiology of arthropod-borne viruses: the role of arthropods as hosts and vectors and of vertebrate hosts in natural transmission cycles. *In*: Schlesinger, R.W. (ed.), The Togaviruses, pp. 175-227 (Academic Press, New York, 1980).

Choo, Q.L.; Kuo, G.; Weiner, A.J.; Overby, L.R.; Bradley, D.W.; Houghton, M.: Isolation of a cDNA clone derived from a blood-borne non-A, non-B viral hepatitis genome. Science *244:* 359-362 (1989).

Collett, M.S.; Larson, R.; Gold, C.; Strick, D.; Anderson, D.K.; Purchio, A.F.: Molecular cloning and nucleotide sequence of the pestivirus bovine viral diarrhea virus. Virology *165:* 191-199 (1988).

Collett, M.S.; Moennig, V.; Horzinek, M.C.: Recent advances in pestivirus research. J. gen. Virol. *70:* 253-266 (1989).

Gorbalenya, A.E.; Donchenko, A.P.; Koonin, E.V.; Blinov, V.M.: N-terminal domains of putative helicases of flavi- and pestiviruses may be serine proteases. Nuc. Acids Res. *17:* 3889-3897 (1989).

Heinz, F.X.: Epitope mapping of flavivirus glycoproteins. Adv. Virus Res. *31:* 103-168 (1986).

Igarashi, A.; Harrap, K.A.; Casals, J.; Stollar, V.: Morphological, biochemical and serological studies on a viral agent (CFA) which replicates in and causes fusion of *Aedes albopictus* (Singh) cells. Virology *74*:174-187 (1976).

Karabatsos, N.: International catalogue of arboviruses, 3rd edn. (American Society of Tropical Medicine and Hygiene, San Antonio, 1985).

Mandl, C.W.; Guirakhoo, F.; Holzmann, H.; Heinz, F.X.; Kunz, C.: Antigenic structure of the flavivirus envelope protein E at the molecular level, using tick-borne encephalitis virus as a model. J. Virol. *63:* 564-571 (1989).

Meyers, G.; Rumenapf, T.; Thiel, H.-J.: Molecular cloning and nucleotide sequence of the genome of hog cholera virus. Virology *171:* 555-567 (1989).

Meyers, G.; Rumenapf, T.; Thiel, H.-J.: Ubiquitin in a togavirus. Nature *341:* 491 (1989).

Murphy, F.A.: Togavirus morphology and morphogenesis. *In* Schlesinger, R.W. (ed.),. The Togaviruses, pp. 241-316 (Academic Press, New York, 1980).

Nowak, T.; Farber, P.M.; Wengler, G.: Analyses of the terminal sequences of West Nile virus structural proteins and of the *in vitro* translation of these proteins allow the proposal of a complete scheme of the proteolytic cleavages involved in their synthesis. Virology *169:* 365-376 (1989).

Rice, C.M.; Grakoui, A.; Galler, R.; Chambers, T.J.: Transcription of infectious yellow fever virus RNA from full-length cDNA templates produced by *in vitro* ligation. New Biol. *1:* 285-296 (1989).

Rice, C.M.; Lenches, E.M.; Eddy, S.R.; Shin, S.J.; Sheets, R.L.; Strauss, J.H.: Nucleotide sequence of yellow fever virus: Implications for flavivirus gene expression and evolution. Science *229*:726-733 (1985).

Roehrig, J.T.: The use of monoclonal antibodies in studies of the structural proteins of togaviruses and flaviviruses. *In*: Schlesinger, S.; Schlesinger, M.J. (eds.), The *Togaviridae* and *Flaviviridae*, pp. 251-278 (Plenum Press, New York, London, 1985).

Ruiz-Linares, A.; Cahour, A.; Despres, P.; Girard, M.; Bouloy, M.: Processing of yellow fever virus polyprotein: Role of cellular proteases in maturation of the structural proteins. J. Virol. *63:* 4199-4209 (1989).

Schlesinger, S.; Schlesinger, M.J.: The *Togaviridae* and *Flaviviridae*. (Plenum Press, New York, 1986).

Shapiro, D.; Brandt, W.E.; Russell, P.K.: Change involving a viral membrane glycoprotein during morphogenesis of group B arboviruses. Virology *50:* 906-911 (1972).

Speight, G.; Coia, G.; Parker, M.D.; Westaway, E.G.: Gene mapping and positive identification of the non-structural proteins NS2A, NS2B, NS3, NS4B and NS5 of the flavivirus Kunjin and their cleavage sites. J. gen. Virol. *69:* 23-34 (1988).

Trousdale, M.D.; Trent, D.W.; Shelokov, A.: Simian hemorrhagic fever virus: a new togavirus. Proc. Soc. Exp. Biol. Med. *150*:707-711 (1975).

Westaway, E.G.: Flavivirus replication strategy. Adv. Virus Res. *33*:45-90 (1987).

Westaway, E.G.; Brinton, M.A.; Gaidamovich, S.Ya.; Horzinek, M.C.; Igarashi, A.; Kaariainen, L.; Lvov, D.K.; Porterfield, J.S.; Russell, P.K.; Trent, D.W.: *Flaviviridae*. Intervirology *24*:183-192 (1985).

Taxonomic status	English vernacular name	International name
FAMILY	CORONAVIRUS GROUP	***CORONAVIRIDAE***

Reported by D. Cavanagh

GENUS	—	***CORONAVIRUS***
TYPE SPECIES	AVIAN INFECTIOUS BRONCHITIS VIRUS (IBV)	—

Properties of the Virus Particle

Morphology Spherical or pleomorphic enveloped particles, 60 to 220 nm in diameter. Club-shaped surface projections, 12-24 nm in length protruding from envelope. Internal RNP structure seen by negative staining as helix of 9-13 nm or strands of 9 nm in diameter.

Physicochemical properties Buoyant density = 1.15-1.18 g/cm^3 in sucrose. Disrupted by ether, chloroform and detergents. Spike but not haemagglutinin-esterase protein of BCV removed by bromelain.

Nucleic acid One molecule of infectious ssRNA; MW = 9.0-11.0 x 10^6 (IBV genome is 27.6 kb; murine hepatitis virus ≈ 33 kb). Polyadenylated at 3'-terminus. MHV genomic RNA known to be capped.

Protein 3 or 4 proteins. All coronaviruses have spike (S), membrane (M) and nucleocapsid (N) proteins and some have haemagglutinin-esterase (HE) protein. HE protein has homology with subunit 1 of haemagglutinin-esterase-fusion protein of influenza C virus; but nature of gene acquisition uncertain (recombination?). S (MW = 170-220 x 10^3) may be cleaved or uncleaved (two subunits: N-terminal S1, C-terminal S2). M present in several differentially glycosylated forms (MW of main species = 23-29 x 10^3). Nucleocapsid (MW = 47-60 x 10^3) phosphorylated and associated with RNA. Membrane fusion and esterase activity associated with S and HE proteins, respectively.

Lipid Present. S protein acylated.

Carbohydrate Present. Spike and haemagglutinin-esterase proteins N-glycosylated. Membrane protein N-glycosylated in IBV, porcine transmissible gastroenteritis and turkey coronaviruses but O-glycosylated in murine hepatitis and bovine coronavirus.

Antigenic properties 3 or 4 major antigens corresponding to each virion protein. Spike and haemagglutinin-esterase predominant antigens involved in neutralization

REPLICATION

Genomic RNA assumed to be mRNA for RNA polymerase responsible for amplification of genome and production of subgenomic mRNAs. One species of negative-stranded RNA is of genome-length that acts as template for the synthesis of a 3'-coterminal set of subgenomic mRNAs which are capped and polyadenylated. Synthesis of mRNA from this template involves a process of discontinuous transcription, probably by a leader-priming mechanism. Apparently, mRNAs serve as templates for their own replication since negative stranded RNAs of mRNA length are also found as part of subgenomic RIs. Another possibility is that the negative stranded subgenomic RNAs may arise by discontinuous transcription on the genome template. Translation of polymerase gene involves ribosomal frame shifting (IBV and murine hepatitis virus). There is a high frequency of recombination (murine hepatitis virus). Number of major subgenomic mRNAs varies from 5-7 depending on virus. The mRNAs encoding structural proteins have been identified for several coronaviruses. Only the 5'-unique regions i.e. those absent from the next smaller RNA, are thought to be translationally active. Structural genes clustered at 3'-end of genome. Virions mature in the cytoplasm by budding through endoplasmic reticulum and golgi membranes. No budding at plasmalemma.

BIOLOGICAL ASPECTS

Host range Infections generally restricted to natural vertebrate host. Often associated with respiratory or gastrointestinal organs.

Transmission Biological vectors not known. Respiratory and faecal-oral transmission. Mechanical transmission common.

OTHER MEMBERS

Human coronavirus
Murine hepatitis virus
Porcine hemagglutinating encephalomyelitis virus
Porcine transmissible gastroenteritis virus
Bovine coronavirus
Canine coronavirus
Feline infectious peritonitis virus

Taxonomic status	English vernacular name	International name
	Turkey coronavirus	
	Probable members	
	Rat coronavirus	
	Porcine epidemic diarrhea virus	
	Possible member	
	Rabbit coronavirus	

Derivation of Name corona: from Latin 'crown', from appearance of surface projections in negatively stained electron micrographs.

REFERENCES

Banner, L.R.; Keck, J.G.; Lai, M.M.C.: A clustering of RNA recombination sites adjacent to a hypervariable region of the peplomer gene of murine coronavirus. Virology *175:* 548-555 (1990).

Boursnell, M.E.G.; Brown, T.D.K.; Foulds, I.J.; Green, P.F.; Tomley, F.M.; Binns, M.M.: Completion of the sequence of the genome of the coronavirus avian infectious bronchitis virus. J. gen. Virol. *68*:57-77 (1987).

Brierley, I.; Digard, P.; Inglis, S.C.: Characterization of an efficient coronavirus ribosomal frameshifting signal: requirement for an RNA pseudoknot. Cell *57:* 537-547 (1989).

Cavanagh, D.; Brian, D.A.; Enjuanes, L.; Holmes, K.V.; Lai, M.M.C.; Laude, H.; Siddell, S.G.; Spaan, W.J.M.; Taguchi, F.; Talbot, P.J.: Recommendations of the Coronavirus Study Group for the nomenclature of the structural proteins, mRNAs and genes of coronaviruses. Virology *176:* 306-307 (1990).

Cavanagh, D.; Brown, T.D.K.: Coronaviruses and Their Diseases. (Plenum Press, New York, 1990).

Luytjes, W.; Bredenbeek, P.J.; Noten, A.F.H.; Horzinek, M.C.; Spaan, W.J.M.: Sequence of mouse hepatitis virus A59 mRNA2: indications for RNA recombination between coronaviruses and influenza C virus. Virology *166:* 415-422 (1988).

Pachuk, C.J.; Bredenbeek, P.J.; Zoltick, P.W.; Spaan, W.J.M.; Weiss, S.R.: Molecular cloning of the gene encoding the putative polymerase of mouse hepatitis coronavirus, strain A59. Virology *171:* 141-148 (1989).

Rasschaert, D.; Gelfi, J.; Laude, H.: Enteric coronavirus TGEV: partial sequence of the genomic RNA, its organization and expression. Biochimie *69:* 591-600 (1987).

Sawicki, S.G.; Sawicki, D.L.: Coronavirus transcription: subgenomic mouse hepatitis virus replicative intermediates function in RNA synthesis. J. Virol. *64:* 1050-1056 (1990).

Sethna, P.B.; Hung, S.L.; Brian, D.A.: Coronavirus subgenomic minus-strand RNAs and the potential for mRNA replicons. Proc. Natl. Acad. Sci. USA *86:* 5626-5630 (1989).

Spaan, W.J.M.; Cavanagh, D.; Horzinek, M.C.: Coronaviruses. *In* van Regenmortel, M.H.V.; Neurath, A.R. (eds.), Immunochemistry of Viruses, Vol. II, pp. 359-379 (Elsevier/North Holland, Amsterdam, 1990).

Spaan, W.J.M.; Cavanagh, D.; Horzinek, M.C.: Coronaviruses: structure and genome expression. J. gen. Virol. *69:* 2939-2952 (1988).

Vlasak, R.; Luytjes, W.; Leider, J.; Spaan, W.; Palese, P.: The E3 protein of bovine coronavirus is a receptor-destroying enzyme with acetylesterase activity. J. Virol. *62:* 4686-4690 (1988).

Taxonomic status	English vernacular name	International name
GENUS	—	*TOROVIRUS*
TYPES SPECIES	BERNE VIRUS	—

Compiled by M.C. Horzinek

PROPERTIES OF THE VIRUS PARTICLE

Morphology Pleomorphic, bioconcave disk-, kidney- and rod-shaped particles 120-140 nm in diameter containing an elongated tubular capsid with helical symmetry. Peplomer-bearing envelope.

Physicochemical properties S_{20w} = 380-400; buoyant density in sucrose 1.16-1.17 g/cm^3; stable between pH 2.5 and 9.7.

Nucleic acid Polyadenylated linear positive-sense ssRNA (infectious) > 20 kb.

Protein Three major proteins in virus particle with MW ≈ 18 x 10^3 (nucleocapsid), 26 x 10^3 (envelope) and 80-100 x 10^3 (peplomer dimer derived from 200 x 10^3 precursor).

Lipid Present.

Carbohydrate Only peplomer protein is glycosylated.

REPLICATION

In cytoplasm, 3'-coterminal nested set of 5 subgenomic mRNAs is detected. The polymerase gene contains two overlapping ORFs; the more downstream one is expressed by ribosomal frame-shifting during translation of genomic RNA. Budding of preformed tubular capsids through Golgi membranes and endoplasmic reticulum; host cell nuclear function required.

BIOLOGICAL ASPECTS

Host range Ungulates, man; probably also carnivores (mustellids).

Transmission Probably via the faecal-oral route.

OTHER MEMBERS

Breda virus (cattle).

Derivation of Name toro: from Latin *torus*, 'lowest convex molding in the base of a column'.

References

Horzinek, M.C.; Flewett, T.H.; Saif, L.J.; Spaan, W.J.M.; Weiss, M.; Woode, G.N.: A new family of vertebrate viruses: *Toroviridae*. Intervirology *27:* 17-24 (1987).

Snijder, E.J.; Ederveen, J.; Spaan, W.J.M.; Weiss, M.; Horzinek, M.C.: Characterization of Berne virus genomic and messenger RNAs. J. gen. Virol. *69:* 2135-2144 (1988).

Snijder, E.J.; Horzinek, M.C.; Spaan, W.J.M.: A 3'-coterminal nested set of independently transcribed messenger RNAs is generated during Berne virus replication. J. Virol. *64:* 331-338 (1990).

Weiss, M.; Steck, F.; Horzinek, M.C.: Purification and partial characterization of a new enveloped RNA virus (Berne virus). J. gen. Virol. *64:* 1849-1858 (1983).

Weiss, M.; Horzinek, M.C.: Morphogenesis of Berne virus (proposed family *Toroviridae*). J. gen. Virol. *67:* 1305-1314 (1986).

Taxonomic status	English vernacular name	International name
ORDER	—	*MONONEGAVIRALES*

Compiled by C.R. Pringle

GENERAL

The order embraces the three families of eukaryotic viruses possessing linear non-segmented negative-strand RNA genomes, i.e. the *Filoviridae, Paramyxoviridae* and *Rhabdoviridae*. Common features include the negative-sense template RNA in the virion, the helical nucleocapsid, the initiation of primary transcription by a virion-associated RNA dependent RNA polymerase, similar gene order (3' NTR - core protein genes - envelope protein genes - polymerase gene - 5' NTR) and single 3' promoter. Maturation is by budding, predominantly from the plasma membrane; rarely from internal membranes (rabies virus) or the inner nuclear membrane (many plant rhabdoviruses). Cytoplasmic, except for some plant rhabdoviruses.

PROPERTIES OF THE VIRUS PARTICLE

Morphology

The virions are large enveloped structures with a prominent fringe of spikes, 5-10 nm long and spaced 7-10 nm apart. The morphologies of the particles are variable but distinguish the three families: Simple, branched, U-shaped, 6-shaped, or circular filaments of uniform diameter (≈ 80 nm) extending up to 14,000 nm are characteristic of the *Filoviridae*, although purified virions are bacilliform and of uniform length (e.g. 790 nm in the case of Marburg virus); filamentous, pleomorphic or spherical structures of variable diameter are characteristic of the *Paramyxoviridae;* and regular bullet-shaped or bacilliform particles are characteristic of the *Rhabdoviridae*. The helical ribonucleoprotein core has a diameter of 13-20 nm which in filoviruses and rhabdoviruses is organised into a helical nucleocapsid of ≈50 nm diameter. The nucleocapsid of VSV is infectious.

Physicochemical properties

MW = 300-1000 x 10^6. S_{20w} = 550- >1000. Buoyant density in sucrose = 1.18-1.20g/cm^3.

Nucleic acid

One molecule of linear non-infectious negative-sense single-stranded RNA, MW = 3.5-7 x 10^6. 0.5-2.0 % of particle by weight. Genome comprises a linear sequence of non-overlapping genes with short terminal untranscribed regions and intergenic regions ranging from

Taxonomic status	English vernacular name	International name

2 to several hundred nucleotides; the only known exceptions are a short overlap of the 9th and 10th genes of respiratory syncytial virus, and encoding of genetic information in all three reading frames in the P genes of paramyxoviruses and morbilliviruses. Genome sizes so far determined range from 11.161 kb (VSV) to 15.892 kb (measles virus). Infectivity sensitive to lipid solvents.

Protein Limited in number in relation to the large particle size; probably no more than 5-7 structural proteins comprising envelope glycoprotein(s), a matrix protein, a major RNA-binding protein, nucleocapsid-associated protein(s) and a large molecular weight polymerase protein, plus in paramyxoviruses several non-structural proteins of unknown function. Enzymes associated with virions may include transcriptase, polyadenylate transferase, mRNA methyl transferase, neuraminidase.

Lipid 15-25 % by weight, composition dependent on host cell.

Carbohydrate 3-6 %, where known.

Antigenic properties Membrane glycoproteins involved in neutralisation; serotypes defined by surface antigens. Filoviruses exceptional in that cannot be neutralised *in vitro*.

Pathogenic potential Variable, but in human hosts tends to be characteristic of family: Haemorrhagic fever (*Filoviridae*); respiratory and neurological disease (*Paramyxoviridae*); mild febrile to fatal neurological disease (*Rhabdoviridae*).

REPLICATION

Discrete unprocessed messenger RNAs are transcribed by sequential interrupted synthesis. Generally genes do not overlap and only 1 ORF utilised; the P genes of paramyxoviruses and morbilliviruses are exceptional in that all 3 ORFs may be utilised; alternate starts, non-AUG starts, and mRNA editing by insertion of non-templated nucleotides to change reading frame are devices uniquely employed in the expression of P gene products. Replication occurs by synthesis of a complete positive-sense RNA anti-genome. Maturation of the independently assembled helical nucleocapsids occurs by budding through host membranes and investment by a host-derived lipid envelope containing transmembrane virus proteins.

Biological Aspects

Host range

Ranging from restricted to unrestricted. Filoviruses have only been isolated from primates. Paramyxoviruses occur only in vertebrates and no vectors are known. Rhabdoviruses infect invertebrates, vertebrates and plants: Some rhabdoviruses multiply in both invertebrates and vertebrates, some in invertebrates and plants, but none in all three hosts.

Families, Sub-Families and Genera

Family	Sub-family	Genus
Filoviridae		*Filovirus*
Paramyxoviridae		
	Paramyxovirinae	*Morbillivirus* *Paramyxovirus*
	Pneumovirinae	*Pneumovirus*
Rhabdoviridae		*Lyssavirus* *Vesiculovirus*

Derivation of Name

mono from Greek *monos* 'single';
nega from *nega*tive strand RNA;
virales from Latin *virales* 'viruses'.

REFERENCES

Kiley, M.P.; Bowen, E.T.A.; Eddy, G.A.; Isaacson, M.; Johnson K.M.; McCormick, J.B.; Murphy, F.A.; Pattyn, S.R.; Peters, D.; Prozesky, O.W.; Regnery, R.L.; Simpson, D.I.H.; Slenczka, W.; Sureau, P.; van der Groen, G.; Webb , P.A.; Wulff, H.: *Filoviridae*, a Taxonomic Home for Marburg and Ebola Viruses: Intervirology *18*: 24-32 (1982).

Kingsbury, D.W.: The Paramyxoviruses. (Plenum Press, New York, 1991).

Pringle, C.R.: The order *Mononegavirales*. Arch. Virol. *117*:137-140 (1991).

Wagner, R.R.: The Rhabdoviruses. (Plenum Press, New York, 1987).

Taxonomic status	English vernacular name	International name
FAMILY	—	***PARAMYXOVIRIDAE***

Reported by C.R. Pringle

PROPERTIES OF THE VIRUS PARTICLE

Morphology

Pleomorphic, usually roughly spherical, 150 nm or more in diameter, but filamentous forms common; envelope derived from cell membrane lipids, incorporating 2 or 3 virus glycoproteins and 1 or 2 unglycosylated proteins. Surface projections 8-12 nm in length, spaced 7-10 nm apart according to genus, contain virus glycoproteins. Nucleocapsid has helical symmetry, 13-18 nm in diameter and 5.5-7 nm pitch according to genus; length up to 1 μm in some genera.

Physicochemical properties

MW > 500 x 10^6, much more for pleomorphic multiploid virions; S_{20w} at least 1000; buoyant density in sucrose = 1.18-1.20 g/cm^3; sensitive to lipid solvents, non-ionic detergents, formaldehyde, and oxidising agents.

Nucleic acid

Single molecule of ssRNA, MW = 5-7 x 10^6. About 0.5% by weight of virus particle. Genomic size fairly uniform: 15.156 kb for Newcastle disease virus, 15.222 kb for human respiratory syncytial virus, 15.285 kb for Sendai virus, 15.463 kb for parainfluenza virus type 3 and 15.892 kb for measles virus. Most particles contain a negative-sense strand, but some contain positive-sense template strands. Thus partial self-annealing of isolated RNA may occur.

Protein

Paramyxoviruses and morbilliviruses have 7-8 ORFs (genes) that encode 10-12 proteins (MW ≈ 5-200 x 10^3), of which 4-5 are derived from 2-3 overlapping ORFs of the P locus. Pneumoviruses have 10 ORFs encoding 10 proteins of MW = 7.5-200 x 10^3, the 9th and the 10th ORFs overlapping in respiratory syncytial virus. Proteins common to all genera are: three nucleocapsid-associated proteins, namely an RNA-binding protein (N or NP), a phosphoprotein (P), a large putative polymerase protein (L); an unglycosylated envelope protein (M); and two glycosylated envelope proteins, comprising a fusion protein (F) and an attachment protein (G, H or HN). Variable proteins include the nonstructural proteins (C, 1C/NS1, and 1B/NS2), a small integral membrane protein (SH/1A), a second inner envelope unglycosylated protein (M2/22 kDa), and a cysteine-rich protein (V). Enzymes (variously represented and reported among genera);

Taxonomic status	English vernacular name	International name
	transcriptase, polyadenylate transferase, mRNA methyl transferase, neuraminidase.	
Lipid	20-25% by weight, host cell derived.	
Carbohydrate	6% by weight, composition dependent on host cell.	
Antigenic properties	One or more surface antigens involved in virus neutralisation; one nucleocapsid antigen described; specificities of antigens vary among genera.	
Effect on cells	Generally cytolytic, but temperate and persistent infections are common; other features are inclusions, syncytium formation, and haemadsorption.	

REPLICATION

Virus entry by fusion of envelope with cell surface membrane at neutral pH; genome transcribed from single promoter into 6-10 separate mRNAs, nucleocapsid is the functional template for transcription of complementary viral mRNA species and for RNA replication. Independently assembled nucleocapsids are enveloped on cell surface at sites containing virus envelope proteins. Paramyxoviruses and morbilliviruses exhibit a novel strategy whereby the variable inclusion of non-templated nucleotides at one site in mRNA from the P locus results in a shift in reading frame and expression of the V protein (and in some of the D protein). C protein(s) are expressed from alternate and non-AUG starts by independent ribosomal initiation. Virions are released by budding. Variable dependence on host nuclear functions.

BIOLOGICAL ASPECTS

Host range Only found in vertebrates. Each virus has its own host range in nature and in the laboratory.

Transmission Horizontal, mainly airborne; no vectors.

SUBFAMILY — *PARAMYXOVIRINAE*

GENERA

Parainfluenza virus group	*Paramyxovirus*
Measles-rinderpest-distemper virus group	*Morbillivirus*

Taxonomic status	English vernacular name	International name
GENUS	PARAINFLUENZA VIRUS GROUP	***PARAMYXOVIRUS***
TYPE SPECIES	NEWCASTLE DISEASE VIRUS AVIAN PARAMYXOVIRUS TYPE 1 (PMV-1)	—

PROPERTIES OF THE VIRUS PARTICLE

All members of the genus possess a neuraminidase, in contrast to members of the other two genera.

OTHER MEMBERS

Avian paramyxovirus 2 (Yucaipa)	(PMV-2)
Avian paramyxovirus 3	(PMV-3)
Avian paramyxovirus 4	(PMV-4)
Avian paramyxovirus 5 (Kunitachi)	(PMV-5)
Avian paramyxovirus 6	(PMV-6)
Avian paramyxovirus 7	(PMV-7)
Avian paramyxovirus 8	(PMV-8)
Avian paramyxovirus 9	(PMV-9)
Parainfluenza virus type 1	human, murine (Sendai)
Parainfluenza virus type 3	human, bovine, ovine, simian (SA10)
Parainfluenza virus type 2	canine, human, simian (SV5, SV41)
Parainfluenza virus type 4	human
Mumps virus	human

GENUS	MEASLES-RINDERPEST-DISTEMPER VIRUS GROUP	***MORBILLIVIRUS***
TYPE SPECIES	MEASLES VIRUS	—

PROPERTIES OF THE VIRUS PARTICLE

All members lack the neuraminidase possessed by the genus *Paramyxovirus*, and differ from the genus *Pneumovirus* in size of the nucleocapsid and other features. All members produce both cytoplasmic and intranuclear inclusion bodies which contain viral RNP. Members of this genus are related antigenically.

OTHER MEMBERS

Canine distemper virus	canine,mustelid
Phocine distemper virus	phocine-phocid, phocoenine

Taxonomic status	English vernacular name	International name
	Peste-des-petits-ruminants virus	caprine,ovine
	Rinderpest virus	bovine,caprine, ovine,porcine

SUBFAMILY	—	*PNEUMOVIRINAE*

GENERA

	Respiratory syncytial virus group	*Pneumovirus*

GENUS	RESPIRATORY SYNCYTIAL VIRUS GROUP	*PNEUMOVIRUS*
TYPE SPECIES	HUMAN RESPIRATORY SYNCYTIAL VIRUS	—

PROPERTIES OF THE VIRUS PARTICLE

Lacks neuraminidase; haemagglutinin absent in bovine and human respiratory syncytial viruses, present in pneumonia virus of mice. Differs from the other two genera in several features: gene number (10 compared with 7/8 transcriptional units), smaller average gene size, possession of one additional unglycosylated membrane-associated protein (M2/22 kDa), inversion of attachment (G) and fusion (F) proteins in the gene order, extensive O-linked glycosylation of the G protein, P locus encodes a single protein. Nucleocapsid diameter (13-14 nm compared with 18 nm), nucleocapsid pitch (7 nm compared with 5.5 nm), length of spike (10-12 nm compared with 8 nm).

OTHER MEMBERS

Bovine respiratory syncytial virus	bovine,caprine,ovine
Pneumonia virus of mice	rodent
Turkey rhinotracheitis virus	avian

Uncharacterised paramyxoviruses

Fer-de-Lance virus	reptillian
La-Piedad-Michoacan-Mexico virus (LPMV)	porcine
Mapuera virus	chiropteran
Nariva virus	rodent
Several viruses from penguins distinct from PMV1-9	avian

Derivation of Name	paramyxo: from Greek *para*, 'by the side of', and *myxa* 'mucus' (relating to activity of haemagglutinin and neuraminidase). morbilli: plural of Latin *morbillus*, diminutive of *morbus*, 'disease'; measles from Germanic *Masemn*. pneumo: from Greek *pneuma*, 'breath'.

REFERENCES

Alexander, D.J.: The classification, host range and distribution of avian paramyxoviruses, *In* McFerran, J.B.; McNulty, M.S. (eds.): Acute Virus Infections in Poultry, pp. 52-66 (Martinus Nijhoff, Dordrecht, 1986).

Bishop, D.H.L.; Compans, R.W.: Nonsegmented negative strand viruses; Paramyxoviruses and Rhabdoviruses (Academic Press, Orlando, 1984).

Choppin, P.W.; Compans, R.W.: Reproduction of paramyxoviruses. *In* Fraenkel-Conrat, H.; Wagner, R.R. (eds.), Comprehensive Virology, Vol. 4, pp. 95-178 (Plenum Press, New York, 1975).

Kingsbury, D.W.: The Paramyxoviruses (Plenum Press, New York, 1991).

Kingsbury, D.W.: *Paramyxoviridae* and their replication. *In* Fields, B.N.; Knight, J.C. (eds.), Virology, Vol. 1, 2nd edn., pp. 945-962 (Raven Press, New York, 1990).

Mahy, B.; Kolakofsky, D.: The Biology of negative strand viruses (Elsevier/North Holland, Amsterdam, Oxford, New York, 1987).

Morrison, T.G.: Structure, function and intracellular processing of paramyxovirus membrane proteins. Virus Res. 10: 113-136 (1988).

Orvell, C.; Norrby, E.: Antigenic structure of paramyxoviruses. *In* van Regenmortel, M.H.V.; Neurath, A.R. (eds.), Immunochemistry of Viruses. The Basis for Serodiagnosis and Vaccines, pp. 241-264 (Elsevier Medical Press, Amsterdam, 1985).

Pringle, C.R.: Paramyxoviruses and Disease. *In* Russell, W.C.; Almond, J.W. (eds.), SGM Symposium 40, Molecular basis of virus disease, pp. 51-90 (Cambridge University Press, Cambridge, 1987).

Stott, E.J.; Taylor, G.: Respiratory syncytial virus; brief review. Arch. Virol. 84: 1-52 (1984).

Taxonomic status	English vernacular name	International name
FAMILY	MARBURG VIRUS GROUP	***FILOVIRIDAE***

Compiled by J.B. McCormick

GENUS	—	***FILOVIRUS***
TYPE SPECIES	MARBURG VIRUS	—

PROPERTIES OF THE VIRUS PARTICLE

Morphology

Pleomorphic, virions appearing as long filamentous forms (sometimes with extensive branching) or as U-shaped, "6"-shaped or circular forms. Particles vary greatly in length (up to 14,000 nm), but of uniform diameter ≈ 80 nm. There are surface projections ≈ 7 nm in length spaced at 10 nm intervals on the particle surface. Virions purified by rate zonal gradient centrifugation are infectious, uniform and bacilliform in shape; Ebola 970 nm and Marbourg 790 nm long. Inside the envelope is a nucleocapsid with a dark central axis ≈ 20 nm in diameter surrounded by a helical tubular capsid ≈ 50 nm in diameter bearing cross-striations with a periodicity ≈ 5 nm. The 20 nm central axis, also seen in infected cells appears to be the virion RNP. A structure with buoyant density ≈ 1.32 g/cm^3 in CsCl, is released from virions by detergent treatment and probably represents the viral RNP. Within the nucleocapsid is an axial channel ≈ 10-15 nm with nucleocapsid proteins (N and VP30); proteins L and VP35, the putative transcriptase are associated with the RNP.

Physicochemical properties

MW = 300-600 x 10^6; S_{20w}.of long particles very high but infectious bacilliform particles ≈ 1,400 S; buoyant density ≈ 1.14 g/cm^3 in potassium tartrate. Infectivity is quite stable at room temperature but is destroyed in 30 min at 60°C. Sensitive to lipid solvents.

Nucleic acid

One molecule of noninfectious (negative strand) linear ss RNA; MW ≈ 4.5 x 10^6 ≈ 1.1% by weight of virus.

Protein

7 proteins designated L, G, N, VP40, VP35, VP30 and VP24. G is very large and 2 are associated with RNA (N and VP30).

Lipid

Present.

Carbohydrate

Associated with surface projections and possibly glycolipid.

Taxonomic status	English vernacular name	International name

Antigenic properties

Virus cannot be neutralized *in vitro*. There is no antigenic cross-reaction between Marburg and Ebola. The two Ebola biotypes, Zaire and Sudan, can be differentiated antigenically. G protein seems to define serotype.

REPLICATION

Seven virion proteins are translated from monocistronic mRNA complementary to virion RNA. Virion transcriptase activity has been detected. Synthesized proteins accumulate in the cytoplasm. Budding of virions appears to be through plasma membrane. Little virion RNA accumulates in infected cells suggesting a very efficient maturation process. Viruses share similar replication signals with both rhabdoviruses and paramyxoviruses.

BIOLOGICAL ASPECTS

Both viruses are indigenous to Africa. Ebola strains have also come from South-east Asia. Some strains cause severe hemorrhagic fevers in humans. Marburg was first isolated from hemorrhagic fever patients in West Germany and Yugoslavia in 1967 by contact with tissues and blood from infected but apparently healthy monkeys (*Ceriopithecus aethiops*) imported from Uganda. Ebola virus was first isolated from two separate outbreaks in northern Zaire and southern Sudan in the fall of 1976.

Host range

The natural reservoir or source of either virus is unknown. In the laboratory, monkey, mouse, guinea pig and hamster have been experimentally infected.

Transmission

In human cases, transmission appears to occur only by close personal contact. Mortality in outbreaks may be as high as 88%.

OTHER MEMBER

Ebola virus (Zaire and Sudan biotypes).

Derivation of Name

filo: from Latin *filo* 'thread-like', for the morphology of the particles.

REFERENCES

Buchmeier, M.J.; DeFries, R.U.; McCormick, J.B.; Kiley, M.P.: Comparative analysis of the structural polypeptides of Ebola viruses from Sudan and Zaire. J. Infect. Dis. *147*:276-281 (1983).

Cox, N.J.; McCormick, J.B.; Johnson, K.M.; Kiley, M.P.: Evidence for two subtypes of Ebola virus based on oligonucleotide mapping of RNA. J. Infect. Dis. *147*:272-275 (1983).

Elliott, L.H.; Kiley, M.P.; McCormick, J.B.: Descriptive analysis of Ebola virus proteins. Virology *147*:169-176 (1985).

Kiley, M.P.; Bowen, E.T.A.; Eddy, G.A.; Isaacson, M.; Johnson, K.M.; McCormick, J.B.; Murphy, F.A.; Pattyn, S.R.; Peters, D.; Prozesky, O.W.; Regnery, R.L.; Simpson, D.I.H.; Slenczka, W.; Sureau, P.; van der Groen, G.; Webb, P.A.; Wulff, H.: *Filoviridae*: a Taxonomic Home for Marburg and Ebola viruses? Intervirology *18*:24-32 (1982).

Kiley, M.P.; Wilusz, J.; McCormick, J.B.; Keene, J.D.: Conservation of the 3' terminal nucleotide sequences of Ebola and Marburg virus. Virology *149*:251-254 (1986).

Martini, G.; Siegert, R.: Marburg Virus Disease. (Springer, New York, 1971).

Pattyn, S.R.: Ebola virus haemorrhagic fever. (Elsevier/North Holland, Amsterdam, 1978).

Regnery, R.L.; Johnson, K.M.; Kiley, M.P.: Virion nucleic acid of Ebola virus. J. Virol. *36*:465-469 (1980).

Regnery, R.L.; Johnson, K.M.; Kiley, M.P.: Marburg and Ebola viruses: possible members of a new group of negative strand viruses. *In* Bishop, D.H.L.; Compans, R.W. (eds.), The replication of negative strand viruses. pp. 971-977 (Elsevier/North Holland, New York, 1981).

Richman, D.D.; Cleveland, P.H.; McCormick, J.B.; Johnson, K.M.: Antigenic analysis of strains of Ebola virus: identification of two Ebola virus serotypes. J. Infect. Dis. *147*:268-271 (1983).

Sanchez, A.; Kiley, M.P.: Identification and analysis of Ebola virus messenger RNA. Virology *157*:414-420 (1987).

Taxonomic status	English vernacular name	International name
FAMILY	BULLET-SHAPED VIRUS GROUP	***RHABDOVIRIDAE***

Revised by W.H. Wunner & D. Peters

PROPERTIES OF THE VIRUS PARTICLE

Morphology

Viruses infecting vertebrates and invertebrates are usually bullet-shaped, and those infecting plants are usually bacilliform; 100-430 nm long and 45-100 nm in diameter, with surface projections (G protein) 5-10 nm long and ≈ 3 nm in diameter. In thin section, a central axial channel is seen. Characteristic cross-striations (spacing 4.5-5.0 nm) are seen in negatively stained and thin-sectioned particles. Truncated particles 0.1-0.5 of the length of the virus may be common except perhaps in members infecting plants. Abnormally long and double-length particles and tandem formations are sometimes observed. A honeycomb pattern is observed on the surface of some members. The inner nucleocapsid, ≈ 50 nm in diameter, with helical symmetry, consists of an RNA+N protein complex together with L and NS proteins, surrounded by an envelope containing M protein. The nucleocapsid contains transcriptase activity and is infectious. It uncoils to a helical structure ≈ 20 x 700 nm.

Physicochemical properties

MW = 300-1,000 x 10^6; S_{20w} = 550-1,000; buoyant density in CsCl = 1.19-1.20 g/cm^3 and in sucrose, 1.17-1.19 g/cm^3. Infectivity; stable at the range pH 5-10; rapidly inactivated at 56°C and by UV- and X-irradiation; sensitive to lipid solvents.

Nucleic acid

One molecule of noninfectious linear (negative-sense) ssRNA; S_{20w} = 38-45; MW = 3.5-4.6 x 10^6; 1-2% by weight of virus.

Protein

Five major polypeptides [designated L,G,N,NS and M for vesicular stomatitis-Indiana (VS-I) virus]; 65-75% by weight of the virus. Other polypeptides may be present in minor amounts. Transcriptase and other enzyme activities are present in virus.

Lipid

15-25% by weight of virus, the composition being dependent on the host cell.

Carbohydrate

3% by weight of virus. Associated with surface projections and glycolipids; minor variation with host cell type.

Antigenic properties

G protein is involved in virus neutralization and defines the serotype. N protein shows cross-reactions between some vesiculoviruses and between some lyssaviruses. N antigen is apparently different in two serotypes of potato yellow dwarf virus.

REPLICATION

Viral proteins accumulate in the cytoplasm except for some plant members. Virus RNA is transcribed by virion transcriptase into several positive-strand RNA species which act as mRNA in polyribosome complexes. Virus RNA replication involves ribonucleoprotein (RNA+N protein, RNP) complex as nucleocapsid template to synthesize full-length positive-strand RNA intermediate. Viral nucleocapsid structures containing negative-strand RNA+N, NS, and L proteins are formed in cytoplasm and virus is assembled from nucleocapsid+M protein complex with envelope produced independently by insertion of viral G protein into pre-existing host cell membranes. Site of formation of mature particles is variable, depending on virus and host cell - e.g. VS-I nucleocapsid is synthesized in cytoplasm and then virus predominantly buds from the plasma membrane in most, but not all, cells; rabies virus buds predominantly from intracytoplasmic membranes; and about half of the plant members bud from the inner nuclear membrane. Complete particles of these viruses accumulate in the perinuclear space.

BIOLOGICAL ASPECTS

Host range

Some members multiply in arthropods as well as vertebrates, others in arthropods and plants. Sigma virus was recognized first as a congenital infection of *Drosophila*. Some vertebrate members have a wide experimental host range. A wide range of vertebrate and invertebrate cells are susceptible to vertebrate viruses *in vitro*. Plant members usually have narrow host range among higher plants; some have been grown in insect cell cultures.

Transmission

Some viruses are transmitted vertically in insects, but none is so transmitted in vertebrates or plants. Some can be transmitted mechanically in plants. Vector transmission by mosquitoes, sandflies, culicoides, mites, aphids, or leafhoppers. Mechanical transmission of viruses infecting vertebrates can be by contact or aerosol, bite or venereal.

Taxonomic status	English vernacular name	International name
	GENERA/GROUPS	
	Vesicular stomatitis virus group	*Vesiculovirus*
	Rabies virus group	*Lyssavirus*
	Plant rhabdovirus group	-

GENUS	VESICULAR STOMATITIS VIRUS GROUP	***VESICULOVIRUS***
TYPE SPECIES	VESICULAR STOMATITIS -INDIANA VIRUS	—

PROPERTIES OF THE VIRUS PARTICLE

Morphology

Virus is ≈ 170 nm long, ≈ 70 nm wide. Helix of the nucleocapsid has an outer diameter of ≈ 49 nm; inner diameter ≈ 29 nm; 35 subunits per turn. The RNP is linear and ≈ 1 µm long.

Physicochemical properties

$S_{20w} \approx 625$.

Nucleic acid

The RNA genome of VS-I virus consists of 5 genes in tandem with no overlaps in the order 3'-N-NS-M-G-L-5'. All but 70 of the 11,161 nucleotides are represented in positive-strand transcripts comprising five monocistronic mRNAs plus an untranslated 3'-leader sequence of 47 nucleotides. The untranscribed regions are a 59 nucleotide 5'-terminal region of the L gene, a 3 nucleotide spacer between leader and N gene, and 4 dinucleotide spacers (CA and GA) at the four inter-cistronic junctions. There is a common nucleotide sequence 3'-AUACUUUUUUU-5' preceding each intercistronic junction, and the sequences complementary to the 5'-end of each mRNA have the general form 3'-UUGUCNNUAG-5'.

Protein

L (large) MW ≈ 150 x 10^3; G (glycoprotein) MW = 70-80 x 10^3; N (nucleoprotein) MW = 50-62 x 10^3; NS (nonstructural and phosphorylated) MW = 40-50 x 10^3; M (matrix, phosphorylated) MW = 20-30 x 10^3. Number of protein subunits in virion: L, 20-50; G, 500-1,500; N, 1,000-2,000; NS, 100-300; M, 1,500-4,000. Enzymes in virion: transcriptase (made up of L + NS proteins); protein kinase (host?); guanyl and methyl transferases; nucleotide triphosphatase; nucleoside diphosphate kinase; 5' capping enzyme.

Antigenic properties

G protein functions as type-specific immunizing antigen; N is cross-reacting CF antigen.

Replication

VS-I virus replicates in enucleate cells. Phenotypic mixing is extensive between VS-I and heterologous lytic viruses (simian virus 5, Newcastle disease virus, fowl plague virus, herpes simplex virus), nonlytic viruses (avian myeloblastosis virus, murine leukemia virus, mouse mammary tumor virus), and partially expressed endogenous viruses.Phenotypic mixing (complementation) also occurs within but not between serological types of vesiculoviruses. Complementation is reported to occur by re-utilization of structural components of UV-irradiated VS-I virus. Complementation shown with VS (Indiana, Cocal, New Jersey) and Chandipura. Five or six non-overlapping groups (identified). Inter-strain complementation only observed with serologically related viruses - e.g. VS-I and Cocal.

Biological Aspects

VS-I serotype isolated from vertebrates and insects.

Other Members

Isolated in nature from vertebrates (V) or invertebrates (I):

BeAn 157575 (V)
Boteke (I)
Calchaqui (I)
Carajas (I)
Chandipura (I, V)
Cocal (I, V)
Eel virus American/Eel virus European (V)
Grasscarp rhabdovirus (V)
Gray Lodge (I)
Isfahan (I)
Jurona (I)
Klamath (V)
Kwatta
La Joya (I)
Malpais Spring (I)
Maraba (I)
Mount Elgon bat (V)
Perinet (I)
Pike fry rhabdovirus (V)
Piry (V)
Porton (I)
Radi (I) (= ISS Ph1 116)
Spring viremia of carp (V) (= Rhabdovirus carpia)

Taxonomic status	English vernacular name	International name
	Tupaia (V) Ulcerative disease rhabdovirus (V) Vesicular stomatitis Alagoas (V) Vesicular stomatitis New Jersey (I, V) Yug Bogdanovac (I)	

GENUS	RABIES VIRUS GROUP	***LYSSAVIRUS***
TYPE SPECIES	RABIES VIRUS	—

PROPERTIES OF THE VIRUS PARTICLE

Physiological properties

Morphologically and physicochemically similar to *Vesiculovirus* but antigenically distinct. Virus is ≈ 180 nm (130-200 nm) long, ≈ 75 nm (60-110 nm) wide. Helical nucleocapsid has 30-35 subunits per turn. Unwound filamentous nucleocapsid = 4.2-4.6 µm. Surface projections (G protein) ≈ 10 nm long.

Nucleic acid

The RNA genome of rabies (PV strain) has the same gene organization as VS-I. Five monocistronic mRNAs plus a 3'-leader sequence of 58 nucleotides are transcribed. The untranscribed regions are a 70 nucleotide 5'-terminal region of the L gene, one dinucleotide spacer between N and NS genes, two pentanucleotide spacers between NS and M genes and between M and G genes, and a 423 nucleotide spacer between the G and L genes. The same poly (U) stretch at the 5'-end of each gene and sequences complementary to the 5'-end of each mRNA are present as in VS-I genome. A sixth mRNA is transcribed from the large G-L intercistronic region of infectious haematopoietic necrosis virus.

Protein

L (large), MW ≈ 190 x 10^3; G (glycoprotein), MW = 65-80 x 10^3; N (nucleoprotein, phosphorylated), MW = 58-62 x 10^3; NS (M1, phosphorylated), MW = 35-40 x 10^3; M (M2, matrix), MW = 22-25 x 10^3. Number of protein subunits in virion: L, 17-150; G, 1,600-1,900; N, 1,750; NS (M1), 900-950; M (M2), 1,650-1,700. Enzymes in virion: transcriptase (L + NS proteins). A sixth protein designated nonviral (NV) is encoded by the sixth gene of infectious haematopoietic necrosis virus; function unknown.

Antigenic properties

On the basis of serum-neutralization tests, some lyssaviruses have been grouped into four serotypes; 1 (rabies), 2 (Lagos bat), 3 (Mokola), 4 (Duvenhage).
The nucleocapsid proteins (N, NS) share common epitopes, however polyclonal anti-RNP as well as monoclonal anti-RNP antibodies make possible the

distinction between groups. G protein provides virus-neutralizing determinants.

REPLICATION

Rabies virus is neurotropic. The virus multiplies in neurons and myotubes of vertebrates. The virus also multiplies in insects. *In vitro*, the virus growth cycle is four times longer than VS-I cycle. Infection does not inhibit cellular macromolecular synthesis.

BIOLOGICAL ASPECTS

The type species (rabies virus) is transmitted through bites and rarely through aerosols or corneal grafts. The virus has been isolated from warm-blooded animals and insects.

OTHER MEMBERS

Isolated in nature from vertebrates (V) or invertebrates (I):

Adelaide River (V)
Berrimah (V)
Bivens Arm (I)
Bovine ephemeral fever (I, V)
Charleville (I, V)
Coastal Plains (V)
Duvenhage (V)
Eel virus B12 (V)
European bat type 1 (V)
European bat type 2 (V)
Hirame rhabdovirus (V)
Humpty Doo (I)
Infectious haematopoietic necrosis (V)
Kimberley (I, V)
Kolongo (V)
Kotonkan (I)
Lagos bat (V)
Malakal (I)
Mokola (V)
Nasoule (V)
Ngaingan (I)
Oak-Vale (I)
Obodhiang (I)
Parry Creek (I)
Puchong (I)
Rochambeau (I)
Sandjimba (V)
Snakehead rhabdovirus (V)
Sweetwater Branch (I)

Taxonomic status	English vernacular name	International name
	Tibrogargan (I) Viral haemorrhagic septicemia (V) (= Egtved)	

Probable members

Bahia Grande serogroup
- Bahia Grande (I)
- Muir Springs (I)
- Reed Ranch (I)

Hart Park serogroup
- Flanders (I, V)
- Hart Park (I, V)
- Kamese (I)
- Mosqueiro (I)
- Mossuril ((I, V)

Kern Canyon serogroup
- Barur (I, V)
- Fukuoka (I)
- Kern Canyon (V)
- Nkolbisson (I)

Le Dantec serogroup
- Keuraliba (V)
- Le Dantec (V)

Sawgrass serogroup
- Connecticut (I)
- New Minto (I)
- Sawgrass (I)

Timbo serogroup
- Chaco (V)
- Sena Madureira (V)
- Timbo (V)

No serogroup assigned
- Almpiwar (V)
- Aruac (I)
- Atlantic cod ulcus syndrome (V)
- Bangoran (I, V)
- Bimbo (V)
- DakArK 7292 (I)
- Gossas (V)
- Joinjakaka (I)
- Kannamangalam (V)
- Landjia (V)
- Marco (V)
- Mn 936-77 (I)
- Navarro (V)
- Oita 296 (V)
- Ouango (V)
- Perch rhabdovirus (V)
- Rhabdovirus of blue crab (I)

Taxonomic status	English vernacular name	International name
	Rhabdovirus of entamoeba (I) Rhabdovirus salmonis (V) Rio Grande cichlid (I) Sigma (I) Sripur (I) Xiburema (I) Yata (I)	
GROUP	PLANT RHABDOVIRUS GROUP (244)	—

PROPERTIES OF THE VIRUS PARTICLE

Morphology

Particles are bacilliform and/or bullet-shaped with a distinct prevalence of the bacilliform. Mature virions are 100-430 nm long and 45-100 nm wide. The nucleocapsid is formed by a helically wound ribonucleoprotein (negative-sense ssRNA plus N protein).

Physiochemical properties

S_{20w} = 774-1045; buoyant density = 1.17-1.20 in sucrose; inactivated by lipid solvents.

Nucleic acid

One molecule of noninfectious ssRNA (MW = 4.2-4.6 x 10^3). Genome of sonchus yellow net virus (subgroup B) consists of 6 ORFs (3'-N-M2-sc4-M1-G-L-5') separated by dinucleotide GG spacers lying within a common "gene junction" consensus sequence (AUUCUUUUUGGU-UGG) with some relatedness to the gene junction regions of vesicular stomatitis and rabies viruses.

Protein

Viruses of subgroup A have one matrix (M) protein (MW = 18-25 x 10^3) and readily detectable *in vitro* transcriptase activity. Protein L (MW = 145-170 x 10^3) is detected in some members of subgroup A. Viruses of subgroup B possess M1 protein (MW = 27-44 x 10^3) and M2 protein (MW = 21-39 x 10^3). Viruses of both groups have G protein (MW = 71-93 x 10^3) and N protein (MW = 55-60 x 10^3).

Antigenic properties

Generally poor immunogens, but polyclonal antisera to several viruses have been prepared, and some shown to contain antibodies to all the structural proteins. Some of the well characterized viruses have been shown to be antigenically related.

REPLICATION

Subgroup A viruses replicate in the cytoplasm in association with masses of thread-like structures

Taxonomic status	English vernacular name	International name

(viroplasms) and morphogenisis occurs in association with vesicles of the endoplasmic reticulum. A nuclear phase appears to be involved in replication of some members (e.g. lettuce necrotic yellows virus) but evidence in others is lacking (e.g. barley yellow striate mosaic virus).

Subgroup B viruses multiply in the nuclei forming large granular inclusions thought to be sites of replication. Viral proteins synthesized from discrete polyadenylated mRNAs accumulate in nucleus and virus morphogenesis occurs at the inner nuclear envelope. Complete virus particles accumulate in perinuclear spaces. In protoplasts treated with tunicamycin, morphogenesis is interrupted and nucleocapsids accumulate in the nucleoplasm.

Biological Aspects

A wide variety of plants are susceptible to rhabdoviruses although each virus usually has a restricted host range. Most are transmitted by leafhoppers, planthoppers or aphids although one mite and one lacebug-transmitted virus have also been identified. Some viruses are also sap-transmissible. In all carefully examined cases, the virus has been shown to replicate in both plant and insect vector.

Subgroups

	Plant rhabdovirus subgroup A	—
	Plant rhabdovirus subgroup B	—

SUBGROUP	PLANT RHABDOVIRUS SUBGROUP A	—
TYPE SPECIES	LETTUCE NECROTIC YELLOWS (APHID) (26,343)	—

Other Members

Barley yellow striate mosaic (leafhopper) (312)
Broccoli necrotic yellows (aphid) (85)
Datura yellow vein
Festuca leaf streak
Maize mosaic (94)
Northern cereal mosaic (leafhopper) (322)
Sonchus
Strawberry crinkle (aphid) (163)
Wheat American striate mosaic (leafhopper) (99)

Taxonomic status	English vernacular name	International name
SUBGROUP	PLANT RHABDOVIRUS SUBGROUP B	—
TYPE SPECIES	POTATO YELLOW DWARF (LEAFHOPPER) (35)	—

OTHER MEMBERS

Eggplant mottled dwarf (115)
(= *Pittosporum* vein yellowing and tomato vein yellowing)
Sonchus yellow net (aphid) (205)
Sowthistle yellow vein (aphid) (62)

Probable members of Plant Rhabdovirus group

Officially ungrouped, but listed according to type of vector (where known). Transmitted experimentally but not characterized physico-chemically.

Aphid
- Carrot latent
- Coriander feathery red vein
- Lucerne enation
- Raspberry vein chlorosis (174)

Leafhopper
- Cereal chlorotic mottle (251)
- *Colocasia* bobone disease
- *Digitaria* striate
- Finger millet mosaic
- Maize sterile stunt
- Oat striate mosaic
- Papaya apical necrosis
- Rice transitory yellowing (100)
- *Sorghum* stunt
- *Sorghum* stunt mosaic
- Wheat chlorotic streak
- Wheat rosette stunt
- Winter wheat Russian mosaic

Lace bug
- Beet leaf curl (268)

Mite
- Coffee ringspot

Taxonomic status	English vernacular name	International name
	Not known	
	Chrysanthemum frutescens	
	Cow parsnip mosaic	
	Cynara	
	Gomphrena	
	Parsley latent	
	Pelargonium vein clearing	
	Pisum	
	Pittosporum vein yellowing	
	Raphanus	

Possible members of Plant Rhabdovirus group

Recognized only as rhabdovirus virus-like particles:

Atropa belladonna
Callistephus chinensis chlorosis
Caper vein yellowing
Carnation bacilliform
Cassava symptomless
Chondrilla stunting
Chrysanthemum vein chlorosis
Clover enation
Cynodon chlorotic streak
Endive
Euonymus fasciation
Gerbera symptomless
Gloriosa fleck
Holcus lanatus yellowing
Honeysuckle vein chlorosis
Iris germanica leaf stripe
Ivy vein clearing
Laburnum yellow vein
Laelia red leafspot
Launea arborescens stunt
Lemon scented thyme leaf chlorosis
Lolium (ryegrass)
Lotus streak
Lupin yellow vein
Malva silvestris
Melilotus latent
Melon leaf variegation
Mentha piperita latent
Passionfruit vein clearing
Patchouli (*Pogostemon patchouli*) mottle
Peanut veinal chlorosis
Pigeon pea (*Cajanus cajan*) proliferation
Pineapple chlorotic leaf streak

Taxonomic status	English vernacular name	International name
	Plantain (*Plantago lanceolata*) mottle	
	Ranunculus repens symptomless	
	Red clover mosaic	
	Saintpaulia leaf necrosis	
	Sambucus vein clearing	
	Sarracenia purpurea	
	Strawberry latent C	
	Tomato vein clearing	
	Triticum aestivum chlorotic spot	
	Vigna sinensis mosaic	
	Zea mays	
	Recognized as nonenveloped rhabdovirus-like particles:	
	Citrus leprosis	
	Orchid fleck	
	Dendrobium leaf streak	
	Phalaenopsis chlorotic spot	

Derivation of Name

rhabdo: from Greek *rhabdos*, 'rod'
vesiculo: from Latin *vesicula*, diminutive of *vesica*, 'bladder, blister'.
lyssa: from Greek 'rage, rabies'

References

Brown, F.; Bishop, D.H.L.; Crick, J.; Francki, R.I.B.; Holland, J.J.; Hull, R.; Johnson, K.M.; Martelli, G.; Murphy, F.A.; Obijeski, J.F.; Peters, D.; Pringle, C.R.; Reichmann, M.E.; Schneider, L.G.; Shope, R.E.; Simpson, D.I.H.; Summers, D.F.; Wagner, R.R.: *Rhabdoviridae*. Intervirology *12*:1-7 (1979).

Calisher, C.H.; Karabatsos, N.; Zeller, H.G.; Digoutte, J.-P.; Tesh, R.B.; Shope, R.E.; Travassos da Rosa, A.P.A.; St. George, T.D.: Antigenic relationships among rhabdoviruses from vertebrates and hematophagous arthropods. Intervirology *30:* 241-257 (1989).

Dietzgen, R.G.; Francki, R.I.B.: Analysis of lettuce necrotic yellows virus structural proteins with monoclonal antibodies and concanavalin A. Virology *156:* 486-494 (1988).

Dietzgen, R.G.; Hunter, B.G.; Francki, R.I.B.; Jackson, A.O.: Cloning of lettuce necrotic yellows virus RNA and identification of virus-specific polyadenylated RNAs in infected *Nicotiana glutinosa* leaves. J. gen. Virol. *70:* 2299-2307 (1989).

Francki, R.I.B.; Milne, R.G.; Hatta, T.: Plant rhabdoviridae, *In* Atlas of Plant Viruses, Vol. 1, pp. 73-100. (CRC Press, Boca Raton, Fl., 1985).

Frerichs, G.N.: Rhabdoviruses of fishes; *In* Ahne, W.; Kurstak, E. (eds.), Viruses of Lower Vertebrates. pp. 316-332. (Springer, Berlin, Heidelberg, New York, Tokyo, 1989).

Heaton, L.A.; Hillman, B.I.; Hunter, B.G.; Zuidema, D.; Jackson, A.O.: Physical map of the genome of sonchus yellow net virus, a plant rhabdovirus with six genes and conserved gene junction sequences. Proc. Natl. Acad. Sci. USA *86:* 8665-8668 (1989).

Jackson, A.O.; Francki, R.I.B.; Zuidema, D.: Biology, structure and replication of plant rhabdoviruses; *In* Wagner, R.R., (ed.), The Viruses, the Rhabdoviruses, pp. 427-508. (Plenum Press, New York, 1987).

Kimura, T.; Yoshimizu, M.; Oseko, N.; Nishizawa, T.: Rhabdovirus Olivaceius (Hirame rhabdovirus); *In* Ahne, W.; Kurstak, E., (eds.), Viruses of Lower Vertebrates, pp. 388-395. (Springer, Berlin, Heidelberg, New York, Tokyo, 1989).

Kurath, G.; Leong, J.C.: Characterization of infectious hematopoietic necrosis virus mRNA species reveals a nonvirion rhabdovirus protein. J. Virol. *53:* 462-468 (1985).

Kurath, G.; Ahern, K.G.; Pearson, G.D.; Leong, J.C.: Molecular cloning of the six mRNA species of infectious hematopoietic necrosis virus, a fish rhabdovirus, and gene order determination by R-loop mapping. J. Virol. *53:* 469-476 (1985).

Milne, R.G.; Masenga, V.; Conti, M.: Serological relationships between the nucleocapsids of some planthopper-borne rhabdoviruses of cereals. Intervirology *25:* 83-87 (1986).

Rose, J.; Schubert, M.: Rhabdovirus genomes and their products; *In* Wagner, R.R., (ed.), The Viruses, the Rhabdoviruses, pp. 129-166. (Plenum Press, New York, 1987).

Tesh, R.B.; Travassos da Rosa, A.P.A.; Travassos da Rosa, J.S.: Antigenic relationship among rhabdoviruses infecting terrestrial vertebrates. J. gen. Virol. *64:* 169-176 (1983).

Tordo, N.; Poch, O.; Ermine, A.; Keith, G.: Primary structure of leader RNA and nucleoprotein genes of the rabies genome: segmented homology with VSV. Nuc. Acids Res. *14:* 2671-2683 (1986).

Tordo, N.; Poch, O.; Ermine, A.; Keith, G.; Rougeon, F.: Walking along the rabies genome: Is the large G-L intergenic region a remnant gene? Proc. Natl. Acad. Sci. USA, *83:* 3914-3918 (1986).

Tordo, N.; Poch, O.; Ermine, A.; Keith, G.; Rougeon, F.: Completion of the rabies virus genome sequence determination: highly conserved domains among the L (polymerase) proteins of unsegmented negative-strand RNA viruses. Virology. *165:* 565-576 (1988).

van Beek, N.A.M.; Lohuis, D.; Dijkstra, J.; Peters, D.: Morphogenesis of sonchus yellow net virus in cowpea protoplasts. J. Ultrastruct. Res. *90:* 294-303 (1985).

Vestergard Jorhensen, P.E.; Olesen, N.J.; Ahne, W.; Lorenzen, N.: SVCV and PFR viruses: Serological examination of 22 isolates indicates close relationship between the two fish rhabdoviruses; *In* Ahne, W., Kurstak, E., (eds.), Viruses of Lower Vertebrates, pp. 349-366. (Springer, Berlin, Heidelberg, New York, Tokyo, 1989).

Wunner, W.H.: The chemical composition and molecular structure of rabies viruses; *In* Baer, L., (ed.), Natural History of Rabies, pp. 31-67. (CRC Press, Boca Raton, Fl., 1991).

Taxonomic status	English vernacular name	International name
FAMILY	—	***ORTHOMYXOVIRIDAE***

Compiled by H.-D. Klenk

GENUS	INFLUENZA VIRUS A AND B	—
TYPE SPECIES	INFLUENZAVIRUS A/PR/8/34 (H1N1)	—

PROPERTIES OF THE VIRUS PARTICLE

Morphology

Nucleocapsid(s) of helical symmetry and diameter 9-15 nm are enclosed within lipoprotein membrane having surface projections. Nucleoproteins of different size classes (50-130 nm length), with loop at each end, are extractable from virions or infected cells. Arrangement within virion uncertain, although coils of about 4-20 turns of a 7 nm thick material are sometimes seen in partially disrupted virus. Virions are pleomorphic, 20-120 nm in diameter, but filamentous forms occur having length up to several micrometers. M_1 protein is believed to form a layer inside the lipid bilayer, with HA and NA glycoproteins projecting about 10-14 nm from the surface. About 500 "spikes" project from the surface of a spherical virion. Most are HA, with NA clusters interposed irregularly. The ratio of HA to NA varies, but is usually about 4 or 5 to 1. The HA "spikes" are rods, 13.5 nm in length and 4 nm diameter. They comprise a coil of α-helices from the three subunits extending from the membrane as a 7.6 nm stalk, with a globular region of antiparallel β-sheets at the distal end that contains the receptor binding site. The NA glycoprotein has a box-shaped head, 10 x 10 x 6 nm, attached to a slender stalk about 100 nm long projecting from the membrane. Each NA subunit is composed of six topologically identical β-sheets arranged in the formation of a "propeller". Cores containing M_1, RNP, and P proteins may be generated by controlled chemical disruption of virions.

Physicochemical properties

MW = 250 x 10^6; S_{20w} of nonfilamentous particles 700-800; density in sucrose/H_2O ≈ 1.19 g/cm^3.
Virus infectivity reduced within minutes by exposure to low pH (5) or heat (56°C). Lipid solvents and detergents (anionic, cationic, or neutral) destroy membrane integrity with resultant reduction in infectivity. Infectivity may be totally destroyed by treatment with formaldehyde, β-propiolactone, UV light or gamma irradiation, without affecting antigenic specificity. Prolonged exposure to

chemicals or radiation inactivates different replicative events at different rates, presumably as a result of induced lesions in individual RNA segments of different sizes. Influenza virus shows multiplicity reactivation.

Nucleic acid

Eight complete segments of linear negative sense ssRNA may be detected by gel electrophoresis. Incomplete RNA segments may be present. Chain lengths are ≈ 900 to 2350 nucleotides for complete segments, total MW ≈ 4.5 x 10^6. The largest three segments code for three polymerase proteins, three intermediate size segments code for surface glycoproteins and nucleoprotein, and the smallest two segments code for matrix protein and several non-structural proteins. Additionally, one of the intermediate size segments (RNA 6) of influenza B viruses codes for a non-structural protein. The exact order of electrophoretic migration of the RNA segments varies with strain and electrophoretic conditions. Conserved nucleotide sequences are present at the 5' and 3' termini (13 and 12 nucleotides respectively in type A; 11 and 9 nucleotides respectively in type B). Type A conserved 5' sequence is 5'-AGUAGAAACAAGG and type B conserved 5' sequence is 5'-AGUAG-AACAA. Type A conserved 3' sequence is 3'-UCGUUUUCGUCC in most segments and 3'-UCGUUUCGUCC in segments 1-3 and in segment 7 of human virus strains. Type B conserved 3' sequence is 3'-UCGUCUUCG.

Protein

Seven virion proteins. Three proteins (PB1, PB2, and PA) and one intermediate size protein (NP) are found in the RNA polymerase complex which has transcriptase and endonuclease activities: PB2 (a basic protein) contains ≈ 760 amino acids and recognizes 5' terminal caps of mRNA and is involved in endonucleolytic cleavage of mRNA primers. PB1 (another basic protein) contains ≈ 760 amino acids and is involved in catalyzing the addition of nucleotides to the nascent mRNA chains. PA (an acidic protein), contains ≈ 720 amino acids (function unknown). The nucleoprotein (NP), which contains ≈ 500 amino acids (MW ≈ 56 x 10^3) is phosphorylated, and is associated with the RNA genome segments in the form of a ribonucleoprotein. NP is a species-specific antigen used to identify type A and B viruses in serological tests.

Hemagglutinin (HA) is a class I membrane protein containing an amino-terminal signal sequence, which is removed by cotranslational cleavage, and a carboxy-terminal transmembrane region, which anchors the glycoprotein in the cell or virion membranes. It initiates infection by binding to sialic acid-containing receptors and by inducing fusion of the viral envelope with cellular membranes. HA is the major surface antigen. The structure of HA, except for side chain coordinates and the C-terminal region of HA_2 (see below), has been resolved for one strain to a resolution of 0.29 nm. HA is composed of three identical subunits, each containing ≈ 550 amino acids. The location and number of most potential N-linked glycosylation sites are not conserved among HAs of difference strains and subtypes. These changes in glycosylation are associated with masking/unmasking antigenic determinants, altered host range, and virulence. Fusion activity requires posttranslational cleavage of HA by cellular proteases into the disulfide-linked fragments HA_1 (≈ 330 amino acids) and HA_2 (≈ 220 amino acids) yielding a highly conserved sequence of 15 amino acids at the amino-terminus of HA_2. Cleavability by a given protease depends, among other factors, on the number of basic amino acids present at the cleavage site. HA is acylated at the membrane-spanning region.

Neuraminidase (NA) is a second surface glycoprotein. It is a class II membrane protein containing an amino-terminal hydrophobic region which serves both as a membrane insertion signal and as a membrane anchor. NA has enzymatic activity which cleaves the alpha-glycosidic bond joining the keto group of sialic acid to D-galactose or D-galactosamine. NA is a minor surface antigen. The structure has been resolved to 0.29 nm, except for side chain coordinates and for the N-terminal region. NA is a tetramer. Each subunit contains 450-470 amino acids. In some cases pairs of subunits are disulfide bonded to each other, depending on the number of cysteine residues and their location relative to proteolytically cleaved sites.

The matrix or membrane (M_1) protein is ≈ 250 amino acids, MW ≈ 28 x 10^3. It is the most abundant virion protein, underlies the lipid bilayer, and is soluble in chloroform/methanol.

Both influenza A and B virus encode small integral membrane proteins of very similar structure, M_2 (97 amino acid residues) and NB (100 amino acids residues) respectively. These proteins are class I integral membrane proteins that contain an uncleaved signal/anchor domain such that they are oriented with a 18-23 residue N-terminal extracellular domain and a C-terminal cytoplasmic domain. Both M_2 and NB are expressed abundantly at the infected cell surface and both proteins are tetramers that can form higher oligomeric forms. NB contains two sides for the

Taxonomic status	English vernacular name	International name

addition of N-linked carbohydrate and both have been found to be modified by the addition of carbohydrate chains which are further modified by the addition of polylactosaminoglycan. The influenza A virus M_2 protein transmembrane domain is linked genetically to the sensitive influenza A virus to the antiviral drug amantadine hydrochloride. Although the M_2 protein is abundantly expressed in influenza A virus infected cells, it is under-represented in purified virions, but it has been found that each virion (A/WSN/33 strain) contains on average 40-63 molecules of M_2. Although the presence of NS in influenza B virus has not been reported, the available evidence does not rule out the presence of a small number of molecules in virions.

Influenza A virus M_2 protein is encoded by a spiced mRNA that is processed from the colinear transcript mRNA that encodes the M_1 protein. M_1 and M_2 proteins share nine N-terminal residues before the sequences diverge. An alternatively spliced mRNA derived from the colinear RNA segment 7 transcript is also found in virus infected cells, but the predicted polypeptide product (9 amino acids which would be the same as the 9 C-terminal residues of the M_1 protein) has not been identified. The influenza B virus NB glycoprotein is encoded in an overlapping reading frame on RNA segment 6 which also encodes NA. The available evidence indicates that the mRNA for NB and NA is bicistronic.

Influenza B virus RNA segment 7, in addition to encoding the M_1 protein, also encodes the BM_2 protein (MW $\approx$ 12,000) that is translated from an overlapping reading frame. The BM_2 protein initiation codon overlaps with the termination codon of the M_1 protein in an overlapping translational stop-start pentanucleotide UAAUG. The available data indicate that expression of the BM_2 protein requires 5'-adjacent termination of M_1 synthesis and that a termination/reinitiation scheme is used in translation of a bicistronic mRNA. BM_2 is predicted to be very different from influenza A virus M_2 protein, as BM_2 is likely to be water soluble, globular protein lacking membrane spanning hydrophobic domains.

Two non-structural proteins are found in influenza virus infected cells, NS_1, NS_2. These proteins are encoded by RNA segment 8. NS is encoded by a mRNA that is encoded by a colinear transcript derived from RNA segment 8. NS_1 is encoded by a spiced mRNA. NS_1 and NS2 share ten N-terminal residues before the sequences diverge. The coding regions for NS_1 and NS_2 overlap by

Taxonomic status	English vernacular name	International name

70 amino acids that are translated from different reading frames. The function of these non-structural proteins in the influenza virusreplicative cycle has not been elucidated but both proteins are localized to the nucleus and nucleolus of infected cells.

Lipid

18-37% by weight of virion. Present in virion envelope. Resembles lipids of plasma membrane of host cell in composition.

Carbohydrate

≈ 5% by weight of virion. Present as oligosaccharide side chains of glycoproteins, as glycolipids, and as mucopolysaccharide. HA (carbohydrate content ≈ 15%) has N-glycosidic side chains of complex and oligomannosidic type. NA (carbohydrate content ≈ 15%) has, in addition, N-linked oligosaccharides containing N-acetylgalactosamine. NB (carbohydrate content 36%) has N-linked polylactosaminoglycan. Composition of viral carbohydrates host- and virus-dependent. Carbohydrates lack sialic acid due to action of virus NA, may contain covalently bound sulphate.

Antigenic properties

The best studied antigens are NP, M_1, HA, and NA. NP and M_1 are species-specific for A and B influenza strains. Variation occurring within HA and NA antigens has been analyzed in great detail. Fourteen subgroups of HA and nine subgroups of NA are recognized for influenza A viruses, with minimal serological crossreaction between subgroups. Additional variation occurs within subgroups, particularly for human viruses isolated in different years, although only a small number of strains of any subgroup are epidemiologically active at any time. Continual evolution of new strains occurs, and older strains apparently disappear from circulation. HA and NA antigens of influenza B viruses exhibit less antigenic variation than for influenza A, and no subgroups are defined. Antibody to HA neutralizes infectivity. Antibody to HA neutralizes infectivity. If NA antibody is present during multicycle replication, it may inhibit virus release and, thus, reduce virus yield. Antibody to N-terminus of M_2 greatly reduces virus yield in tissue culture.

Effect on cells

Erythrocytes of many species are agglutinated by virions. Sialic acid-containing virus receptors of erythrocytes may be destroyed by NA of attached virions, resulting in elution of virus. Hemolysis of erythrocytes may be produced at pH of about 5.

REPLICATION

Attachment of virions occurs by binding of the hemagglutinin (HA) to N-acetylneuraminic acid-containing receptors on the plasma membrane. Specificity of strains may be for 2-3 or 2-6 glycosidic linkages, depending on sequence of receptor site in HA. Entry is by endocytosis into endosomal vesicles. Fusion between the virus envelope and the endosomal membrane is apparently triggered by a conformational change that occurs only in cleaved HA proteins when the pH is reduced to about 5. This leads to release of the transcription complex into the cytoplasm.

Transcriptase complex synthesises messenger RNA transcripts in the cell nucleus; this process is primed by 5'-methyl-guanosine (capped) RNA fragments 8-15 nucleotides in length. These primers are generated from host heterogeneous nuclear RNA by a viral endonuclease activity associated with the viral PB2 protein. Virus-specific messenger RNA synthesis is inhibited by actinomycin D or α-amanitin due to blockage of host DNA-dependent transcription and a presumed lack of newly synthesized substrate for viral endonuclease to generate primers. Viral-specific mRNA is polyadenylated at the 3' termini, and lacks sequences corresponding to the 5'-terminal 16 nucleotides of the corresponding vRNA segment. The mechanism for early termination during transcription of mRNA is unknown.

Complementary RNA molecules which act as templates for new vRNA synthesis are complete transcripts of vRNA, and are neither capped nor polyadenylated. These RNAs are also probably synthesized in the nucleus of infected cells.

Protein synthesis occurs in the cytoplasm. Nucleoprotein and NS_1 protein antigens accumulate in the nucleus during the first hours of infection, then migrate to the cytoplasm. Inclusions of NS_1 may form. M_1 has also been observed in nucleus. HA and NA proteins migrate through the Golgi apparatus to localized regions of the plasma membrane where new virions form by budding, incorporating M protein and RNP's which have aligned below regions of plasma membrane containing HA and NA on their surface. M_1 protein of influenza A, and NB protein of influenza B, also accumulate after intracellular transport by the exocytotic pathway on plasma membranes. Budding is from the apical surface in polarized cells. Gene reassortment occurs during mixed infections with virus of the same species, but not between virus species. True recombination of RNA has also been detected.

BIOLOGICAL ASPECTS

Host range

Influenza A viruses naturally infect man, and several other mammalian species and a wide variety of avian species. Some interspecies transmission believed to occur. Epidemics of respiratory disease in man have been caused by influenza A viruses having antigenic composition H1N1, H2N2, H3N2, and possibly H3N8. Influenza A viruses of subtype H7N7 and H3N8 (previously designated equine 1 and equine 2 viruses) cause outbreaks of respiratory diseases in horses. Type A (H1N1) viruses, and type A (H3N2) viruses have been frequently isolated from swine. The H1N1 viruses isolated from swine in recent years appear to be of three general categories: those closely related to classical "swine influenza" and which cause occasional human cases (e.g., A/New Jersey/8/76-like strains), those first recognized in avian specimens (e.g., A/Alberta/35/76-like strains), but which have caused outbreaks among swine in France, and those resembling viruses isolated from epidemics in man since 1977 (e.g., A/USSR/90/77-like strains). H3N2 viruses from swine all appear to contain HA and NA genes closely related to those from human epidemic strains. Type A (H7N7 and H4N5) viruses have caused outbreaks in seals, with virus spread to nonrespiratory tissue in this host. Such virus has accidentally infected the conjunctiva of one laboratory worker. Pacific Ocean whales were reportedly infected with type A (H1N1) virus. Other influenza subtypes have also been isolated from lungs of Atlantic Ocean whales in North America. Type A (H10N4) virus has caused outbreaks in mink. All subtypes of HA and NA, in many different combinations, have been identified in isolates from avian species, particularly chickens, turkeys, and ducks. Pathology in avian species varies from unapparent infection (often involving replication in, and probable transmission via, the intestinal tract), to virulent infections (only observed with subtypes H5 and H7) with spread to many tissues and high mortality rates. Structure of the HA protein, in particular the specificity of its receptor binding site and its cleavability by naturally occurring tissue proteases, appears critical in determining the host range of the virus. In addition, interactions between gene products determine the outcome of infection. Thus, host range of influenza viruses is generally unpredictable. Interspecies transmission has apparently occurred in some instances

Taxonomic status	English vernacular name	International name

without genetic reassortment (e.g., H1N1 virus from swine to man and vice versa, or H3N2 virus from man to swine), but in other cases of interspecies transmission it is proposed that reassortment in hosts infected with more than one strain may have resulted in viruses with new constellations of genes having altered host ranges or epidemic properties (e.g., H3N2 viruses probably derived in 1968 by reassortment of human H2N2 viruses and an unknown H3-containing virus; seal H7N7 virus probably derived by reassortment of two or more avian influenza viruses; and reassortment of human H1N1 and H3N2 viruses in 1978 led to outbreaks of virus with H1N1 surface antigens but 4 or 5 genes of H3N2 origin). Laboratory animals that may be artificially infected with influenza A viruses include ferrets, mice, hamsters, and guinea pigs as well as some small primates such as squirrel monkeys.

Influenza B strains appear to naturally infect only man and cause epidemics every few years. They also artificially infect laboratory rodents. Most type A and B strains grow in the amniotic cavity of embryonated hen's eggs, and after adaptation type A and B viruses grow in the allantoic cavity. Primary kidney cells from monkeys, humans, calves, pigs, and chickens support replication of many virus strains. Host range may be extended by addition of trypsin to growth medium, so that replication also can be obtained in some continuous cell lines. Clinical specimens from influenza-infected hosts sometimes contain subpopulations of virus with minor sequence differences in at least their HA protein. These subpopulations may differ in their receptor specificity or their propensity for growth in different host cells.

Transmission Aerosol (human and most non-aquatic hosts) or waterborne (ducks).

GENUS	INFLUENZA VIRUS C	—
TYPE SPECIES	INFLUENZA VIRUS C/TAYLOR/1233/47	—

PROPERTIES OF THE VIRUS PARTICLE

Morphology Size generally similar to influenza A and B viruses with reticular structure often, but not always, observed on virion surface.

Nucleic acid

Seven molecules of negative sense ssRNA. Size = 975-2,350 nucleotides, with total molecular weight of RNA = 4-5 x 10^6. RNA segments 1-3 code for 3 polymerase proteins, segments 4, 5 and 6 code for envelope glycoprotein, nucleoprotein, and membrane protein, respectively, and segment 7 codes for 2 non-structural proteins. Nucleotide sequences at the 5' and 3' termini conserved between segments, and are 5'AGCAGUAGCAA and 3'UCGUUUCGUC, respectively. These sequences closely resemble those of the influenzavirus A and B genus.

Protein

Six virion proteins. Nucleocapsid contains polymerase proteins PB_2 (774 amino acids), PB_1 (754 amino acids), PB_3 (709 amino acids) and nucleoprotein (NP) (565 amino acids). The single glycoprotein (HEF) present in the viral envelope has 3 functions. (1) it hemagglutinates and initiates infection by binding to 9-0-acetyl-N-acetyl-neuraminic acid as the essential receptor compound, (2) it has neuraminate 0-acetyl esterase activity which functions as receptor destroying enzyme, (3) it induces membrane fusion. HEF, which is about 100 amino acids longer than HA of influenza A and B viruses, is synthesized as a precursor polypeptide ≈ 655 amino acids long (≈ 72 x 10^3) including a cotranslationally cleaved hydrophobic leader sequence. Posttranslational cleavage produces a large fragment (HEF_1) of ≈ 48 x 10^3 and a small fragment (HEF_2) of 22.5 x 10^3 with an N-terminus resembling F_1 polypeptide of paramyxoviruses. N- and C-termini of HEF_2 are hydrophobic, similar to HA_2 of influenza A and B viruses. 8 potential N-glycosylation sites have been identified, 6 in HEF_1 and 2 in HEF_2. Homologies with influenza A and B HA are largely confined to the N- and C-termini, and to 6 of the cysteines. Virions contain also large amounts of internal membrane protein which, unlike influenza A and B M_1, is translated from spliced mRNA.

Non structural proteins. A colinear and a spliced mRNA are derived from RNA 7 encoding the non-structural proteins NS_1 (286 amino acids) and NS_2 (122 amino acids), respectively.

REPLICATION

Like influenzaviruses A and B, replication can be inhibited by agents that interfere with host cell DNA-dependent RNA synthesis.

Biological Aspects

Host range

Infection of man is common in childhood. Occasional outbreaks, but not epidemics, have been detected. Swine in China reported to be infected by viruses similar to contemporary human strains.

Other Members of the Family

D, comprising tick borne viruses (e.g. Dhori and Thogoto viruses) occasionally infecting man. Such viruses, morphologically resembling influenza viruses, contain 6 or 7 ss RNA segments of negative sense, which have 3' and 5' ends similar to those of other orthomyxoviridae. Based on nucleotide sequences that have been compared to those of influenza A, B and C viruses, segments 2, 4, 5, and 6 of Dhori virus have been predicted to code for PB1, the glycoprotein, the nucleoprotein, and the matrix protein, respectively. The sequenced segments 3 and 4 of Thogoto virus show evolutionary relatedness to PA and to the major surface glycoproteins of orthomyxoviridae, respectively.

Derivation of Name

ortho: from Greek *orthos* "straight, correct"
myxo: from Greek *myxa* "mucus" (relating to activity of hemagglutinin and neuraminidase).
influenza: Italian form of Latin influentia, "epidemic". So used because epidemics were thought to be due to astrological or other occult "influences".

References

Fuller, F.J., Clay, W.C., Freedman, E.Z., McEntree, M., and Barnes, J.A.: Nucleotide sequence of the major structural protein genes of the tick-borne, othomyxo-like Dhori/Indian/1313/61 virus. *In* Kolakovsky, D., Mahy, B.W.J. (eds.), Genetics and pathogenicity of negative strand viruses pp. 279-286 (Elsevier, Amsterdam, 1989).

Kingsbury, E.D.: Influenza. (Plenum Medical Book Company, New York, London, 1987).

Krug, R.M.: The Influenza Viruses. *In* Fraenkel-Conrat, H.; Wagner, R.R. (eds.), The viruses (Plenum Press, New York, 1990).

Palese, P.; Kingsbury, D.W.: Genetics of influenza viruses. (Springer, Wien, New York, 1983).

Taxonomic status	English vernacular name	International name
FAMILY	—	*BUNYAVIRIDAE*

Reported by C.H. Calisher

PROPERTIES OF THE VIRUS PARTICLE

Morphology Spherical or pleomorphic enveloped particles (80-100 nm in diameter) with glycoprotein surface projections; ribonucleocapsids composed of 3 circular, helical strands, 2-2.5 nm diameter, sometimes supercoiled, 0.2-3 µm in length depending on arrangement.

Physicochemical properties MW = 300-400 x 10^6; S_{20w} = 350-500; buoyant density in CsCl ≈ 1.2 g/cm^3. Sensitive to lipid solvents and detergents.

Nucleic acid Three molecules (large [L], medium [M], and small [S]) of negative or ambisense ssRNA. Ends are hydrogen-bonded, RNA and nucleocapsids circular. Differences exist between terminal nucleotide sequences of gene segments of viruses of different genera. MW = 2.2-4.9 (6.5-14.4 kb), 1.0-2.3 (3.2-6.3 kb) and 0.28-0.8 x 10^6 (0.8-2.0 kb), respectively; 1-2% by weight.

Proteins Usually 4 consisting of 2 external glycoproteins (G1, G2), a nucleocapsid protein (N), and a large protein (L) which is presumably a transcriptase. Transcriptase activity present in virion.

Lipid 20-30% by weight; forms lipoprotein envelope, which is cell-derived.

Carbohydrate 2-7% by weight; components of the glycoproteins and glycolipids.

Antigenic properties Hemagglutinin and neutralizing antigenic determinants present on viral glycoproteins. CF antigenic determinants principally associated with nucleocapsid protein.

Effect of virus on vertebrate cells Some induce cell fusion at low pH. Most cause cytopathic effects; hantaviruses do not cause cytopathic effects. Some members have ion-dependent hemagglutinating activity.

REPLICATION

Replicate in cytoplasm. Host RNA sequence shown to prime viral mRNA synthesis. Some code for non-structural (NS) protein(s). Genetic reassortment demonstrated for certain members. Virion RNA segments

Taxonomic status	English vernacular name	International name

are transcribed in the cytoplasm to complementary mRNAs by the virion transcriptase. The L RNA encodes the L protein, a single open reading frame in the M RNA encodes the glycoproteins, which are cotranslationally cleaved to G1 and G2. The S RNA encodes the N protein and, in some instances, a nonstructural protein NS_S. Mature by budding into smooth-surfaced vesicles in or near the Golgi region but maturation at the plasma membrane has also been observed.

BIOLOGICAL ASPECTS

Host range Various arthropods and/or warm or cold-blooded vertebrates.

Transmission Mosquitoes, ticks, phlebotomine flies and other arthropod vectors. Transovarial and venereal transmission demonstrated for some mosquito-borne viruses. Aerosol infection occurs in certain situations or is the principal means of transmission for some viruses. In some instances, avian host and/or vector movements may result in virus dissemination. No arthropod vector demonstrated in *Hantavirus* transmission.

GENERA

Bunyamwera supergroup	*Bunyavirus*
Sandfly fever and Uukuniemi group	*Phlebovirus*
Nairobi sheep disease group	*Nairovirus*
Hantaan group	*Hantavirus*
Tomato spotted wilt group	*Tospovirus*

GENUS	BUNYAMWERA SUPERGROUP	***BUNYAVIRUS***
TYPE SPECIES	BUNYAMWERA VIRUS	—

PROPERTIES OF THE VIRUS PARTICLE

Nucleic acid L RNA = $2.7\text{-}3.1 \times 10^6$ (≈ 7 kb); M RNA = $1.8\text{-}2.3 \times 10^6$ (4.45-4.54 kb); S RNA = $0.28\text{-}0.50 \times 10^6$ (0.85-0.99 kb); 3'-terminal nucleotide sequences of L, M and S gene segments = UCAUCACAUGA..., 5'- terminal nucleotide sequences of M and S gene segments = AGUAGUGUGCU...

Protein G1 = $108\text{-}120 \times 10^3$; G2 = $29\text{-}41 \times 10^3$; N = $19\text{-}25 \times 10^3$; L ≈ 200×10^3. Both glycoproteins and $15\text{-}18 \times 10^3$ NS_M derived from M RNA; N and NS_S coded in overlapping reading frames by S RNA. L protein coded by L RNA.

REPLICATION

Viral induced mRNA species (1 per RNA segment) are subgenomic, viral-complementary in sequence and have host mRNA-derived 5' terminal sequences.

BIOLOGICAL ASPECTS

Host range

Various vertebrate species; also insects, primarily mosquitoes but occasionally other arthropods, e.g. ceratopogonids in the genus *Culicoides*, phlebotomines and ticks.

Virulence

Primarily determined by viral M RNA gene products (glycoproteins).

OTHER MEMBERS

There are 18 antigenic groups of the genus *Bunyavirus* (at least 162 viruses) and 4 ungrouped viruses. Serologically unrelated to members of other genera. Mostly mosquito-transmitted; some (Tete group) tick-transmitted. Some transmitted transovarially in arthropods.

The groups are:

Anopheles A (12): Anopheles A, Las Maloyas, Lukuni, Tacaiuma, Trombetas, Virgin River, CoAr3624, CoAr1071, CoAr3627, ColAn57389, SPAr2317, H32580

Anopheles B (2): Anopheles B, Boraceia

Bakau (5): Bakau, Ketapang, Nola, Tanjong Rabok, Telok Forest

Bunyamwera (32): Anhembi, Batai, Birao, Bozo, Bunyamwera, Cache Valley, Fort Sherman, Germiston, Guaroa, Iaco, Ilesha, Kairi, Lokern, Macaua, Maguari, Main Drain, Mboke, Ngari, Northway, Playas, Santa Rosa, Shokwe, Sororoca, Taiassui, Tensaw, Tlacotalpan, Tucunduba, Wyeomyia, Xingu, Ag83-1746, BeAr328208, CbaAr426

Bwamba (2): Bwamba, Pongola

C (14): Apeu, Bruconha, Caraparu, Gumbo Limbo, Itaqui, Madrid, Marituba, Murutucu, Nepuyo, Oriboca, Ossa, Restan, Vinces, 63U11

Taxonomic status	English vernacular name	International name
	California (13): California encephalitis, Inkoo, Jamestown Canyon, Keystone, La Crosse, Melao, San Angelo, Serra do Navio, snowshoe hare, South River, Tahyna, trivittatus, AG83-497	
	Capim (10): Acara, Benevides, Benfica, Bushbush, Capim, Guajara, Juan Diaz, Moriche, GU71u344, GU71u350	
	Gamboa (8): Alajuela, Brus Laguna, Gamboa, Pueblo Viejo, San Juan, 75V-2374, 75V-2621, 78V-2441	
	Guama (12): Ananindeua, Bertioga, Bimiti, Cananeia, Catu, Guama, Guaratuba, Itimirim, Mahogany Hammock, Mirim, Moju, Timboteua	
	Koongol (2): Koongol, Wongal	
	Minatitlan (2): Minatitlan, Palestina	
	Nyando (2): Nyando, Eret-147	
	Olifantsvlei (5): Bobia, Botambi, Dabakala, Olifantsvlei, Oubi	
	Patois (7): Abras, Babahoyo, Estero Real, Pahayokee, Patois, Shark River, Zegla	
	Simbu (24): Aino, Akabane, Buttonwillow, Douglas, Facey's Paddock, Ingwavuma, Inini, Kaikalur, Manzanilla, Mermet, Oropouche, Para, Peaton, Sabo, Sango, Sathuperi, Shamonda, Shuni, Simbu, Thimiri, Tinaroo, Utinga, Utive, Yaba-7	
	Tete (6): Bahig, Batama, Matruh, Tete, Tsuruse, Weldona	
	Turlock (4): Lednice, Turlock, Umbre, Yaba-1	
	Ungrouped (4): Kaeng Khoi, Leanyer, Mojui dos Campos, Termeil	

Taxonomic status	English vernacular name	International name
GENUS	SANDFLY FEVER AND UUKUNIEMI GROUP VIRUSES	***PHLEBOVIRUS***
TYPE SPECIES	SANDFLY FEVER (SF) SICILIAN VIRUS	—

PROPERTIES OF THE VIRUS PARTICLE

Nucleic acid L RNA = 2-2.8 x 10^6 (6.5-8.5 kb); M RNA = 1.1-2.2 x 10^6 (3.2-4.3 kb); S RNA = 0.4-0.8 x 10^6 (1.7-1.9 kb); 3'-terminal nucleotide sequences of L, M and S gene segments = UGUGUUUC..., 5'-terminal nucleotide sequences of M gene segment = ACACAAAGAC...

Protein G1 = 55-75 x 10^3; G2 = 50-70 x 10^3; N = 20-30 x 10^3; L = 145-200 x 10^3. Both glycoproteins coded by M RNA; the N protein coded by S RNA.

REPLICATION

Virion M and S RNA segments are transcribed into complementary mRNAs by virion RNA transcriptase. The S RNA exhibits an ambisense coding strategy, i.e. it is transcribed by virion RNA polymerase to a subsegmental viral complementary mRNA that encodes the N protein and to a subsegmental viral-sense mRNA that encodes a nonstructural (NS_S) protein (MW ≈ 30,000), the function of which is unknown. At least the M and S mRNA contain host mRNA-derived 5' primer sequences. Viruses of the sandfly fever group but not of the Uukuniemi group have a pre-glycoprotein coding region (NS_M) of unknown function.

BIOLOGICAL ASPECTS

Host range Sandfly fever group viruses have been isolated from various vertebrate species and from phlebotomines and occasional alternate arthropods, e.g. mosquitoes or ceratopogonids in the genus *Culicoides*. Uukuniemi serogroup viruses are isolated from various vertebrate species and from ticks. Virulence factors are coded by genes on each of the RNA species.

OTHER MEMBERS

There are 8 antigenic complexes (at least 23 viruses) within the sandfly fever group; 16 viruses related to sandfly fever Sicilian virus have not been assigned to an antigenic complex. Uukuniemi group viruses belong to a single serogroup (12 viruses). Low-level antigenic cross-

Taxonomic status	English vernacular name	International name

reactivity occurs between certain sandfly fever group and certain Uukuniemi group viruses but sandfly fever and Uukuniemi group viruses are antigenically unrelated to members of other genera. Sandfly fever group viruses are transmitted by phlebotomines, mosquitoes or ceratopogonids of the genus *Culicoides*; Uukuniemi group viruses are transmitted by ticks.

Uukuniemi and sandfly fever group viruses are related in that (i) they share the same ambisense coding strategy for the S RNA segment, (ii) they have identical 5' and 3' terminal nucleotide sequences, (iii) they display low, but significant homology between the glycoproteins, (iv) the N proteins show a high degree of homology, and (v) certain members of each group are antigenically related to certain members of the other group.

The complexes are (sandfly fever group):

Sandfly fever Naples (4): Karimabad, Sandfly fever Naples, Tehran, Toscana

Bujaru (2): Bujaru, Munguba

Candiru (6): Alenquer, Candiru, Itaituba, Nique, Oriximina, Turuna

Chilibre (2): Cacao, Chilibre

Frijoles (2): Frijoles, Joa

Punta Toro (2): Buenaventura, Punta Toro

Rift Valley fever (3): Belterra, Icoaraci, Rift Valley fever

Salehabad (2): Arbia, Salehabad

No complex assigned (16): Aguacate, Anhanga, Arboledas, Arumowot, Caimito, Chagres, Corfou, Gabek Forest, Gordil, Itaporanga, Odrenisrou, Pacui, Rio Grande, Saint-Floris, Sandfly fever Sicilian, Urucuri

Uukuniemi group (12): Grand Arbaud, Manawa, Murre, Oceanside, Ponteves, Precarious Point, St. Abbs Head, Uukuniemi, Zaliv Terpeniya, EgAn1825-61, Fin V-707, RML 105355.

Taxonomic status	English vernacular name	International name
GENUS	NAIROBI SHEEP DISEASE AND RELATED VIRUSES	*NAIROVIRUS*
TYPE SPECIES	CRIMEAN-CONGO HEMORRHAGIC FEVER (CCHF) VIRUS	—

PROPERTIES OF THE VIRUS PARTICLE

Nucleic acid

L RNA = 4.1-4.9 x 10^6 (11.0-14.4 kb); M RNA = 1.5-2.3 x 10^6 (4.4-6.3 kb); S RNA = 0.6-0.7 x 10^6 (1.7-2.1 kb); 3'-terminal nucleotide sequences of L, M and S gene segments = AGAGAUUCU...

Protein

G1 = 72-84 x 10^3; G2 = 30-40 x 10^3; N = 48-54 x 10^3; L = 145-200 x 10^3.

REPLICATION

At least two non-structural glycoprotein precursors synthesized in infected cells. Nucleocapsid protein coded by S RNA in viral-complementary sequences.

BIOLOGICAL ASPECTS

Host range

Various vertebrate species; primarily ticks but occasional alternate arthropod species, mosquitoes and ceratopogonids of the genus *Culicoides*.

OTHER MEMBERS

There are 7 antigenic groups of the genus *Nairovirus* (at least 33 viruses). Serologically unrelated to members of other genera. Most are tick-transmitted.

The groups are:

Crimean-Congo hemorrhagic fever (3): Crimean-Congo hemorrhagic fever, Hazara, Khasan

Dera Ghazi Khan (6): Abu Hammad, Abu Mina, Dera Ghazi Khan, Kao Shuan, Pathum Thani, Pretoria

Hughes (10): Farallon, Fraser Point, Great Saltee, Hughes, Puffin Island, Punta Salinas, Raza, Sapphire II, Soldado, Zirqa

Nairobi sheep disease (2): Dugbe, Nairobi sheep disease

Qalyub (3): Bandia, Omo, Qalyub

Taxonomic status	English vernacular name	International name
Other possible members of family	At least 7 groups (19 viruses) and 22 ungrouped viruses. Not shown to be antigenically related to members of other Bunyaviridae genera. For most, no biochemical characterization of the viruses has been reported to confirm their family or generic status. The groups are: Bhanja (3): Bhanja, Forecariah, Kismayo Kaisodi (3): Kaisodi, Lanjan, Silverwater Mapputta (4): Gan Gan, Mapputta, Maprik, Trubanaman Okola (2): Okola, Tanga Resistencia (3): Antequera, Barranqueras, Resistencia Upolu (2): Aransas Bay, Upolu Yogue (2): Yogue, Kasokero Ungrouped viruses (22): Bangui, Batken, Belem, Belmont, Bobaya, Caddo Canyon, Chim, Enseada, Issyk-Kul (Keterah), Kowanyama, Lone Star, Pacora, Razdan, Salanga, Santarem, Sunday Canyon, Tai, Tamdy, Tataguine, Wanowrie, Witwatersrand, Yacaaba.	
Derivation of Name	bunya: from *Bunya*mwera; place in Uganda, where type virus was isolated. nairo: from *Nairo*bi sheep disease; first reported disease caused by member virus. phlebo: refers to *phlebo*tomine vectors of sandfly fever group viruses; Greek *phlebos*, "vein". hanta: from *Hanta*an; river in South Korea near where type virus was isolated. tospo: derived from To (*To*mato) spo (*spo*tted wilt), the type member of the genus.	

REFERENCES

Arikawa, J.; Lapenotiere, H.F.; Iacono-Connors, L.; Wang, M.; Schmaljohn, C.S.: Coding properties of the S and M genome segments of Sapporo Rat virus: comparison to other causative agents of Hemorrhagic fever with renal syndrome. Virology *176:* 114-125 (1990).

Bishop, D.H.L.: *Bunyaviridae* and their replication, Part I. *Bunyaviridae*; *In* Fields, B.N.; Knight, J.C. (eds.), Virology, Vol. 1, 2nd edn., pp. 1155-1173 (Raven Press, New York, 1990).

Bishop, D.H.L.; Shope, R.E.: *Bunyaviridae*; *In* Fraenkel-Conrat, H.; Wagner, R.R. (eds.), Comprehensive Virology, Vol. 14, pp. 1-156 (Plenum Press, New York, 1979).

de Haan, P.; Wagemakers, L.; Peters, D.; Goldbach, R.: Molecular cloning and terminal sequence determination of the S and M RNAs of tomato spotted wilt virus. J. gen. Virol. *70:* 3469-3473 (1989).

de Haan, P.; Wagemakers, L.; Peters, D.; Goldbach, R.: The S RNA segment of tomato spotted wilt virus has an ambisense character. J. gen. Virol. *71:* 1001-1007 (1990).

Elliott, R.M.: Molecular biology of the *Bunyaviridae*. J. gen. Virol. *71:* 501-522 (1990).

Francki, R.I.B.; Hatta, T.: Tomato spotted wilt virus; *In* Kurstak, E. (ed.), Handbook of Plant Virus Infections and Comparative Diagnosis, pp. 491-512 (Elsevier/North-Holland, Amsterdam, 1981).

Gonzalez-Scarano, F.; Nathanson, N.: Bunyaviruses. *In* Fields, B.N.; Knight, J.C. (eds.), Virology, Vol.1, 2nd edn., pp. 1195-1228 (Raven Press, New York, 1990).

Karabatsos, N.: International catalogue of arboviruses including certain other viruses of vertebrates. (American Society of Tropical Medicine and Hygiene, San Antonio, Tx, USA, 1985).

Marriott, A.C.; Ward, V.K.; Nuttall, P.A.: The S RNA segment of sandfly fever Sicilian virus: evidence for an ambisense genome. Virology. *169:* 341-345 (1989).

Milne, R.G.; Francki, R.I.B.: Should tomato spotted wilt virus be considered as a possible member of the family *Bunyaviridae*? Intervirology *22:* 72-76 (1984).

Pettersson, R.F.; Gahmberg, N.; Kuismanen, E.; Kääriäinen, L.; Ronnholm, R.; Saraste, J.: Bunyavirus membrane glycoproteins as models for Golgi-specific proteins. *In* Satir, B. (ed.), Modern Cell Biology, pp. 65-96 (Alan R. Liss, New York, 1988).

Schmaljohn, C.S.; Dalrymple, J.M.: Analysis of Hantaan virus RNA: Evidence for a new genus of *Bunyaviridae*. Virology *131*:482-491 (1983).

Schmaljohn, C.S.; Patterson, J.L.: *Bunyaviridae* and their replication, Part II. Replication of *Bunyaviridae*; *In* Fields, B.N.; Knight, J.C. (eds.), Virology, Vol.1, 2nd edn., pp. 1175-1194 (Raven Press, New York, 1990).

Simons, J.F.; Hellmann, U.; Pettersson, R.F.: Uukuniemi virus S RNA segment: ambisense coding strategy, packaging of complementary strands into virions, and homology to members of the genus Phlebovirus. J. Virol. *64:* 247-255 (1990).

Tesh, R.B.; Peters, C.J.; Meegan, J.M.: Studies on the antigenic relationship among phleboviruses. Am. J. Trop. Med. Hyg. *31:* 149-155 (1982).

Ward, V.K.; Marriott, A.C.; El-Ghorr, A.A.; Nuttall, P.A.: Coding strategy of the S RNA segment of Dugbe virus (Nairovirus; *Bunyaviridae*). Virology *175:* 518-524 (1990).

Zeller, H.G.; Karabatsos, N.; Calisher, C.H.; Digoutte, J.P.; Cropp, C.B.; Murphy, F.A.; Shope, R.E.: Electron microscopic and antigenic studies of uncharacterized viruses. II. Evidence suggesting the placement of viruses in the family *Bunyaviridae*. Arch. Virol. *108:* 211-227 (1989).

Taxonomic status	English vernacular name	International name
FAMILY	ARENAVIRUS GROUP	***ARENAVIRIDAE***

Reported by M.J. Buchmeier

GENUS	LCM VIRUS GROUP	***ARENAVIRUS***
TYPE SPECIES	LYMPHOCYTIC CHORIOMENINGITIS VIRUS (LCM)	—

PROPERTIES OF THE VIRUS PARTICLE

Morphology

Enveloped, spherical to pleomorphic particles, 50-300 nm diameter (usually 110-130 nm). The dense lipid bilayer envelope has surface projections 10 nm long and club-shaped. Varying numbers of ribosome-like particles (20-25 nm diameter) appear free within the envelope. Isolated nucleocapsids, free of contaminating host ribosomes, display a linear array of nucleosomal subunits organized in closed circles varying in length from 450-1300 nm.

Physicochemical properties

S_{20w} = 325-500; buoyant density in sucrose = 1.17-1.18 g/cm^3; in CsCl = 1.19-1.20 g/cm^3; in amidotrizoate compounds ≈ 1.14 g/cm^3. Relatively unstable in vitro. Rapidly inactivated below pH 5.5 and above pH 8.5. Inactivated rapidly at 56°C and by solvents. Highly sensitive to UV and gamma radiation.

Nucleic acid

Two virus specific ssRNA molecules, L and S (MWs = 2.2-2.8 x 10^6 and ≈ 1.1 x 10^6), and three RNAs of cell origin, ≈ 28S, 18S and 4-6S. The 4-6S RNA also contains a subgenomic viral mRNA encoding the Z gene. Proportions of S to L RNA are not equimolar due to frequent packaging of multiple S RNA strands.

The S RNAs of Pichinde, LCM, Lassa and Tacaribe viruses have been sequenced and consist of 3375-3432 nucleotides. These RNAs are similarly organized and share considerable sequence homology. A 3' region (19-30 nucleotides) of conserved sequences is shared by the different viruses and is complementary to a similar region found at the 5' end. Similar regions are found at the termini of the L RNAs. Two genes encoded in an ambisense are associated with S RNA. The gene encoding the nucleoprotein is found in the 3' half of the molecule (in message complementary sense) while the gene for the glycoproteins is encoded (in message sense) in the 5' half. The intergenic regions contain nucleotide sequences with

Taxonomic status	English vernacular name	International name

the potential of forming hairpin configurations which may regulate transcription.

The L RNAs of LCM and Tacaribe viruses have also been sequenced and contain 7220 and 7102 nucleotides, respectively. Viral L RNA encodes a large protein (L) which may function as an RNA dependent, RNA polymerase and a small protein (p11, Z), which has a zinc-binding domain. The L RNA genes also have an ambiense arrangement with the L protein encoded on the 3' end (in message complementary sense) and the Z protein encoded (in message sense) on the 5' portion (ca. 0.5 kb) of the L RNA segment.

Protein

One nonglycosylated polypeptide (MW = 63-72 x 10^3) associated with the RNA as part of RNP complex. One glycosylated polypeptide with MW = 34-44 x 10^3 found in all members of the family and a second glycosylated polypeptide of MW = 44-72 x 10^3 noted in some but not other members. These envelope glycoproteins are synthesized in the cell as a single mannose-rich precursor molecule which is trimmed and proteolytically cleaved during transport to the plasma membrane. L protein, (MW ≈ 2 x 10^5) has been identified in virions as well as infected cells is an RNA-dependent RNA polymerase. Other minor proteins of unknown significance have also been detected. Enzymes found in purified virions include the transcriptase associated with the RNP of Pichinde virus, poly(U) and poly(A) polymerases which appear to be associated with the packaged host cell ribosomes and a protein kinase capable of phosphorylating the nucleoprotein.

Lipid

Present; phospliolipid composition is similar to that of the host cell plasma membrane.

Carbohydrate

Glucosamine, fucose and galactose are incorporated into numerous asparagine-linked branched chain complex carbohydrates of the viral glycoproteins.

Antigenic properties

At least 3 distinct antigenic molecules. Antigens on the surface glycoprotein (MW = 34-44 x 10^3) are involved in virus neutralization. These antigens are type-specific although cross-neutralization tests have demonstrated partial shared antigenicity between Tacaribe and Junin viruses, and cross-protection between Junin and Lassa viruses following prior infection by Tacaribe and Mopeia viruses has been demonstrated. CF antigens are used to define the Tacaribe complex. Major CF antigens are associated with the nucleoprotein. Monoclonal antibodies react with common epitopes on nucleocapsid proteins of all members. Fluorescent antibody techniques show that antisera against all Tacaribe complex viruses, as well as Lassa virus, react with LCM virus. No haemagglutinin has been identified. By monoclonal and polyclonal antibodies, LCM-Lassa complex viruses are distinguishable from Tacaribe complex viruses. Cytotoxic T-lympocyte epitopes are well characterized on the nucleoprotein and glycoprotein of LCM virus. Number and location of epitopes vary depending on virus strain and host MHC class molecules.

Replication

Limited data support the concept of differential transcription of the ambisense S RNA segment. Early events post infection include the synthesis of mRNA for the nucleoprotein which is required for the synthesis of complementary S RNA and progeny RNA. Messenger RNA for glycoprotein precursor is synthesized from the complementary RNA. Infected cells synthesise a protein (MW $\approx$ 64 x 10^3) to yield RNP, and two other proteins (MW $\approx$ 42 x 10^3 and 200 x 10^3), the smaller giving rise to a fully glycosylated precursor (MW = 79 x 10^3) which in turn is cleaved to yield the envelope glycoproteins. High-frequency genetic recombination is found as expected for viruses with segmented genomes. Reassortment studies of LCM suggest that genetic information in L RNA controls plaque morphology and virulence in guinea pigs while tissue tropism and virulence in mice are associated with S RNA. The synthesis of LCM DI virus has been observed in vivo as well as in vitro. DNA synthesis inhibitors have no effect on arenavirus multiplication, but a functional host-cell nucleus is required. Replication *in vitro* of a number of arenaviruses is inhibited by amantadine, α-amanitin, benzimidazoles, glucosamine, and thio-semicarbazones. Ribavirin appears to inhibit the replication of several arenaviruses *in vitro* and spares monkeys infected with Machupo and Lassa viruses.

Most, if not all, arenaviruses probably have limited cell killing potential. However, virus replication commonly occurs in the absence of overt cytopathic effects and carrier cultures are readily established in tissue culture. DI particles are readily produced and interference may play a role in preventing cell destruction.

Intracytoplasmic inclusion bodies are prominent in cells infected with arenaviruses; they consist of ribosome masses in a moderately electron-dense matrix. The relative proportion of ribosomes and matrix may vary widely in different inclusions, but as infection progresses a condensation of inclusion material results in rather uniformly marginated, large masses.

BIOLOGICAL ASPECTS

Host range

Natural: Virus isolates from the Old World appear to be restricted to the family *Muridae* and those from the New World to the family *Cricetidae* with the two exceptions of Amapari virus which was isolated from a *Muridae* (Neacomys) and TAC virus which was isolated from a fruit-eating bat (*Artibeus*). Most viruses are found as a chronic infection in a single rodent host (*Mus, Calomys, Mastomys, Oryzomys, Sigmodon, Praomys,* and the fruit-eating bat *Artibeus*) in which persistent infection with viremia and/or viruria occur or are suspected. Such infections may be caused by a slow and/or insufficient immune response of the host. Natural spread to other mammals and humans is unusual except for Lassa virus, a common infection of humans in West Africa, and Junin virus, a less common but significant infection of humans in Argentina. LCMV infection of humans has been significant in some urban areas with high rodent populations. It has also been reported to be acquired from pet hamsters. Disease outcome in experimentally infected laboratory animals (mouse, hamster, guinea pig, rhesus monkey, marmoset, rat) vary with the type of virus used. In general, viruses of the Tacaribe complex are pathogenic for suckling but not weaned mice; LCM and Lassa viruses produce the opposite effect. Cross-protection is seen against Junin and Lassa with prior infection by Tacaribe and Mozambique viruses, respectively. LCM virus has been found to grow in murine lymphocytes. Vero and BHK21 infected cells are most commonly used for virus isolation and growth, but the viruses grow moderately well in many other mammalian cells.

Transmission

Vertical - transuterine, transovarian and postpartum (most likely by milk-, saliva- or urine-borne routes) in natural hosts. Horizontal - important as a mechanism for viruses to escape from their natural host. Venereal transmission suspected as an important mode for intra-species spread. Vectors - a few arthropod isolations which have never been shown to have any place in transmission cycles in nature. Biological - unknown. Mechanical - unknown.

Taxonomic status	English vernacular name	International name

OTHER MEMBERS

LCM-Lassa Complex:

Lymphocytic choriomeningitis (LCM)
Lassa
Mobala
Mopeia
Ippy

Tacaribe Complex:

Tacaribe
Junin
Macupo
Amapari
Parana
Tamiami
Pichinde
Latino
Flexal

Derivation of Name arena: from Latin *arenosus*, 'sandy', from appearance of particles in electron microscope sections.

REFERENCES

Auperin, D.D.; Romanowski, V.; Galinski, M.; Bishop, D.H.L.: Sequencing studies of Pichinde arenavirus S RNA indicate a novel coding strategy, an ambisense viral S RNA. J. Virol. *52*:897-904 (1984).

Auperin, D.D.; Sasso, D.R.; McCormick, J.B.: Nucleotide sequence of the glycoprotein gene and intergenic region of the Lassa virus S genome RNA. Virology *154*:155-167 (1986).

Bruns, M.; Cihak, J.; Muller, G.; Lehmann-Grube, F.: Lymphocytic choriomeningitis virus. VI. Isolation of a glycoprotein mediating neutralization. Virology *130*:247-251 (1983).

Buchmeier, M.J.; Southern, P.J.; Parekh, B.S.; Wooddell, M.K.; Oldstone, M.B.A.: Site-specific antibodies define a cleavage site conserved among Arenavirus GP-C glycoproteins. J. Virol. *61*:982-985 (1987).

Casals, J.; Buckley, S.M.; Cedeno, R.: Antigenic properties of the arenaviruses. Bull WHO *52:* 421-427 (1975).

Franze-Fernandez, M.T.; Zetina, C.; Iapalucci, S.; Lucero, M.A.; Bouissou, C.; Lopez, R.; Rey, O.; Daheli, M.; Cohen, G.N.; Zakin, M.M.: Molecular structure and early events in the replication of Tacaribe arenavirus S RNA. Virus Res. *7*:309-324 (1987).

Fuller-Pace, F.V.; Southern, P.J.: Detection of virus-specific RNA-dependent RNA polymerase activity in extracts from cells infected with lymphocytic choriomeningitis virus: In vitro synthesis of full-length viral RNA species. J. Virol. *63:* 1938-1944 (1989).

Gonzalez, J.P.: Les arenavirus D'Afrique: un nouveau paradigme d'evolution. Bull. Inst. Pasteur *84*:67-85 (1986).

Harnish, D.G.; Dimock, K.; Bishop, D.H.L; Rawls, W.E.: Gene mapping in Pichinde virus: assignment of viral polypeptides to genomic L and S RNAs. J. Virol. *46*:638-641 (1983).

Howard, C.R.: Arenaviruses, *In* Perspectives in Medical Virology, Vol. 2 (Elsevier, New York, 1986).

Howard, C.R.; Buchmeier, M.J.: A protein kinase activity in lymphocytic choriomeningitis virus and identification of the phosphorylated product using monoclonal antibody. Virology *126*:538-547 (1983).

Iapalucci, S.; Lopez, N.; Rey, O.; Zakin, M.M.; Cohen, G.N.; Franze-Fernandez, M.T.: The 5' region of Tacaribe virus L RNA encodes a protein with a potential metal binding domain. Virology *173:* 357-361 (1989).

Iapalucci, S.; Lopez, R.; Rey, O.; Lopez, N.; Franze-Fernandez, M.T.; Cohen, G.N.; Lucero, M.A.; Ochoa, A.; Zakin, M.M.: Tacaribe virus L gene encodes a protein of 2210 amino acid residues. Virology *170:* 40-47 (1989).

Parekh, B.S.; Buchmeier, M.J.: Proteins of lymphocytic choriomeningitis virus: Antigenic topography of the viral glycoproteins. Virology *153*:168-178 (1986).

Romanowski, V.; Matsuura, Y.; Bishop, D.H.L.: Complete sequence of the S RNA of lymphocytic choriomeningitis virus (WE strain) compared to that of Pichinde arenavirus. Virus Res. *3*:101-114 (1985).

Rowe, W.P.; Murphy, F.A.; Bergold, G.H.; Casals, J.; Hotchin, J.; Johnson, K.M.; Lehmann-Grube, F.; Mims, C.A.; Traub, E.; Webb, P.A.: Arenoviruses: proposed name for a newly defined virus group. J. Virol. *5:* 651-652 (1970).

Southern, P.J.; Singh, M.K.; Riviere, Y.; Jacoby, D.R.; Buchmeier, M.J.; Oldstone, M.B.A.: Molecular characterization of the genomic S RNA segment from lymphocytic choriomeningitis virus. Virology *157*:145-155 (1987).

Young, P.R.; Howard, C.R.: Fine structure analysis of Pichinde virus nucleocapsids. J. gen. Virol. *64*:833-842 (1983).

Taxonomic status	English vernacular name	International name
FAMILY	RNA TUMOR VIRUSES (AND RELATED AGENTS)	***RETROVIRIDAE***

Reported by J.M. Coffin

PROPERTIES OF THE VIRUS PARTICLE

Morphology

Spherical, enveloped virions 80-100 nm in diameter. Glycoprotein surface projections of approximately 8 nm diameter. Internal structure: spherical to rod-shaped capsid containing a possibly helical RNP. Special features in thin sections: outer envelope, inner membrane (shell) and central nucleoid. The central nucleoid is located acentrically in type B virions, concentrically in type C virions, and is in the shape of a rod or truncated cone in lentiviruses.

Physicochemical properties

Density between 1.16 and 1.18 g/cm^3 in sucrose gradients. Disrupted by lipid solvents and detergents. Surface glycoproteins partially removable by proteolytic enzymes. Relatively resistant to UV light.

Nucleic acid

Dimer of linear positive-sense ssRNA 7-10 kbp in length (about 2% by weight). Monomers held together by hydrogen bonds. Polyadenylated at the 3' end, with a cap structure ($m^7G^5ppp^{5'}GmpNp$) at the 5' end of the genome. The virion RNA is not infectious.

Protein

About 60% by weight. Three to four internal nonglycosylated structural proteins (encoded by the *gag* gene): MA (matrix); CA (capsid); and NC (nucleocapsid); and (in some genera) one more protein of undetermined function. The MA protein is often acylated with a fatty acid (e.g. myristate) group at its NH_2-terminus. A protease (PR) is encoded by the *pro* gene. Reverse transcriptase (RT) and integrase (IN) encoded by the *pol* gene. Two envelope (*env* gene encoded) proteins SU (surface) and TM (transmembrane).

Lipid

About 35% by weight. Derived from the plasma membrane of the host cell.

Carbohydrate

About 3.5% by weight. At least one of the two *env* proteins (SU) is glycosylated; in most viruses both are. Cellular carbohydrates and glycolipids are found in the viral envelope.

Antigenic properties

Virion proteins contain type-specific and group-specific determinants, the latter sometimes shared among members of a genus. The type-specific determinants of the envelope glycoproteins are involved in antibody neutralization.

Genetic structure

Although virions carry two copies of the genome, it is not known whether both are functional. Basic genetic information for the production of infectious progeny virus consists of 4 genes: *gag*, coding for internal structural proteins of the virion; *pro*, encoding the virion protease; *pol*, coding for reverse transcriptase; and *env*, coding for envelope glycoproteins of the virion. The order of these genes is invariably 5' *gag, pro, pol, env* 3'. Some retroviruses also contain genes encoding non virion proteins which are important for the regulation of expression. Others carry cell-derived genetic information for nonstructural proteins that are important in pathogenesis. These cellular sequences are either inserted in a complete retrovirus genome (some strains of Rous sarcoma virus) or they form substitutions for deleted viral replicative sequences (most other rapidly oncogenic retroviruses). Such deletions render the virus replication-defective and dependent on nontransforming helper virus for production of infectious progeny. In many cases the cell-derived sequences form a fused gene with viral structural information that is then translated into one protein (e.g., '*gag*-onc' protein).

REPLICATION

Entry into the host cell is mediated by interaction between an envelope glycoprotein of the virion and specific receptors at the cell surface, possibly resulting in fusion of the viral envelope to the plasma membrane either directly or following endocytosis. Receptors are cell surface proteins of which two have been identified to date: one (the CD4 protein recognized by HIV) has a single transmembrane region; the other (the receptor for ecotropic MLV) has a more complex structure with multiple transmembrane domains. The further process of intracellular uncoating of the viral particle is not understood, but subsequent early events are carried out in the context of a nucleoprotein complex derived from the capsid. Replication starts with reverse transcription of virion RNA into DNA. The linear dsDNA transcripts of the viral genome contain long terminal repeats (LTR's) composed of sequences from the 3' (U3) and 5' (U5) ends of the viral RNA flanking a sequence (R) found near both ends of the viral RNA. Retroviral DNA becomes integrated into the chromosomal DNA of the host to form a *provirus* by a mechanism involving the viral IN protein. The ends of virus DNA are joined to cell DNA, removing one or two bases from the ends of the linear viral DNA and generating a short duplication of cell sequences at the integration site. Viral DNA can integrate at many sites in the cellular genome, and once integrated is apparently incapable of further "transposition" within the same cell. The map of the integrated provirus is coextensive with that of unintegrated linear viral DNA. Integration appears to be a prerequisite of virus replication. The integrated provirus is transcribed by cellular RNA polymerase II into virion RNA and mRNA in response to transcriptional signals in the LTR's. There are several classes of mRNA reflecting the genetic map of retroviruses. An mRNA comprising the whole genome serves for the translation of the *gag, pro,* and *pol* genes positioned at the 5' portion of this RNA into polyprotein precursors which are cleaved by the PR protein to yield the structural proteins, protease and reverse transcriptase, respectively. A smaller mRNA consisting of the 3' sequences of the genome, including the *env* gene and the U3 and R regions, is translated into the precursor of the envelope proteins. In viruses that contain additional genes, other forms of spliced mRNA are also found. All mRNAs share a common sequence at their 5' ends. In the less-than-genome size mRNAs this sequence is acquired by RNA splicing. Most primary translational products in retrovirus infection are polyproteins which require proteolytic cleavage before becoming functional. The gag, pro and pol products are produced from a nested set of primary products whose translation is mediated by partial readthrough of translational terminal signals (usually by ribosomal frame shifting) at the gag-pro and/or the pro-pol boundaries. Virions mature either at the plasma membrane (type C and most other viruses) or as intracytoplasmic (type A) particles and are released from the cell by a budding process.

Biological Aspects

Host range

Retroviruses are widely distributed as exogenous infectious agents of vertebrates, particularly mammals and birds. Endogenous proviruses that have resulted from infection of the germ line and are inherited as Mendelian genes occur widely among vertebrates.

Association with disease

Retroviruses are associated with a large variety of diseases, including malignancies (leukemias, lymphomas, sarcomas and other tumors of mesodermal origin, mammary carcinomas, carcinomas of liver and kidney);

Taxonomic status	English vernacular name	International name
	immunodeficiencies, such as AIDS; autoimmune disease; lower motor neuron disease; and several acute diseases with tissue damage. Some retroviruses are nonpathogenic.	
Transmission	Transmission is horizontal via a number of routes, including blood, saliva, intimate contact, insects, and others and vertical via direct infection of the developing embryo, via milk, or other perinatal routes. Endogenous retroviruses are also transmitted by inheritance of proviruses.	

GENERA

In view of current knowledge of retroviruses, the "previous" classification into subfamilies (*oncovirinae*, *lentivirinae*, *spumavirinae*) is no longer appropriate, since the genera that made up, for example, *oncovirinae* are no more closely related (or similar) to one another than they are to members of other previously designed subfamilies. Retroviruses are currently classified into 7 genera as follows:

Mammalian type B oncovirus group	-
MLV-related viruses (Mammalian type C retrovirus group)	-
Type D retrovirus group	-
Avian type C retrovirus group (ALV-related viruses)	-
Foamy virus group	*Spumavirus*
HTLV-BLV group	-
Lentivirus group	*Lentivirus*

GENUS	MAMMALIAN TYPE B ONCOVIRUS GROUP	—
TYPE SPECIES	MOUSE MAMMARY TUMOR VIRUSES	—

DISTINGUISHING CHARACTERISTICS

Virion: B-type morphology (prominent surface spikes, eccentric condensed core, assembly occurs within the cytoplasm as A-type particles prior to budding). Proteins: MA $\approx$ 10 kDa; p21;CA $\approx$ 27 kDa; NC $\approx$ 14 kDa; PR $\approx$ 13 kDa; SU $\approx$ 52 kDa; TM $\approx$ 36 kDa.

Genome: $\approx$ 10 kb. One additional gene (*orf*- function unknown) 3' of *gag-pro-pol* and *env*. Primer $tRNA^{Lys\text{-}3}$. LTR $\approx$ 1300 bp. (U3 1200, R 15, U5 120).

Taxonomic status	English vernacular name	International name
	Distribution: Limited to a few exogenous, vertically-transmitted (via milk) and endogenous viruses of mice. Associated with mammary carcinoma and T-lymphoma. Related endogenous sequences have been found in other rodents and primates. No oncogene-containing members are known.	
GENUS	MLV-RELATED VIRUSES (MAMMALIAN TYPE C RETROVIRUS GROUP)	—
TYPE SPECIES	MURINE LEUKEMIA VIRUS	—

DISTINGUISHING CHARACTERISTICS

Virion: C-type morphology (barely visible surface spikes, central condensed core, assembly occurs at the inner surface of the membrane at the same time as budding). Proteins: MA ≈ 15 kDa; p12; CA ≈ 30 kDa; NC ≈ 10 kDa; PR ≈ 14 kDa; SU ≈ 70 kDa; TM ≈ 15 kDa.

Genome: ≈ 8.3 kb. No known additional genes to *gag-pro-pol* and *env*. Primer $tRNA^{Pro}$ ($tRNA^{Glu}$ is found in a few endogenous mouse viruses). LTR ≈ 600 bp. (U3 500, R 60, U5 75).

Distribution: Widespread exogenous vertically and horizontally transmitted and endogenous viruses found in many groups of mammals. The reticuloendotheliosis group comprises a few isolates from birds, with no known corresponding endogenous relatives. Related endogenous sequences are found in mammals. Associated with a variety of diseases including malignancies, immunosuppression, neurological disorders and others. Many oncogene-containing members of the mammalian and reticuloendotheliosis virus groups have been isolated.

SUBGENERA

Mammalian type C viruses

species:
- Murine sarcoma and leukemia viruses
- Feline sarcoma and leukemia viruses
- Gibbon ape leukemia virus
- Guinea pig type C virus
- Porcine type C virus
- Woolly monkey sarcoma virus

Reticuloendotheliosis viruses

Taxonomic status	English vernacular name	International name
	species: Avian reticuloendotheliosis virus	
	Reptilian type C viruses	
	species: Viper retrovirus	
GENUS	TYPE D RETROVIRUS GROUP	—
TYPE SPECIES	MASON-PFIZER MONKEY VIRUS	—

DISTINGUISHING CHARACTERISTICS

Virion: D-type morphology (resembling B-type except for less prominent surface spikes). Proteins: MA ≈ 10 kDa; p18; CA ≈ 27 kDa; NC ≈ 14 kDa; PR unknown; SU ≈ 70 kDa; TM ≈ 22 kDa.

Genome: ≈ 8.0 kb. No known additional genes to *gag-pro-pol* and *env*. Primer $tRNA^{Lys\ 1,2}$. LTR ≈ 350 bp. (U3 240, R 15, U5 95).

Distribution: Several isolates of exogenous, horizontally transmitted and endogenous viruses of new and old world primate species. Exogenous virus isolates associated with immunodeficiency diseases. No oncogene-containing members are known.

OTHER MEMBERS

Squirrel monkey retrovirus
Langur virus (PO-1-Lu)

Taxonomic status	English vernacular name	International name
GENUS	AVIAN TYPE C RETROVIRUS GROUP (ALV-RELATED VIRUSES)	—
TYPE SPECIES	AVIAN LEUKOSIS VIRUS	—

DISTINGUISHING CHARACTERISTICS

Virion: C-type morphology. Proteins: MA ≈ 19 kDa; p10; CA ≈ 27 kDa; NC ≈ 12 kDa; PR ≈ 15 kDa; SU ≈ 85 kDa; TM ≈ 37 kDa.

Genome: ≈ 7.2 kb. No known additional genes to *gag-pro-pol* and *env*. Primer $tRNA^{Trp}$. LTR ≈ 350 bp. (U3 250, R 20, U5 80).

Distribution: Widespread exogenous vertically and horizontally transmitted and endogenous viruses found in

Taxonomic status	English vernacular name	International name

chickens and some other birds. Distantly related endogenous sequences are found in birds and mammals. Associated with malignancies and some other diseases such as wasting, osteopetrosis. Many oncogene-containing members of this group have been isolated.

Other Members

Avian sarcoma and leukemia viruses

GENUS	FOAMY VIRUS GROUP	***SPUMAVIRUS***
TYPE SPECIES	HUMAN FOAMY VIRUS	—

Distinguishing characteristics

Virion: Distinctive (but unnamed) morphology (prominent surface spikes, central condensed core, assembly occurs in the cytoplasm prior to budding). Proteins are not yet well defined.

Genome: ≈ 11 kb. Several additional open reading frames (tentatively designated "bel 1,2,3,4" of unknown coding capacity and function) 3' to *gag-pro-pol* and *env*. Primer $tRNA^{Lys\ 1,2}$. LTR ≈ 1150 bp. (U3 800, R 200, U5 150).

Distribution: Widespread exogenous viruses found in many groups of mammals. No related endogenous viruses are known. Although many isolates cause characteristic "foamy" cytopathology in cell culture, no associated diseases have been described. No oncogene-containing members of this group have been isolated.

Other Members

Simian foamy virus
Feline syncytial virus
Bovine syncytial virus

Taxonomic status	English vernacular name	International name
GENUS	HTLV-BLV GROUP	—
TYPE SPECIES	HUMAN T-CELL LYMPHOTROPIC VIRUS TYPE 1	—

DISTINGUISHING CHARACTERISTICS

Virion: Similar to C-type in morphology and assembly. Proteins: MA ≈ 19 kDa; CA ≈ 24 kDa; NC ≈ 12,15 kDa; PR ≈ 14 kDa; SU ≈ 60 kDa; TM ≈ 21 kDa.

Genome: ≈ 8.3 kb. Two additional genes (*tax* and *rex*) whose products are involved in regulation of synthesis and processing of virus RNA 3' to *gag-pro-pol* and *env*. Primer $tRNA^{Pro}$. LTR ≈ 550-750 bp. (U3 200-300, R 135-235, U5 100-200).

Distribution: Exogenous horizontally-transmitted viruses found in a few groups of mammals. No related endogenous viruses are known. Associated with B or adult T cell leukemia/lymphoma with a very long latency and less than 100% incidence. No oncogene-containing members of this group have been isolated.

OTHER MEMBERS

Human T-cell lymphotropic virus type 2
Simian T-cell lymphotropic virus
Bovine leukemia virus

Taxonomic status	English vernacular name	International name
GENUS	LENTIVIRUS GROUP	***LENTIVIRUS***
TYPE SPECIES	HUMAN IMMUNO-DEFICIENCY VIRUS	—

DISTINGUISHING CHARACTERISTICS

Virion: Distinctive (but unnamed) morphology with a bar (or truncated cone)-shaped core. Assembly at the cell membrane. Proteins: MA ≈ 17 kDa; CA ≈ 24 kDa; NC ≈ 7-11 kDa; PR ≈ 14 kDa; SU ≈ 120 kDa; TM ≈ 41 kDa.

Genome: ≈ 9.2 kb. Several additional genes varying somewhat among the groups (e.g. *vif, vpr, tat, rev, vpu* in HIV-1) whose products are involved in regulation of synthesis and processing of virus RNA and possibly other functions, located 3' to *gag-pro-pol* and 5' to *env*, as well as one (*nef* in HIV) 3' of *env*. Primer $tRNA^{Lys\ 1,2}$. LTR ≈ 600 bp. (U3 450, R 100, U5 70).

Taxonomic status	English vernacular name	International name
FAMILY	CALICIVIRUS FAMILY	*CALICIVIRIDAE*

Compiled by F.L. Schaffer

GENUS	CALICIVIRUS GROUP	*CALICIVIRUS*
TYPES SPECIES	VESICULAR EXANTHEMA OF SWINE VIRUS (VESV) (SEROTYPE A)	—

PROPERTIES OF THE VIRUS PARTICLE

Morphology Roughly spherical, 35-39 nm diameter, with 32 cup-shaped surface depressions arranged in icosahedral symmetry. Capsid probably consists of 180 polypeptides.

Physicochemical properties MW ≈ 15 x 10^6; S_{20w} = 170-183; buoyant density in CsCl = 1.33-1.39 g/cm^3 depending upon virus and strain. Insensitive to ether, chlorolform or mild detergents. Inactivated at pH values between 3 and 5. Thermal inactivation is accelerated in high concentrations of Mg^{++}. Some members inactivated by trypsin.

Nucleic acid One molecule of infectious positive-sense ssRNA, MW = 2.6-2.8 x 10^6 (≈ 8.2 kb). Polyadenylated at 3'-terminus: no methylated cap at 5'-terminus.

Protein One major capsid polypeptide, MW = 60-71 x 10^3 with blocked N-terminal end. A protein with apparent MW = 10-15 x 10^3, essential for infectivity, is covalently linked to virion RNA, presumably a Vpg at the 5'-end. A minor polypeptide, MW = 15-19 x 10^3 and < 2% of total protein possibly noncovalently associated with RNA, has also been reported.

Lipid None.

Carbohydrate None reported.

Antigenic properties Neutralization indicates distinct serotypes of vesicular exanthema of swine and San Miguel sealion viruses, but considerable cross-reactivity among feline calicivirus strains. Precipitin reactions indicate antigenic relationships among swine vesicular exanthema, San Miguel sealion and feline caliciviruses but no relationship of these to canine calicivirus.

Effect on cells Lysis.

Replication

Two major virus-specific ssRNA species in infected cells are a genome-sized RNA and a smaller RNA (≈ 2.4 kb for feline calicivirus, apparent size may differ depending on virus and method of analysis). Genome-sized RNA presumably serves as mRNA coding for non-structural polypeptides, and the smaller RNA is presumably a subgenomic mRNA coding for the major capsid polypeptide (probably via cleavage of a precursor). Two dsRNAs corresponding to the major ssRNAs, and RNA partially resistant to RNase (presumptive RI) are found in infected cells. Minor ssRNAs, two of intermediate length, 4.8 and 4.2 kb have also been observed; functions of these have not been established. The viral RNAs all appear to be polyadenylated and represent nested coterminal transcripts with common 3'-ends. Capsid polypeptide is the major protein synthetic product; an uncertain number of additional polypeptides is also synthesized; precursor-product relationships among them are not fully established. Virions mature in the cytoplasm.

Biological Aspects

Host range

Natural - vesicular exanthema of swine virus: swine (pinnipeds ?); San Miguel sea lion virus: pinnipeds, fish, swine; feline calicivirus: felines, dogs; canine calicivirus: dogs. Experimental *in vivo* - vesicular exanthema of swine virus: horse (some strains); SMSV: monkey. Cell culture-vesicular exanthema of swine and San Miguel sealion viruses: porcine, primate (feline?); feline calicivirus: feline (primate?); canine calicivirus: canine, dolphin.

Transmission

Biological vectors not known. Mechanical via contaminated food (vesicular exanthema of swine virus), contact, airborne (feline calicivirus). Marine/terrestrial transmission likely with vesicular exanthema of swine and San Miguel sealion viruses.

Other Members

Feline calicivirus (numerous antigenically related strains)
San Miguel sea lion (8 or more serotypes)
Canine calicivirus.

Probable members

Viruses with typical calicivirus morphology have been identified in other animal species including humans, other

Taxonomic status	English vernacular name	International name

primates, cattle, mink, swine, walruses, dolphins, dogs, rabbits, chickens, reptiles, amphibians and insects, but none of these have been fully characterized. Those from humans and some other species cause gastroenteritis, and are difficult to propagate in cell culture. Other viruses that cause gastroenteritis in humans, generally designated "small round structured viruses", including Norwalk virus and Snow Mountain agent, lack typical calicivirus morphology, but have buoyant density and a single capsid polypeptide typical of caliciviruses.

Limited serological relationships have been found among strains of viruses from humans: little or no serological relationships have been detected among viruses from other species.

Derivation of Name calici: from Latin *calix*, 'cup' or 'goblet', from cup-shaped depressions observed by electron microscopy.

REFERENCES

Carter, M.J.: Feline calicivirus protein synthesis investigated by Western blotting. Arch. Virol. *108:* 69-79 (1989).

Crandell, R.A.: Isolation and characterization of caliciviruses from dogs with vesicular genital disease. Arch. Virol. *98:* 65-71 (1988).

Cubitt, W.D.: Diagnosis, occurrence and clinical significance of the human 'candidate' caliciviruses. Progr. Med. virol. *36:* 103-119 (1989).

Hayashi, Y.; Ando, T.; Utagawa, E.; Sekine, S.; Okada, S.; Yabuuchi, K.; Miki, T.; Ohashi, M.: Western blot (immunoblot) assay of small, round-structured virus associated with an acute gastroenteritis outbreak in Tokyo. J. Clin. Microbiol. *27:* 1728-1733 (1989).

Neill, J.D.; Mengeling, W.L.: Further characterization of the virus-specific RNAs in feline calicivirus infected cells. Virus Res. *11:* 59-72 (1988).

Ohlinger, V.F.; Haas, B.; Ahl, R.; Weiland, F.: Die infektiose hamorrhagische Krankheit der Kaninchen-eine durch ein Calicivirus verursachte Tierseuche. Z. Geb. Veterinarmed. *44:* 284-294 (1989).

Schaffer, F.L.: Caliciviruses, *In* Fraenkel-Conrat, H.; Wagner, R.R. (eds.), Comprehensive Virology, Vol. 14, pp. 249-284 (Plenum Press, New York, 1979).

Schaffer, F.L.; Bachrach, H.L.; Brown, F.; Gillespie, J.H.; Burroughs, J.N.; Madin, S.H.; Madeley, C.R.; Povey, R.C.; Scott, F.; Smith, A.W.; Studdert, M.J.: *Caliciviridae.* Intervirology *14*:1-6 (1980).

Smith, A.W.; Boyt, P.M.: Caliciviruses of Ocean Origin: A Review. J. Zoo Wildlife Med. *21:* 3-23 (1990).

Studdert, M.J.: Caliciviruses: a brief review. Arch. Virol. *58*:157-191 (1978).

Taxonomic status	English vernacular name	International name
GROUP	CARNATION MOTTLE VIRUS GROUP	*CARMOVIRUS*

Compiled by T.J. Morris

TYPE MEMBER	CARNATION MOTTLE VIRUS (CARMV) (7)	—

PROPERTIES OF THE VIRUS PARTICLE

Morphology Isometric particles with rounded outline, 33 nm in diameter. The X-ray crystallographic structure of the closely related turnip crinkle virus has been resolved at 3.2Å. The virions are composed of 180 protein subunits that are structurally similar to those of tomato bushy stunt virus with two globular domains (P and S) and an N terminal basic domain.

Physicochemical properties MW ≈ 8.2 x 10^6; S_{20w} ≈ 122; buoyant density in CsCl ≈ 1.35 g/cm^3.

Nucleic acid One molecule of positive sense ssRNA, MW ≈ 1.3 x 10^6 (4003 nt); 17% by weight of virus. The genomes of both CarMV and turnip crinkle virus (4051 nt) have been sequenced and infectious transcripts have been produced from full-length genomic clones of turnip crinkle virus. Satellite RNAs and defective interfering RNAs have been reported for turnip crinkle virus.

Protein One major coat polypeptide, MW ≈ 38 x 10^3.

Lipid None reported.

Carbohydrate None reported.

Antigenic properties Good immunogens. Single precipitin line in gel-diffusion tests; no serological cross reactivity between members.

REPLICATION

The 4 kb viral genomic RNA encodes a gene product, MW ≈ 27 x 10^3 and two potential readthrough polypeptides of 86 and 98 x 10^3. Two viral-specific 3'-coterminal, subgenomic RNAs of 1.7 and 1.5 kb code for a putative MW = 7 x 10^3 protein and the coat protein, respectively. Three viral-specific dsRNA species (4.0, 1.7 and 1.5 kbp) corresponding to each of the viral specific ssRNA species have been detected in infected tissues. Virus particles have been located in the cytoplasm, vascular tissues and within the nucleus, and are associated with cytoplasmic membranes. The genome features of turnip crinkle virus are very similar to those of CarMV including open reading frames corresponding to a MW = 28×10^3 gene product and an MW = 88×10^3 readthrough product, both of which are necessary for replication in protoplasts. Encoded gene products of MW = 8×10^3 and the 38×10^3 coat protein are required for systemic infection in plants.

BIOLOGICAL ASPECTS

Host range Wide among angiosperms.

Transmission Readily transmitted by mechanical inoculation. Acquisition through soil is possible. Melon necrotic spot virus is transmitted in nature by the chytrid fungus *Olpidium radicale.*

SEQUENCE SIMILARITIES WITH OTHER VIRUS GROUPS OR FAMILIES

The non-structural gene sequences of carmoviruses (CarMV and turnip crinkle virus) show striking similarity to analogous regions of maize chlorotic mottle virus and members of the tombusviruses, dianthoviruses and luteoviruses. The sequence conservation typical of the putative RNA polymerase domains (GDD motif) among members of the Sindbis virus superfamily is evident although sequence motifs for other conserved regions are not present.

OTHER MEMBERS

Cucumber soil-borne
Galinsoga mosaic (252)
Hibiscus chlorotic ringspot (227)
Melon necrotic spot (302)
Pelargonium flower break (130)
Saguaro cactus (148)
Turnip crinkle (109)

Possible members

Bean mild mosaic (231)
Blackgram mottle (237)
Cowpea mottle (212)
Cucumber leaf spot (319)
Elderberry latent (127)
Glycine mottle
Narcissus tip necrosis (166)
Plantain 6
Tephrosia symptomless

Derivation of Name carmo: sigla from *car*nation *mo*ttle virus.

References

Carrington, J.C.; Heaton, L.A.; Zuidema, D.; Hillman, B.I.; Morris, T.J.: The genome structure of turnip crinkle virus. Virology *170:* 219-226 (1989).

Guilley, H.; Carrington, J.C.; Balazs, E.; Jonard, G.; Richards, K.; Morris, T.J.: Nucleotide sequence and genome organization of carnation mottle virus RNA. Nuc. Acids Res. *13:* 6663-6677 (1985).

Heaton, L.A.; Carrington, J.C.; Morris, T.J.: Turnip crinkle virus infection from RNA synthesized *in vitro*. Virology *170:* 214-218 (1989).

Hogle, J.M.; Maeda, A.; Harrison, S.C.: Structure and assembly of turnip crinkle virus. I. X-ray crystallographic structure analysis at 3.2Å resolution. J. Mol. Biol. *191:* 625-638 (1986).

Li, X.H.; Heaton, L.A.; Morris, T.J.; Simon, A.E.: Turnip crinkle virus defective interfering RNAs intensify viral symptoms and are generated *de novo*. Proc. Nat. Acad. Sci., USA *86:* 9173-9177 (1989).

Morris, T.J.; Carrington, J.C.: Carnation mottle virus and viruses with similar properties. *In* Koenig, R. (ed.), The Plant Viruses, Polyhedral virions with monopartite RNA genomes, Vol. 3, pp. 73-112. (Plenum Press, New York , 1988).

Nutter, R.C.; Scheets, K.; Panganiban, L.C.; Lommel, S.A.: The complete nucleotide sequence of the maize chlorotic mottle virus genome. Nuc.Acids Res. *17:* 3163-3177 (1989).

Taxonomic status	English vernacular name	International name
FAMILY	ssRNA PHAGES	***LEVIVIRIDAE***

Revised by H.-W. Ackermann

PROPERTIES OF THE VIRUS PARTICLE

Morphology Quasi-icosahedral, ≈ 24 nm in diameter, probably 32 capsomers (T = 3), no envelope.

Physicochemical properties MW ≈ 4.0 x 10^6; S_{20w} = 79-84; buoyant density in CsCl = 1.42-1.47 g/cm^3. Infectivity is ether-, chloroform- and detergent-resistant.

Nucleic acid One molecule of linear positive-sense ssRNA; MW ≈ 1.2 x 10^6; 31% by weight of particle, G+C = 51-52%; 3-4 partly overlapping genes.

Protein 180 copies of capsid protein (MW = 12-17 x 10^3) and one copy of A protein (MW = 35-44 x 10^3), which is required for maturation and infectivity. Capsid proteins may lack histidine or methionine.

Lipid None.

Carbohydrates None.

REPLICATION

Adsorption to sides of pili determined by a wide variety of different plasmids. Infecting phage RNA is transcribed into a negative strand which acts as a template for positive strand synthesis. Viral RNA acts as template and as messenger for A protein, coat protein, lysis protein and RNA polymerase. Capsids assemble in cytoplasm around phage RNA. Crystalline arrays in infected bacteria. Virulent, lysis with release of sometimes thousands of particles for each bacterial cell.

BIOLOGICAL ASPECTS

Host range Enterobacteria, *Caulobacter, Pseudomonas*.

GENERA

Coliphage MS2-GA group	*Levivirus*
Coliphage Qβ-SP group	*Allolevivirus*

Taxonomic status	English vernacular name	International name
GENUS	COLIPHAGE MS2-GA GROUP	***LEVIVIRUS***
TYPE SPECIES	COLIPHAGE MS2 GROUP	—

PROPERTIES OF THE VIRUS PARTICLE

Physicochemical properties Buoyant density in CsCl = 1.44-1.46 g/cm^3. Infectivity is relatively UV-resistant.

Nucleic acid MW ≈ 1.2 x 10^6; number of nucleotides is 3,466 for GA and 3,569 for MS2. Four genes; lysis protein gene overlaps coat protein and replicase genes.

Protein Coat protein MW ≈ 13-14 x 10^3, maturation protein MW ≈ 45 x 10^3; no read-through protein.

Antogenic properties Distinct from members of *Allolevivirus* genus.

REPLICATION

Optimal temperature is 37°C (MS2 subgroup) or 30°C (GA subgroup).

BIOLOGICAL ASPECTS

Host range Enterobacteria.

OTHER MEMBERS

a. MS2 subgroup: FH5, fr, f2, M12, R17, μ2; many others of unknown morphology.
b. GA subgroup: at least 16 others, many of them of unknown morphology.

GENUS	COLIPHAGE Qβ-SP GROUP	***ALLOLEVIVIRUS***
TYPE SPECIES	COLIPHAGE Qβ GROUP	—

PROPERTIES OF THE VIRUS PARTICLE

Physicochemical properties Buoyant density in CsCl = 1.47 g/cm^3. Infectivity is relatively UV-sensitive.

Nucleic acid MW ≈ 1.4 x 10^6; number of nucleotides is 4,218 for Qβ and 3,569 for SP. Four genes; read-through protein gene overlaps coat protein and replicase genes.

Protein Coat protein MW ≈ 16.9-17.3 x 10^3, maturation protein MW ≈ 44-48 x 10^3; read-through protein MW ≈ 39 x 10^3.

Taxonomic status	English vernacular name	International name

Antogenic properties Distinct from members of *Levivirus* genus.

REPLICATION

Optimal temperature is 37°C.

BIOLOGICAL ASPECTS

Host range Enterobacteria.

OTHER MEMBERS

a. Qβ subgroup: at least 12 others, mostly of unknown morphology.
b. SP subgroup: at least 5 others, unknown morphology.

OTHER MEMBERS OF THE FAMILY

Other members of family not yet allocated to genera.

a. B6, B7, C-1, C2, fcan, Fo*lac*, Iα, M, pilHα, R23, R34, ZG/1, ZIK/1, ZJ/1, ZL/3, ZS/3, α15, β, μ2, τ, and others (enterobacteria; many plasmid specificities).

b. φCb2, φCb4, φCb5, φCb8r, φCb9, φCb12r, φCb23r, φCP2, φCP18, φCr14, φCr28 (*Caulobacter*).

c. PRR1, PP7, 7s (*Pseudomonas*).

Derivation of Name levi: from Latin *levis,* 'light'.

REFERENCES

Ackermann, H.W.; DuBow, M.S.: Viruses of Prokaryotes, Vol. II, pp 171-218 (CRC Press, Boca Raton, Fl., 1987).
Furuse, K.: Distribution of coliphages in the environment. *In* Goyal, S.M.; Gerber, C.P.; Bitton, G. (eds.), Phage Ecology, pp. 87-124 (John Wiley & Sons, New York, 1987).
van Duin, J.: Single-stranded RNA bacteriophages. *In* Calendar, R. (ed.), The Bacteriophages, Vol. 1, pp 117-167 (Plenum Press, New York, 1988).
Zinder, N.D.: RNA phages (Cold Spring Harbor Laboratory, Cold Spring Harbor, 1975).

Taxonomic status	English vernacular name	International name
GROUP	BARLEY YELLOW DWARF VIRUS GROUP (339)	***LUTEOVIRUS***

Revised by J.W. Randles

TYPE MEMBER	BARLEY YELLOW DWARF VIRUS (BYDV) - MAV ISOLATE (32)	—

PROPERTIES OF THE VIRUS PARTICLE

Morphology Isometric particles, 25-30 nm in diameter.

Physicochemical properties MW ≈ 6.5 x 10^6; S_{20w} = 104-118; buoyant density ≈ 1.40 g/cm^3 in CsCl. Moderately stable.

Nucleic acid One molecule of positive-sense ssRNA MW ≈ 2.0 x 10^6 with Vpg at 5'-end and containing 6 ORFs. 'Satellite' RNAs are associated with some barley yellow dwarf virus (RPV) isolates.

Protein One coat polypeptide, MW ≈ 24 x 10^3. 180 protein subunits arranged in a T = 3 icosahedral lattice.

Lipid None reported.

Carbohydrate None reported.

Antigenic properties Strongly immunogenic. Many members are serologically related.

REPLICATION

Confined to phloem tissues of infected plants. Details of ultra-structural changes vary among members.

BIOLOGICAL ASPECTS

Host range Varies with member - some infect wide range of monocotyledonous plants, others infect many dicotyledonous plants, and some are restricted to smaller plant groups.

Transmission Persistent transmission by aphid vectors; virus apparently does not replicate in vector. Pronounced vector specificity among some virus isolates. Not transmitted by mechanical inoculation to plants, but aphids are rendered inoculative by injection. Not transmitted through seed. Several members are associated with systems of dependent virus transmission by aphids from mixed infections in the host .

Taxonomic status	English vernacular name	International name

Other Members

Characterized isolates of BYDV fall into two groups on the basis of serological properties and cytological effects:

I. MAV, PAV, and SGV
II. RPV, RMV and rice giallume (RGV)

Bean leaf roll (286)
(= legume yellows, Michigan alfalfa, pea leaf roll)
Beet western yellows (89)
(= beet mild yellowing, *Malva* yellows, turnip yellows)
Carrot red leaf (249)
Groundnut rosette assistor
Indonesian soybean dwarf
Potato leaf roll (36; 291)
(= *Solanum* yellows, tomato yellow top)
Soybean dwarf (179)
(= subterranean clover red leaf, strawberry mild yellow edge)
Tobacco necrotic dwarf (234)

Possible members

Beet yellow net
Celery yellow spot
Cotton anthocyanosis
Filaree red leaf
Milk vetch dwarf
Millet red leaf
Physalis mild chlorosis
Physalis vein blotch
Raspberry leaf curl
Tobacco vein distorting
Tobacco yellow net
Tobacco yellow vein assistor

Derivation of Name from Latin *luteus*, 'yellow', from yellowing symptoms shown by infected hosts.

REFERENCES

Diaco, R.; Lister, R.M.; Hill, J.H.; Durand, D.P.: Demonstration of serological relationships among isolates of barley yellow dwarf virus by using polyclonal and monoclonal antibodies. J. gen. Virol. *67:*353-362 (1986).

Duffus, J.E.; Falk, B.W.; Johnstone, G.R.: Luteoviruses - one system, many variations; in Burnett, World Perspectives on Barley Yellow Dwarf, pp. 86-104. (CIMMYT, Mexico, 1990).

Francki, R.I.B.; Milne, R.G.; Hatta, T.: Luteovirus group, *In* Atlas of Plant Viruses, Vol. I, pp 137-152 (CRC Press, Boca Raton, Fl., 1985).

Iwaki, M.; Roechan, M.; Hibino, H.; Tochihara, H.; Tantera, D.M.: A persistent aphid-borne virus of soybean, Indonesian soybean dwarf virus. Plant Dis. *64:*1027-1030 (1980).

Johnstone, G.R.; Ashby, J.W.; Gibbs, A.J.; Duffus, J.E.; Thottapilly, G.; Fletcher, J.D.: The host ranges, classification and identification of eight persistent aphid-transmitted viruses causing diseases in legumes. Neth. J. Pl. Path. *90:*225-245 (1984).

Martin, R.R.; D'Arcy, C.J.: Relationships among luteoviruses based on nucleic acid hybridization and serological studies. Intervirology *31:* 23-30 (1990).

Mayo, M.A.; Robinson, D.J.; Jolly, C.A.; Hyman, L.: Nucleotide sequence of potato leafroll luteovirus RNA. J. gen. Virol. *70:* 1037-1051 (1989).

Miller, W.A.; Hercus, T.; Waterhouse, P.M.; Gerlach, W.L.: Characterization of a satellite RNA associated with a luteovirus. 7th Intl. Congr. Virology, Edmonton. Abstracts, p. 299 (1987).

Miller, W.A.; Waterhouse, P.M.; Gerlach, W.L.: Sequence and organization of barley yellow dwarf virus genomic RNA. Nuc. Acids Res. *16:* 6097-6111 (1988).

Murphy, J.F.; D'Arcy, C.J.; Clark, J.M.: Barley yellow dwarf virus RNA has a 5'-terminal genome-linked protein. J. gen. Virol. *70:* 2253-2256 (1989).

Thomas, J.E.: Characterisation of an Australian isolate of tomato yellow top virus. Ann. appl. Biol. *104:*79-86 (1984).

Yu, T.F.; Hsu, H.-K.; Pei, M.-Y.: Studies on the red-leaf disease of the foxtail millet [*Setaria italia*,(L) Beauv.]. II. Cultivated and wild hosts of millet red leaf virus. Acta Phytopath. Sin. *4:*1-7 (1958).

Taxonomic status	English vernacular name	International name
GROUP	MAIZE CHLOROTIC DWARF VIRUS GROUP	—

Revised by R.I. Hamilton

TYPE MEMBER	MAIZE CHLOROTIC DWARF VIRUS (MCDV) (194)	—

PROPERTIES OF THE VIRUS PARTICLE

Morphology Polyhedral particles ≈ 30 nm in diameter.

Physicochemical properties MW ≈ 8.8 x 10^6; S_{20w} ≈ 183; density in CsCl ≈ 1.51 g/cm^3.

Nucleic acid One molecule of positive-sense ssRNA, MW = 3.2 x 10^6.

Protein Three proteins, MW ≈ 34, 25 and 22.5 x 10^3 have been isolated from purified virus.

Lipid None reported.

Carbohydrate None reported.

Antigenic properties Efficient immunogens.

REPLICATION

No subgenomic RNAs found. Probably translated as polyprotein that is cleaved to produce functional proteins.

BIOLOGICAL ASPECTS

Host range Narrow, limited to members of Gramineae.

Transmission Only by leafhoppers in a semi-persistent manner. A virus-encoded helper component is required for transmission. The principal vector is *Graminella nigrifons*.

OTHER MEMBERS

Possible members

Rice tungro spherical (67)
Anthriscus yellows

References

Choudhury, M.M.; Rosenkranz, E.: Vector relationship of *Graminella nigrifons* to maize chlorotic dwarf virus. Phytopathology *73*:685-690 (1983).

Gingery, R.E.: Maize chlorotic dwarf and related viruses; *In* Koenig, R. (ed.), The plant viruses, Polyhedral virions with monopartite RNA genomes, Vol. 3, pp. 259-272 (Plenum Press, New York, 1988).

Hemida, S.K.; Murant, A.F.; Duncan, G.H.: Purification and some particle properties of anthriscus yellows virus, a phloem-limited semi-persistent aphid-borne virus. Ann. appl. Biol. *114:* 71-86 (1989).

Hunt, R.E.; Nault, L.R.; Gingery, R.E.: Evidence for infectivity of maize chlorotic dwarf virus and for a helper component in its leafhopper transmission. Phytopathology *78:* 499-504 (1988).

Taxonomic status	English vernacular name	International name
GROUP	MAIZE RAYADO FINO VIRUS GROUP	***MARAFIVIRUS***

Revised by K. Tomaru

TYPE MEMBER	MAIZE RAYADO FINO VIRUS (MRFV) (200)	—

PROPERTIES OF THE VIRUS PARTICLE

Morphology Polyhedral particles ≈ 31 nm diameter.

Physicochemical properties Two major classes of particles (B and T); S_{20w} ≈ 120 and 54, respectively; buoyant densities in CsCl ≈ 1.46 and 1.28.

Nucleic acid One molecule of linear ssRNA with MW = 2.0-2.4 x 10^6, accounting for 25-30% of the B particle weight.

Protein One major protein (MW ≈ 22 x 10^3) and a minor one (MW ≈ 28 x 10^3) reported for different isolates of MRFV; both proteins contain common peptide sequences. Single protein (MW ≈ 27 x 10^3) detected in bermuda grass etched-line virus.

Lipid None reported.

Carbohydrate None reported.

Antigenic properties Moderately immunogenic. Serological relationships among members.

REPLICATION

Virions are found in vacuoles of collenchyma and phloem parenchyma cells but also occur in the cytoplasm either dispersed, in single rows in long tubules, or in crystal-like arrays. Viral RNA is translated *in vitro* in rabbit reticulocyte lysates to yield polypeptides of MW ranging from 15-165 x 10^3. No polypeptides with electrophoretic or serological properties of coat protein are obtained by *in vitro* translation.

BIOLOGICAL ASPECTS

Host range Individual members may have broad host ranges; hosts are restricted to the *Gramineae.*

Taxonomic status	English vernacular name	International name
Transmission	Transmitted naturally by leafhoppers; manual transmission is difficult. Replication of MRFV in its vector is suggested by serial passage experiments.	

MEMBERS

Bermuda grass etched-line
Oat blue dwarf (123)

Derivation of Name marafi: Sigla from *ma*ize *ra*yado *fi*no

REFERENCES

Banttari, E.E.; Moore, M.B: Virus cause of blue dwarf of oats and its transmission to barley and flax. Phytopathology *52:*897-902 (1962).

Banttari, E.E.; Zeyen, R.J.: Chromatographic purification of the oat blue dwarf virus. Phytopathology *59:*183-186 (1969).

Espinoza, A.M.; Ramirez, P.; Leon, P.: Cell-free translation of maize rayado fino virus genomic RNA. J. gen. Virol. (in press).

Falk, B.W.; Tsai, J.H.: The two capsid proteins of maize rayado fino virus contain common peptide sequences. Intervirology *25:*111-116 (1986).

Gamez, R.: Transmission of rayado fino virus of maize (*Zea mays*) by *Dalbulus maidis* . Ann. appl. Biol. *73:*285-292 (1973).

Gingery, R.E.; Gordon, D.T.; Nault, L.R.: Purification and properties of an isolate of maize rayado fino virus from the United States. Phytopathology *72:*1313-1318 (1982).

Leon, P.; Gamez, R.: Biologia molecular del virus del rayado fino del maiz. Rev. Biol. Trop. *34:*111-114 (1986).

Leon, P.; Gamez, R.: Some physicochemical properties of maize rayado fino virus. J. gen. Virol. *56:*67-75 (1981).

Lockhart, B.E.L.; Khaless, N.; Lennon, A.M.; El Maatauoi, M.: Properties of Bermuda grass etched-line virus, a new leafhopper-transmitted virus related to maize rayado fino and oat blue dwarf viruses. Phytopathology *75:*1258-1262 (1985).

Pring, D.R.; Zeyen, R.J.; Banttari, E.E.: Isolation and characterization of oat blue dwarf virus ribonucleic acid. Phytopathology *63:*393-396 (1973).

Rivera, C.; Gamez, R.: Multiplication of maize rayado fino virus in the leafhopper vector *Dalbulus maidis*. Intervirology *25:*76-82 (1986).

Taxonomic status	English vernacular name	International name
GROUP	TOBACCO NECROSIS VIRUS GROUP	*NECROVIRUS*

Revised by J.W. Randles

TYPE MEMBER	TOBACCO NECROSIS VIRUS (TNV)(A STRAIN) (14)	—

PROPERTIES OF THE VIRUS PARTICLE

Morphology Polyhedral particles ≈ 28 nm in diameter consisting of 180 protein subunits arranged in a T = 3 icosahedral lattice. Isolates may be associated with a satellite virus (satellite TNV, 17 nm) which depends on TNV for replication of its RNA (1239 nt) but which codes for its own coat protein (195 residues).

Physicochemical properties MW ≈ 7.6 x 10^6; S_{20w} ≈ 118; buoyant density in CsCl ≈ 1.40 g/cm^3.

Nucleic acid One molecule of linear positive-sense ssRNA, MW = 1.3-1.6 x 10^6; 5'- terminus has the sequence ppApGpUp...

Protein Single polypeptide, MW = 22.6-33.3 x 10^3.

Lipid None reported.

Carbohydrate None reported.

Antigenic properties Moderately immunogenic. Single precipitin line in gel diffusion tests.

REPLICATION

A virus-induced RNA-dependent polymerase occurs in infected plants. Three dsRNAs have been detected in infected tissue. One (MW ≈ 2.6 x 10^6) appears to be the replicative form (RF) for genomic RNA; the others (1.05 and 0.94 x 10^6) may be RFs of subgenomic RNAs. Crystal-like aggregates of virus particles often seen in cytoplasm of infected cells.

BIOLOGICAL ASPECTS

Host range Wide among angiosperms: usually restricted to roots in natural infections.

Taxonomic status	English vernacular name	International name
Transmission	Transmitted naturally by the chytrid fungus *Olpidium brassicae*, and experimentally by mechanical inoculation of sap.	

OTHER MEMBERS

Chenopodium necrosis

Possible members

Carnation yellow stripe
Lisianthus necrosis

Derivation of Name necro: from Greek *nekros*, "dead body".

REFERENCES

Condit, C.; Fraenkel-Conrat, H.: Isolation of replicative forms of 3' terminal subgenomic RNAs of tobacco necrosis virus. Virology *97:*122-130 (1979).

Francki, R.I.B.; Milne, R.G.; Hatta, T.: Tobacco necrosis virus group, *In* Atlas of Plant Viruses, Vol. I, pp 171-180 (CRC Press, Boca Raton, Fl., 1985).

Gallitelli, D.; Castellano, M.A.; Di Franco, A.; Rana, G.I.L.: Properties of carnation yellow stripe virus, a member of the tobacco necrosis virus group. Phytopathol. Medit. *18:*31-40 (1979).

Gama, M.I.C.S.; Kitajima, E.W.; Lin, M.T.: Properties of a tobacco necrosis virus isolate from *Pogostemum patchuli* in Brazil. Phytopathology *72:*529-532 (1982).

Iwaki, M.; Hanada, K.; Maria, E.R.A.; Onogi, S.: Lisianthus necrosis virus, a new necrovirus from *Eustoma russellianum.* Phytopathology *77:* 867-870 (1987).

Lesnaw, J.A.; Reichmann, M.E.: The structure of tobacco necrosis virus. I. The protein subunit and the nature of the nucleic acid. Virology *39:*729-737 (1969).

Stussi-Garaud, C.; Lemius, J.: Fraenkel-Conrat, H.: RNA polymerase from tobacco necrosis virus-infected and uninfected tobacco. Virology *81:*224-236 (1977).

Tomlinson, J.A.; Faithfull, E.M.; Webb, M.J.W.; Fraser, R.S.S.; Seeley, N.D.: *Chenopodium* necrosis: a distinctive strain of tobacco necrosis virus isolated from river water. Ann. appl. Biol. *102:* 135-147 (1983).

Uyemoto, J.K.: Tobacco necrosis and satellite viruses; *In* Kurstak, E. (ed.), Handbook of Plant Virus Infections and Comparative Diagnosis, pp. 123-146. (Elsevier/North Holland, Amsterdam, 1981).

Taxonomic status	English vernacular name	International name
GROUP	PARSNIP YELLOW FLECK VIRUS GROUP	—

Compiled by A.F. Murant

TYPE MEMBER	PARSNIP YELLOW FLECK VIRUS (PYFV) (PARSNIP STRAIN) (129)	—

PROPERTIES OF THE VIRUS PARTICLE

Morphology Isometric particles ≈ 30 nm in diameter. Particles of T component are penetrated by negative stain.

Physicochemical properties Particles sediment as two components, T and B, respectively containing ≈ 0 and 42% RNA and with S_{20w} of 60 and 153; buoyant density in CsCl ≈ 1.3 (T) and 1.5 (B).

Nucleic acid One molecule of infective positive-sense linear ss-RNA of MW ≈ 3.5 x 10^6. The molecule is polyadenylated at the 3'-, and a Vpg at the 5'-end.

Protein Three major polypeptides, MW ≈ 31, 26 and 23 x 10^3.

Lipid None reported.

Carbohydrate None reported.

Antigenic properties Efficient immunogens.

REPLICATION

Large inclusion bodies occur in infected cells adjacent to the nucleus. They contain vesicular structures, granular bodies, amorphous matrix material and straight tubules ≈ 30 nm in diameter; mitochondria occur around the periphery. Only the amorphous matrix material is labelled with gold conjugate to PYFV antibody. Virus particles in the cytoplasm often occur within tubules ≈ 45 nm in diameter, which may pass through plasmodesmata.

BIOLOGICAL ASPECTS

Host range Natural host range restricted. Experimental host range moderate to narrow. Symptoms are mottles and mosaics; in some species, wilting and necrosis.

Taxonomic status	English vernacular name	International name
Transmission	Transmitted by aphids in a semi-persistent manner but only in association with a helper virus. No evidence for multiplication of virus in the vector. No seed-transmission reported. Transmissible experimentally by mechanical inoculation.	

OTHER MEMBERS

Parsnip yellow fleck, *Anthriscus* strain (129)

Possible member

Dandelion yellow mosaic

REFERENCES

Bos, L.; Huijberts, N.; Huttinga, H.; Maat, D.Z.: Further characterization of dandelion yellow mosaic virus from lettuce and dandelion. Neth. J. Pl. Path. *89:*207-222 (1983).

Elnagar, S.; Murant, A.F.: Relations of the semi-persistent viruses, parsnip yellow fleck and anthriscus yellows, with their vector, *Cavariella aegopodii*. Ann. appl. Biol. *84:*153-167 (1976).

Elnagar, S.; Murant, A.F.: The role of the helper virus, anthriscus yellows, in the transmission of parsnip yellow fleck virus by the aphid *Cavariella aegopodii*. Ann. appl. Biol. *84:*169-181 (1976).

Fasseas, C.; Roberts, I.M.; Murant, A.F.: Immunogold localization of parsnip yellow fleck virus particle antigen in thin sections of plant cells. J. gen. Virol. *70:* 2741-2749 (1989).

Hemida, S.K.; Murant, A.F.: Particle properties of parsnip yellow fleck virus. Ann. appl. Biol. *114:* 87-100 (1989).

Hemida, S.K.; Murant, A.F.: Host ranges and serological properties of eight isolates of parsnip yellow fleck virus belonging to the two major serotypes. Ann. appl. Biol. *114:* 101-109 (1989).

Murant, A.F.: Parsnip yellow fleck virus, type member of a proposed new plant virus group, and a possible second member, dandelion yellow mosaic virus; *In* Koenig, R. (ed.), The plant viruses. Polyhedral virions with monopartite RNA genomes, Vol. 3, pp. 273-288 (Plenum Press, New York, London, 1988).

Murant, A.F.; Goold, R.A.: Purification, properties and transmission of parsnip yellow fleck, a semi-persistent, aphid-borne virus. Ann. appl. Biol. *62:*123-137 (1968).

Murant, A.F.; Roberts, I.M.; Elnagar, S.: Association of virus-like particles with the foregut of the aphid *Cavariella aegopodii* transmitting the semi-persistent viruses anthriscus yellows and parsnip yellow fleck. J. gen. Virol. *31:*47-57 (1976).

Murant, A.F.; Roberts, I.M.; Hutcheson, A.M.: Effects of parsnip yellow fleck virus on plant cells. J. gen. Virol. *26:*277-285 (1975).

Taxonomic status	English vernacular name	International name
FAMILY	PICORNAVIRUS GROUP	***PICORNAVIRIDAE***

Reported by P. Minor

PROPERTIES OF THE VIRUS PARTICLE

Morphology

Virions are icosahedra (T = 1) with no envelope; the core consists of RNA and a small protein 3B VPg covalently linked to its 5'-end. Electron micrographs (EM) reveal no projections, the surface being almost featureless. Hydrated native particles are 30 nm in diameter but vary from 22-30 nm by EM due to drying and flattening during preparation. Sometimes form long ribonucleoprotein strands upon heating at slightly alkaline pH.

Physicochemical properties

MW = 8-9 x 10^6; S_{20w} = 140-165; buoyant density in CsCl = 1.33-1.45 g/cm^3 depending mainly on genus. Some species are unstable below pH 6; many are less stable at low ionic strength than at high. Insensitive to ether, chloroform or non-ionic detergents. Inactivated by light when grown with, or in the presence of, dyes such as neutral red and proflavin.

Nucleic acid

One molecule of infectious positive-sense ssRNA, MW = 2.4-2.7 x 10^6. A poly A tract, heterogeneous in length, is transcribed onto the 3'-terminus. A protein, VPg (MW ≈ 2,400), is linked covalently to the 5'-terminus.

Protein

Capsid of 60 protein subunits (protomers), each consisting of four polypeptides (three of MW = 24-41 x 10^3 and one of MW = 5.5-13.5 x 10^3) derived by cleavage of a single polyprotein. Protomers vary from 80 kDa for aphthovirus to 97 for poliovirus and some may be incompletely cleaved. The atomic structures of representatives of four of the picornavirus genera have been solved and are very similar to each other and to the icosahedral plant viruses.

Lipid

None. Some strains of poliovirus may carry 60 molecules each of a sphingosine-like molecule. The inner capsid polypeptide 1A (VP4) has a molecule of myristic acid covalently attached to the amino terminal end.

Carbohydrate

None.

Antigenic properties

Native virions are antigenically specific (designated "N" or "D"), but after gentle heating are converted to group specificity (designated "H"). Neutralization by antibody follows first-order inactivation kinetics. Species (equivalent to serotypes) are classified by neutralization of infectivity, complement-fixation or immunodiffusion; some species can be identified by hemagglutination. Neutralization epitopes defined by resistance mutations to monoclonal antibodies typically number 3 or 4 per protomer.

REPLICATION

Replication of viral RNA occurs in complexes associated with cytoplasmic membranes apparently via two distinct RIs. One complex uses positive- strand RNA and the other uses negative-strand RNA as template. Functional proteins are mainly produced from a single large (MW = 240-250 x 10^3) polyprotein by post-translational cleavage. The precursor protein is cleaved during translation and thus not normally detectable. Coat protein is encoded by the 5' half; VPg, proteases and polymerases or polymerase factors are encoded downstream. Many compounds specifically inhibiting replication have been described. Mutants resistant to and dependent on mutants drugs are often easily obtained. Genetic recombination, complementation and phenotypic mixing occur. DI particles have been produced experimentally but are probably not very important in nature because they appear only under extreme selection pressure.

BIOLOGICAL ASPECTS

Host range

Nature: Most species are host-specific. Exceptions include coxsackie B5 virus, EMC virus and aphthoviruses; serologic tests suggest they pass occasionally between man and domestic (cloven-footed) animals.

Laboratory: Most species can be grown in cell cultures. Resistant host cells can often be infected (single round) by transfection with naked infective RNA. Rhinoviruses and many enteroviruses grow poorly or not at all in laboratory animals.

Transmission

Horizontal, mainly mechanically.

GENERA

Enterovirus group	*Enterovirus*
Hepatitis A virus group	*Hepatovirus*
EMC virus group	*Cardiovirus*
Common cold virus group	*Rhinovirus*
Foot-and-mouth disease virus group	*Aphthovirus*

Taxonomic status	English vernacular name	International name
GENUS	ENTEROVIRUS GROUP	***ENTEROVIRUS***
TYPE SPECIES	HUMAN POLIOVIRUS 1	—

PROPERTIES OF THE VIRUS PARTICLE

Stable at acid pH; buoyant density in CsCl = 1.30-1.34 g/cm^3; empty shells often observed with virus; very small amounts (1%) of high density particles (1.43) sometimes observed. Primarily viruses of gastrointestinal tract, but also multiply in other tissues such as nerve, muscle, etc.

BIOLOGICAL ASPECTS

Infection may frequently be asymptomatic. Clinical manifestations may include mild gastrointestinal symptoms, meningitis, paralysis, cardiac symptoms, conjunctivitis and hand, foot and mouth disease.

OTHER MEMBERS

Human polioviruses 2-3
Human coxsackieviruses A1-22, 24 (A23 = echovirus 9) 1
Human coxsackieviruses B1-6 (swine vesicular disease virus is very similar to coxsackievirus B5)
Human echoviruses 1-9, 11-27, 29-34
Human enteroviruses 68-71
Vilyuisk virus
Murine poliovirus (Theiler's encephalomyelitis virus, TO, FA, GD7)
Simian enteroviruses 1-18
Porcine enteroviruses 1-8
Bovine enteroviruses 1-2

GENUS	HEPATITIS A VIRUS GROUP	***HEPATOVIRUS***
TYPE SPECIES	HUMAN HEPATITIS A VIRUS (STRAIN HM 175)	—

PROPERTIES OF THE VIRUS PARTICLE

Very stable, resistant to acid pH and elevated temperature (60°C for 10 min). Buoyant density in CsCl = 1.32-1.34 g/cm^3. Primarily viruses of liver, found in faeces at high titre shortly before clinical signs of hepatitis develop. Strongly conserved antigenic properties and tendency to establish persistent virus infections *in vitro*. VP4 is small. Genomic sequences show no detectable similarity with entero or rhinoviruses.

Taxonomic status	English vernacular name	International name

BIOLOGICAL ASPECTS

Clinical manifestations are hepatitis.

OTHER MEMBERS

Simian hepatitis A virus.

GENUS	EMC VIRUS GROUP	*CARDIOVIRUS*
TYPE SPECIES	ENCEPHALOMYOCARDITIS (EMC) VIRUS	—

PROPERTIES OF THE VIRUS PARTICLE

Unstable at pH 5-6 in presence of 0.1 M halide; buoyant density in CsCl = 1.33-1.34 g/cm^3; single serotype. Poly(C) tract of variable length (80-250 bases) about 150 bases from 5' terminus of RNA. Empty shells seen rarely, if ever.

BIOLOGICAL ASPECTS

Clinical manifestations include encephalitis and myocarditis in mice.

OTHER MEMBERS

Mengovirus
Murine encephalomyelitis (ME) virus
Columbia SK virus
MM virus

GENUS	COMMON COLD VIRUS GROUP	*RHINOVIRUS*
TYPE SPECIES	HUMAN RHINOVIRUS 1A	—

PROPERTIES OF THE VIRUS PARTICLE

Unstable below pH 5-6; buoyant density in CsCl = 1.38-1.42 g/cm^3.

BIOLOGICAL ASPECTS

Clinical manifestations include the common cold in humans.

OTHER MEMBERS

Human rhinoviruses 1B-100
Bovine rhinoviruses 1 and 2

Taxonomic status	English vernacular name	International name
GENUS	FOOT-AND-MOUTH DISEASE VIRUS GROUP	***APHTHOVIRUS***
TYPE SPECIES	APHTHOVIRUS O	—

PROPERTIES OF THE VIRUS PARTICLE

Unstable below pH 5-6; buoyant density in CsCl = 1.43-1.45 g/cm^3; clinical manifestations. Poly(C) tract of variable length (100-170 bases), about 400 bases from 5' terminus of RNA. The genome encodes 3 species of VPg.

BIOLOGICAL ASPECTS

Clinical manifestations include foot and mouth disease of cloven hoofed animals.

OTHER MEMBERS

A
C
SAT1
SAT2
SAT3
Asia 1

OTHER MEMBERS OF FAMILY *PICORNAVIRIDAE* NOT YET ASSIGNED TO GENERA

Equine rhinoviruses types 1 and 2
Cricket paralysis virus
Drosophila C virus
Gonometa virus

UNCLASSIFIED SMALL RNA VIRUSES

About 30 small RNA viruses of unknown affinities have been described. These include: bee acute paralysis, bee slow paralysis, bee virus X, *Drosophila* P and A, sacbrood, Queensland fruitfly virus and *Triatoma* virus and aphid lethal paralysis virus. Parsnip yellow fleck virus, the type member of the parsnip fleck virus group, has many properties in common with picornaviruses.

Derivation of Name

picorna: from the prefix 'pico' (= 'micro-micro') and RNA (= the sigla for ribonucleic acid).
entero: from Greek *enteron*, 'intestine'.
rhino: from Greek *rhis, rhinos*, 'nose'.
cardio: from Greek *kardia*, 'heart'.
aphtho: from Greek *aphtha*, 'vesicles in the mouth'; English aphtho, 'thrush'; French *fievre aphteuse*.

REFERENCES

Acharya, R.; Fry, E.; Stuart, D.; Fox, G.; Rowlands, D.; Brown, F.: The three dimensional structure of foot-and-mouth disease virus at 2.9Å resolution. Nature *337:* 709-716 (1989).

Adair, B.M.; Kennedy, S.; McKillop, E.R.; McNulty, M.S.; McFerran, J.B.: Bovine, porcine and ovine picornaviruses: identification of viruses with properties similar to human coxsackieviruses. Arch. Virol. *97:*49-60 (1987).

Bashiruddin, J.B.; Martin, J.L.; Reinganum, C.: Queensland fruit fly virus, a probable member of the *Picornaviridae*. Arch. Virol. *100:*61-74 (1988).

Cooper, P.D.; Agol, V.I.; Bachrach, H.L.; Brown, F.; Ghendon, Y.; Gibbs, A.J.; Gillespie, J.H.; Lonberg-Holm, K.; Mandel, B.; Melnick, J.L.; Mohanty, S.B.; Povey, R.C.; Rueckert, R.R.; Schaffer, F.L.; Tyrrell, D.A.J.: *Picornaviridae*: Second Report. Intervirology *10:*165- 180 (1978).

Gust, I.D.; Feinstone, S.M.: In Hepatitis A, p. 56 (CRC Press, Boca Raton, Fl., 1988).

Hamparian, V.V.; et al.: A collaborative report: Rhinoviruses - Extension of the numbering system from 89 to 100. Virology *159:* 191-192 (1987).

Knowles, N.J.; Barnett, I.T.R.: A serological classification of bovine enteroviruses. Arch. Virol. *83:*141-155 (1985).

Lipton, H.; Friedmann, A.; Sethi, P.; Crowther, R.R.: Characterization of Vilyuisk virus as a picornavirus. J. Med. Virol. *12:*195-204 (1983).

Melnick, J.L.; Agol, V.I.; Bachrach, H.L.; Brown, F.; Cooper, P.D.; Fiers, W.; Gard, S.; Gear, J.H.S.; Ghendon, Y.; Kasza, L.; LaPlaca, M.; Mandel, B.; McGregor, S.; Mohanty, S.B.; Plummer, G.; Rueckert, R.R.; Schaffer, F.L.; Tagaya, I.; Tyrrell, D.A.J.; Voroshilova, M.; Wenner, H.A: *Picornaviridae*. Intervirology *4:*303-316 (1974).

Moore, N.F.; King, L.A.; Pullin, J.S.K.: Insect Picornaviruses. *In* Rowlands, D.; Mayo, M.A.; Mahy, B. (eds.), The Molecular Biology of the Positive Strand RNA Viruses, pp. 67-74 (Academic Press, New York, 1987).

Murant, A.F.: Parsnip yellow fleck virus, type member of a proposed new plant virus group, and a possible second member, Dandelion Yellow Mosaic Virus, *In* Koenig, R. (ed.), The Plant Viruses, Polyhedral Virions with Monopartite RNA Genomes, Vol. 3, pp. 273-288 (Plenum Press, New York, 1988).

Muscio, O.A.; La Torre, J.L.; Scodeller, E.A.: Characterization of *Triatoma* virus, a picorna-like virus isolated from the triatomine bug *Triatoma infestans*. J. gen. Virol. *69:*2929-2934 (1988).

Ozden, S.; Tangy, F.; Chamorro, M.; Brahic, M.: Theiler's virus genome is closely related to that of Encephalomyocarditis virus, the prototype cardiovirus. J. Virol. *60:*1163-1165 (1986).

Palmenberg, A.: C. Sequence alignments of picornaviral capsid proteins. *In* Semler, B.L.; Ehrenfeld, E. (eds.), Molecular Aspects of picornavirus Infection and Detection, pp. 211-241 (ASM Press, Washington DC, 1989).

Rueckert, R.R.: Picornaviruses and their multiplication. *In* Fields: Virology, pp. in press (Raven Press, New York, 1989).

Tinsley, T.W.; MacCallum, F.O.; Robertson, J.S.; Brown, F.: Relationship of encephalomyocarditis virus to cricket paralysis virus of insects. Intervirology *21*:181-186 (1984).

Williamson, C.; Rybicki, E.P.; Kasdorf, G.G.F.; von Wechmar, M.B.: Characterization of a new picorna-like virus isolated from aphids. J. gen. Virol. *69*:787-795 (1988).

Yamashita, H.; Akashi, H.; Inaba, Y.: Isolation of a new serotype of bovine rhinovirus from cattle. Arch. Virol. *83*:1138-116 (1985).

Taxonomic status	English vernacular name	International name
GROUP	SOUTHERN BEAN MOSAIC VIRUS GROUP	***SOBEMOVIRUS***

Revised by E.P. Rybicki

TYPE MEMBER	SOUTHERN BEAN MOSAIC VIRUS (SBMV) (57;274)	—

PROPERTIES OF THE VIRUS PARTICLE

Morphology Particles ≈ 30 nm diameter with 180 subunits in a T = 3 icosahedral structure stabilized by divalent cations. Each protein subunit has two domains. One forms parts of the icosahedral shell about 3.5 nm thick and the other forms a partially ordered 'arm' in the interior of the virus.

Physicochemical properties MW ≈ 6.6×10^6; S_{20w} ≈ 115; density ≈ 1.36 g/cm^3 in CsCl (but virus forms two or more bands in Cs_2SO_4); particles swell reversibly in EDTA and/or pH increase with concomitant changes in capsid conformation and partial loss of stability.

Nucleic acid One molecule of positive-sense ssRNA MW = 1.4×10^6 (≈ 4.2 kb); Vpg, essential for infectivity of RNA is associated with 5'-end; 3'-end does not contain poly(A) or a tRNA-like structure. A subgenomic, 3'-coterminal RNA (MW ≈ 0.38×10^6) is also found in SBMV particles. Satellite viroid-like RNAs are associated with some members.

Protein One coat polypeptide with MW ≈ 30×10^3.

Lipid None.

Carbohydrate None.

Antigenic properties Efficient immunogens. Single precipitin line in gel diffusion tests. Serological relationships between strains and some members of the group.

REPLICATION

Genomic RNA remains associated with swollen virions during cell-free translation in wheat germ extract. Genome sequencing of SBMV shows four possible ORFs, with coding capacity for proteins of MW ≈ 21×10^3 (ORF 1, 49-603), 105×10^3 (ORF 2, 570-3437), 18×10^3 (ORF 3, 1895-2380) and 31×10^3 (ORF 4, 3217-4053). *In vitro*

Taxonomic status	English vernacular name	International name

translation of full-length SBMV genomic RNA in wheat germ, or of turnip rosette virus RNA in rabbit reticulocyte lysate, yields three proteins (P1, 105 x 10^3; P2, 60 x 10^3; P4, 14-25 x 10^3); however, coat protein (P3, 28 x 10^3) is only translated from 0.3-0.4 x 10^6 virion-associated RNA 2, indicating that this is a subgenomic mRNA. It is suggested that ORF 1 encodes P4(s); ORF 2 encodes P1; P2 is derived by proteolysis from P1; ORF 4 encodes P3. No protein or subgenomic mRNA has been associated with ORF 3. Genome homologies suggest mechanisms of expression of other proteins and of replication are similar to picorna- and potyviruses. Virions are found in both nuclei and cytoplasm; sometimes in crystalline arrays in the latter. The viruses do not appear to be tissue-specific.

Biological Aspects

Host range Each virus has relatively narrow host range.

Transmission Seed transmission in several host plants. Transmitted by beetles or a myrid in the case of velvet tobacco mottle virus. Readily transmitted mechanically.

Sequence Similarities with Other Virus Groups or Families

The predicted amino acid sequence from ORF 2 contains motifs with significant homology to (in order, NH_2-end to COOH-end): the putative ATP-binding domain of picorna- and Sindbis-like viruses; the VPg of picornaviruses; the cysteine protease of picornaviruses; the putative + strand RNA virus polymerase domain. This is similar to the core organisation of picorna-, poty-, como- and nepoviruses and puts the sobemoviruses in the picorna-like virus "superfamily". Other regions of the genome show no similarity to other viruses; this, together with the unique genome organisation, indicates that these viruses should be a distinct taxonomic group.

Other Members

Blueberry shoestring (204)
Cocksfoot mottle (23)
Lucerne transient streak (224)
Rice yellow mottle (149)
Solanum nodiflorum mottle (318)
Sowbane mosaic (64)
Subterranean clover mottle (329)
Turnip rosette virus (125)
Velvet tobacco mottle (317)

Taxonomic status	English vernacular name	International name
	Possible members	
	Cocksfoot mild mosaic	
	Cynosurus mottle	
	Ginger chlorotic fleck (328)	
	Maize chlorotic mottle (284)	
	Olive latent virus-1	
	Panicum mosaic (177)	

Derivation of Name

sobemo: sigla derived from the name of type member *so*uthern *be*an *mo*saic.

REFERENCES

Abad-Zapatero, C: Abdel-Meguid, S.S.; Johnson, J.E.; Leslie, A.G.W.; Rayment, I.; Rossmann, M.G.; Suck, D.; Tsukihara, T.: Structure of southern bean mosaic at 2.8 Å resolution. Nature, Lond. *286:*33-39 (1980).

Francki, R.I.B.; Milne, R.G.; Hatta, T.: *Sobemovirus* group, *In* Atlas of Plant Viruses, Vol. I, pp 153-169 (CRC Press, Boca Raton, Fl., 1985).

Francki, R.I.B.; Randles, J.W.; Hatta, T.; Davies, C.; Chu, P.W.G.; McLean, G.D.: Subterranean clover mottle virus: another virus from Australia with encapsidated viroid-like RNA. Plant Pathol. *32:*47-59 (1983).

Ghosh, A.; Dasgupta, R.; Salerno-Rife, T.; Rutgers, T.; Kaesberg, P.: Southern bean mosaic viral RNA has a 5'-linked protein but lacks 3' terminal poly (A). Nuc. Acids Res. *7:*2137-2146 (1979).

Goldbach, R.: Genome similarities between plant and animal RNA viruses. Microbiol. Sci. *4:* 197-202 (1987).

Gorbalenya, A.E.; Koonin, E.V.; Blinov, V.M.; Donchenko, A.P.: *Sobemovirus* genome appears to encode a serine protease related to cysteine proteases of picornaviruses. FEBS Lett. *236:* 287-290 (1988).

Hull, R.: The *sobemovirus* group. *In* Koenig, R. (ed.), The Plant Viruses, Polyhedral virions with monopartite genomes, Vol. 3, pp 113-146 (Plenum Press, New York, 1988).

Jones, A.T.; Mayo, M.A.: Satellite nature of the viroid-like RNA-2 of *Solanum nodiflorum* mottle virus and the ability of other plant viruses to support the replication of viroid-like RNA molecules. J. gen. Virol. *65:*1713-1721 (1984).

Rossman, M.G.; Abad-Zapatero, C.; Hermodson, M.A.; Erickson, J.W.: Subunit interactions in southern bean mosaic virus. J. Mol. Biol. *166:* 37-83 (1983).

Salerno-Rife, T.; Rutgers, T.; Kaesberg, P.: Translation of southern bean mosaic virus RNA in wheat embryo and rabbit reticulocyte extracts. J. Virol. *34:*51-58 (1980).

Wu, S.; Rinehart, C.A.; Kaesberg, P.: Sequence and organization of southern bean mosaic virus genomic RNA. Virology *161:* 73-80 (1987).

Taxonomic status	English vernacular name	International name
FAMILY	*NUDAURELIA* ß VIRUS GROUP	***TETRAVIRIDAE***

Revised by J.E. Johnson

GENUS	—	—
TYPE SPECIES	*NUDAURELIA* ß VIRUS (ISOLATED FROM *NUDAURELIA CYTHEREA CAPENSIS*)	—

Properties of the Virus Particle

Morphology Virions are icosahedra (probably T = 4).

Physicochemical properties MW = 16.3 x 10^6; S_{20w} = 194-210; buoyant density in CsCl = 1.29 g/cm^3. Stable at pH 3.0.

Nucleic acid One molecule of positive-sense ssRNA. MW ≈ 1.8 x 10^6; 10-11% of particle by weight. RNA is not polyadenylated.

Protein One major polypeptide of MW = 60-70 x 10^3. There are small differences in MW with different isolates.

Lipid Not determined; probably none.

Carbohydrate None detectable.

Antigenic properties Most of the members of the group are serologically interrelated but distinguishable. The majority of the isolates were identified on the basis of their serological reaction with antiserum raised against *Nudaurelia* ß virus.

Replication

The viruses replicate primarily in the cytoplasm of gut cells of several Lepidoptera. Crystalline arrays of virus particles are often seen within cytoplasmic vesicles.

Biological Aspects

Host range Natural - All species were isolated from Lepidoptera, principally from Saturniid, Limacodid and Noctuid moths. There is a considerable range of pathogenicity with different isolates. Effects of infections range from rapid death to growth retardation of larval stages. Artificial - No infections have yet been achieved in cultured invertebrate cells.

OTHER MEMBERS

Probable members

Isolated from:

Antheraea eucalypti
Darna trima
Thosea asiona
Philosamia ricini
Trichoplusia ni
Dasychira pudibunda

Possible members

Isolated from:

Saturnia pavonia
Acherontia atropas
Setora nitens
Eucocytis meeki
Hypocrita jacobeae
Agraulis vanillae
Lymantria ninayi
Euploea corea

Derivation of Name tetra: from Greek *tettara* 'four' as T=4

REFERENCES

Finch, J.T.; Crowther, R.A.; Hendry, D.A.; Struthers, J.K.: The structure of *Nudaurelia capensis* β virus; the first example of a capsid with icosahedral surface symmetry T = 4. J. gen. Virol. *24*:191-200 (1974).

Greenwood, L.K.; Moore, N.F.: The *Nudaurelia* β group of small RNA-containing viruses of insects: serological identification of several new isolates. J. Invertebr. Pathol. *39*:407-409 (1982).

King, L.A.; Merryweather, A.T.; Moore, N.F.: Proteins expressed *in vitro* by two members of the *Nudaurelia* β family of viruses, *Trichoplusia ni* and *Dasychira pudibunda* viruses. Annales de Virologie (Inst. Pasteur) *135E*:335-342 (1984).

King, L.A.; Moore, N.F.: The RNAs of two viruses of the *Nudaurelia* β family share little homology and have no terminal poly (A) tracts. FEMS Microbiology Letts. *26*:41-43 (1985).

Morris, T.J.; Hess, R.T.; Pinnock, D.E.: Physicochemical characterisation of a small RNA virus associated with baculovirus infection in *Trichoplusia ni*. Intervirol. *11*:238-247 (1979).

Moore, N.F.; Reavy, B.; King, L.A.: General characteristics, gene organization and expression of small RNA viruses of insects. J. gen. Virol. *66*:647-659 (1985).

Reinganum, C.; Robertson, J.S.; Tinsley, T.W.: A new group of RNA viruses from insects. J. gen. Virol. *40*:195-202 (1978).

Struthers, J.K.; Hendry, D.A.: Studies of the protein and nucleic acid components of *Nudaurelia capensis* β virus. J. gen. Virol. *22*:355-362 (1974).

Taxonomic status	English vernacular name	International name
GROUP	TOMATO BUSHY STUNT VIRUS GROUP (352)	*TOMBUSVIRUS*

Revised by G.P. Martelli

TYPE MEMBER	TOMATO BUSHY STUNT VIRUS (TBSV) (69)	—

PROPERTIES OF THE VIRUS PARTICLE

Morphology Isometric particles with rounded outline, ≈ 30 nm in diameter. 180 protein subunits are arranged in a T = 3 icosahedral lattice. In TBSV, each protein subunit folds into two distinct major globular domains (P, S), connected by a flexible hinge and a flexibly linked N-terminal arm. Each domain P forms one-half of a dimer-clustered surface protrusion. Domain S forms the icosahedral shell. The inward projecting N-terminal arms (domain R) may have an RNA-binding function. Virions also encapsidate 'satellite' and subgenomic RNAs.

Physicochemical properties MW ≈ 8.9 x 10^6; S_{20w} = 131-140; buoyant density in CsCl ≈ 1.35 g/cm^3.

Nucleic acid One molecule of linear positive-sense ssRNA, MW ≈ 1.5 x 10^6 (4701-4771 nt); 17% by weight of virus. Satellite ssRNA MW ≈ 0.21 x 10^6 (621 nt), defective interfering (DI) ssRNA MW = 0.14-0.24 x 10^6 (0.4-0.7 kb), and two subgenomic ssRNAs MW ≈ 0.7 x 10^6 (2.1 kb) and ≈ 0.3 x 10^6 (0.9 kb) respectively are also encapsidated. 3'-ends of genomic, DI and satellite RNAs do not contain poly (A) tracts; 5'-ends do not have a covalently bound VPg and are probably capped. Extensive sequence homology exists between members, in nucleotide and amino acids of both structural and putative non-structural proteins.

Protein One major coat polypeptide, MW ≈ 43 x 10^3.

Lipid None reported.

Carbohydrate None reported.

Antigenic properties Good immunogens. Single precipitin line in gel-diffusion tests. Serological relationship from close to very distant among members.

REPLICATION

Cytoplasmic, compact membranous inclusions ('multivesicular bodies') are induced by all members during early stages of infection. These bodies develop from modified peroxisomes, or more rarely, mitochondria, and contain dsRNA possibly representing RF or RI. Some members induce peripheral vesiculation of chloroplasts. Excess coat protein may accumulate in the cytoplasm in amorphous electron-dense aggregates. Virus particles are located in the cytoplasm, nuclei or with some members, in the mitochondria, sometimes in crystalline arrays. Cytoplasmic accumulations of virus particles often protrude into the vacuole.

A 4.7 kb genomic RNA and two 3'-coterminal RNA species of ≈ 2.1 and 0.9 kb have been identified both in infected tissues and virions. The genomic RNA has four ORFs. ORF1 encodes a protein MW ≈ 33 x 10^3 and terminates with an amber stop codon. Readthrough of this stop codon would produce a polypeptide MW ≈ 92 x 10^3 resulting from continuous reading of ORFs 1 and 2. ORF3 is translated via subgenomic RNA_1 (2.1 kb) into a polypeptide MW ≈ 41 x 10^3 (coat protein), and ORF4 via subgenomic RNA_2 (0.9 kb) into a polypeptide MW ≈ 22 x 10^3. An additional ORF nested into ORF4 codes for a polypeptide MW ≈ 19 x 10^3. The function of the 22 kDa and 19 kDa proteins is unknown, whereas the 92 kDa protein may be a part of the viral replicase. Three virus-specific dsRNAs corresponding to genomic and subgenomic RNAs have been detected in infected tissues. Satellite RNA has no detectable messenger activity and is present in linear monomers and dimers in single- and double-stranded forms. DI RNAs occur both as single- and double-stranded forms.

BIOLOGICAL ASPECTS

Host range Wide among angiosperms.

Transmission Readily transmitted by mechanical inoculation. Seed transmission is reported for some members. Acquisition through soil is possible. Cucumber necrosis virus is transmitted by the chytrid fungus *Olpidium radicale*.

SIMILARITIES WITH OTHER VIRUS GROUPS

Tombusviruses share with members of the carmovirus group, significant structural similarities in the capsid

Taxonomic status	English vernacular name	International name

protein with respect to polypeptide folding topology and subunit interactions. Physico-chemical properties are also similar but the genome organization is quite different.

OTHER MEMBERS

Artichoke mottle crinkle (69)
Carnation Italian ringspot (69)
Cucumber necrosis (82)
Cymbidium ringspot (178)
Eggplant mottled crinkle
Grapevine Algerian latent
Moroccan pepper
Lato river
Neckar river
Pelargonium leaf curl (69)
Petunia asteroid mosaic (69)

Derivation of Name tombus: sigla from *tom*ato *bu*shy *s*tunt.

REFERENCES

Burgyan, J.; Grieco, F.; Russo, M.: A defective interfering RNA molecule in cymbidium ringspot virus infections. J. gen. Virol. *70:* 235-239 (1989).

Burgyan, J.; Nagy, P.D.; Russo, M.: Synthesis of infectious RNA from full-length cloned cDNA to RNA of cymbidium ringspot tombusvirus. J. Gen. Virol. *71:* 1857-1860 (1990).

Burgyan, J.; Russo, M.; Gallitelli, D.: Translation of cymbidium ringspot virus RNA in cowpea protoplasts and rabbit reticulocyte lysates. J. gen. Virol. *67:*1149-1160 (1986).

Gallitelli, D.; Russo, M.: Some properties of Moroccan pepper virus and tombusvirus Neckar. J. Phytopathol. *119:*106-110 (1987).

Gallitelli, D.; Hull, R.: Characterization of satellite RNAs associated with tomato bushy stunt virus and five other definitive tombusviruses. J. gen. Virol. *66:*1533-1543 (1985).

Grieco, F.; Burgyan, J.; Russo, M.: The nucleotide sequence of cymbidium ringspot virus RNA. Nuc. Acids Res. *17:* 6383 (1989).

Hearne, P.Q.; Knorr, D.A.; Hillman, B.I.; Morris, T.J.: The complete genome structure and synthesis of infectious RNA from clones of tomato bushy stunt virus. Virology *177:* 141-151 (1990).

Hillman, B.I.; Hearne, P.; Rochon, D.; Morris, T.J.: Organization of tomato bushy stunt virus genome: Characterization of the coat protein gene and the 3' terminus. Virology *169:* 42-50 (1989).

Koenig, R.; Gibbs, A.: Serological relationships among tombusviruses. J. gen. Virol. *67:*75-82 (1986).

Makkouk, K.M.; Koenig, R.; Lesemann, D.-E.: Characterization of a tombusvirus isolated from eggplant. Phytopathology *71:*572-577 (1981).

Martelli, G.P.: Tombusviruses. *In* Kurstak, E. (ed.), Handbook of Plant Virus Infections and Comparative Diagnosis, pp. 61-90. (Elsevier/North Holland, Amsterdam, 1981).

Martelli, G.P.; Gallitelli, D.; Russo, M.: Tombusviruses. *In* Koenig, R. (ed.), The Plant Viruses, Polyhedral Virions With Monopartite RNA Genomes, Vol. 3, pp. 13-72. (Plenum Press, New York, 1988).

Olson, A.J.; Bricogne, G.; Harrison, S.C.: Structure of tomato bushy stunt virus. IV. The virus particle at 2.9 Å resolution. J. Mol. Biol. *171*:61-93 (1983).

Rochon, D.M.; Tremaine, J.H.: Cucumber necrosis virus is a member of the tombusvirus group. J. gen. Virol. *69*:395-400 (1988).

Rochon, D.M.; Tremaine, J.H.: Complete nucleotide sequence of the cucumber necrosis virus genome. Virology *169:* 251-259 (1989).

Rubino, L.; Burgyan, J.; Grieco, F.; Russo, M.: Sequence analysis of cymbidium ringspot virus satellite and defective interfering RNAs. J. gen. Virol. *71:* 1655-1660 (1990).

Russo, M.; Di Franco, A.; Martelli, G.P.: Cytopathology in the identification and classification of tombusviruses. Intervirology *28:* 134-143 (1990).

Russo, M.; Burgyan, J.; Carrington, J.C.; Hillman, B.I.; Morris, T.J.: Complementary DNA cloning and characterization of cymbidium ringspot virus RNA. J. gen. Virol. *69*:401-406 (1988).

Taxonomic status	English vernacular name	International name
GROUP	TURNIP YELLOW MOSAIC VIRUS GROUP (214)	***TYMOVIRUS***

Revised by R. Koenig

TYPE MEMBER	TURNIP YELLOW MOSAIC VIRUS (TYMV) (2; 230)	—

PROPERTIES OF THE VIRUS PARTICLE

Morphology

Particles are T = 3 icosahedral structures, ≈ 29 nm in diameter. They are stabilized by protein-protein interactions of the 180 subunits, which are clustered into 12 pentamers and 20 hexamers.

Physicochemical properties

Two major classes of stable particles (B and T) with MWs of 5.6 and 3.6 x 10^6; buoyant densities in CsCl ≈ 1.42 and 1.29 g/cm^3, and S_{20w} = 115 and 54, respectively. Only the B component containing the genome RNA is infectious. Partial specific volume = 0.661. The isoelectric point of TYMV is pH 3.75; those of other members cover a wide pH range. Virus is stable at neutral pH. Several minor nucleoproteins can be isolated with densities intermediate between those of the two major particle types. For TYMV, these contain subgenomic coat protein mRNA and less than full-length pieces of the genome RNA.

Nucleic acid

One molecule of linear positive-sense ssRNA containing 3 ORFs; MW ≈ 2 x 10^6, accounting for ≈ 35% of the weight of the B component. The 5' terminus of TYMV RNA has the sequence $m^7G^{5'}ppp^{5'}Gp$...; the 3' terminus has a tRNA-like structure which accepts valine. Small amounts of subgenomic coat protein mRNA (695 nt; MW ≈ 0.25 x 10^6) are found in several classes of virus particles. Both RNAs have a high content of cytidylic acid (32-41%). Particles of some members may also contain small amounts of transfer RNAs of plant origin. The RNAs of several tymoviruses have been sequenced.

Protein

One coat protein MW ≈ 20 x 10^3. 180 molecules per particle.

Lipid

None.

Carbohydrate

None.

Antigenic properties

Serological relationships between members of the group range from very close, to distant, to not detectable.

Taxonomic status	English vernacular name	International name

REPLICATION

Genomic RNA of TYMV is translated *in vitro* into 2 proteins of 150 and 195 x 10^3, the latter by read through of the 150 x 10^3 gene. A subgenomic RNA (695 nt) corresponding to the 3' region of the genomic RNA is translated *in vitro* into coat protein. Post-translational processing of the 195 x 10^3 protein *in vitro* has been reported. Tymoviruses induce at the periphery of the chloroplasts small flask-shaped double-membrane bounded vesicles which contain membrane-bound viral polymerase. They are probably the main site of production of viral positive-sense RNA. Pentamers and hexamers of the protein are probably produced in the cytoplasm, and virions assembled at the necks of the vesicles. Empty protein shells accumulate in nuclei. Most members cause clumping of chloroplasts in infected cells.

BIOLOGICAL ASPECTS

Host range Possibly restricted to dicotyledonous hosts. Individual viruses often have narrow host range.

Transmission By beetles and mechanical inoculation.

OTHER MEMBERS

Belladonna mottle (52)
Cacao yellow mosaic (11)
Clitoria yellow vein (171)
Desmodium yellow mottle (168)
Dulcamara mottle
Eggplant mosaic (124)
Erysimum latent(222)
Kennedya yellow mosaic (193)
Okra mosaic (128)
Ononis yellow mosaic
Passion fruit yellow mosaic
Peanut yellow mosaic
Physalis mosaic
Plantago mottle
Scrophularia mottle (113)
Voandzeia necrotic mosaic (279)
Wild cucumber mosaic (105)

Possible member

Poinsettia mosaic virus (311)

Derivation of Name tymo: sigla from *t*urnip *y*ellow *mo*saic virus

REFERENCES

Crestani, O.A.; Kitajima, E.W.; Lin, M.T.; Marinho, V.L.A.: Passion fruit yellow mosaic virus, a new tymovirus found in Brazil. Phytopathology *76:*951-955 (1986).

Fauquet, C.; Monsarrat, A.; Thouvenel, J.-C.: *Voandzeia* necrotic mosaic virus, a new tymovirus. 5th Int'l Congr. Virology (Strasbourg) Abstract. P23/03 p. 237 (1981).

Francki, R.I.B.; Milne, R.G.; Hatta, T.: Tymovirus group, *In* Atlas of Plant Viruses, Vol. I, pp 117-136 (CRC Press, Boca Raton, Fl., 1985).

Fulton, R.W.; Fulton, J.L.: Characterization of a tymo-like virus common in *poinsettia.* Phytopathology *70:*321-324 (1980).

Hirth, L.; Givord, L.: Tymoviruses, *In* Koenig, R. (ed.), The Plant Viruses, Polyhedral Plant Virions With Monopartite RNA Genomes, Vol. 3, pp. 163-212 (Plenum Press, New York, 1988).

Koenig, R.: A loop-structure in the serological classification system of tymoviruses. Virology *72:*1-5 (1976).

Koenig, R.; Lesemann, D.-E.: Tymoviruses; *In* Kurstak, E. (ed.), Handbook of Plant Virus Infections and Comparative Diagnosis, pp. 33-60. (Elsevier/North Holland, Amsterdam, 1981).

Lana, A.F.: Properties of a virus occurring in *Arachis hypogea* in Nigeria. Phytopath. Z. *97:*169-178 (1980).

Lesemann, D.-E.: Virus group-specific and virus-specific cytological alterations induced by members of the tymovirus group. Phytopath. Z. *90:*315-336 (1977).

Osorio-Keese, M.E.; Keese, P.; Gibbs, A.J.: Nucleotide sequence of the genome of eggplant mosaic tymovirus. Virology *172:* 547-554 (1989).

Szybiak, U.; Bouley, J.P.; Fritsch, C.: Evidence for the existence of a coat protein messenger RNA associated with the top component of each of three tymoviruses. Nuc. Acids Res. *5:*1821-1831 (1978).

Zagorski, W.; Morch, M.-D.; Haenni, A.-L.: Comparison of three different cell-free systems for turnip yellow mosaic virus RNA translation. Biochimie *65:*127-133 (1983).

Taxonomic status	English vernacular name	International name
GROUP	APPLE STEM GROOVING VIRUS GROUP	***CAPILLOVIRUS***

Revised by M. Bar-Joseph & G.P. Martelli

TYPE MEMBER	APPLE STEM GROOVING VIRUS (ASGV) (31)	—

PROPERTIES OF THE VIRUS PARTICLE

Morphology Flexuous filamentous particles ≈ 640 x 12 nm, with obvious cross-banding (helical symmetry).

Physicochemical properties S_{20w} ≈ 100S.

Nucleic acid One molecule of linear, plus sense ssRNA, MW ≈ 2.5 x 10^6; ≈ 5% by weight of virion. RNA is polyadenylated at 3'-end.

Protein Single polypeptide, MW ≈ 27 x 10^3.

Lipid None reported

Carbohydrate None reported

Antigenic properties Moderately immunogenic; serological relationship between members.

REPLICATION

Not studied.

BIOLOGICAL ASPECTS

Host range Natural host range restricted, experimental host range moderate.

Transmission Vector unknown. Transmitted through seed and by mechanical inoculation of sap.

OTHER MEMBERS

Potato virus T (187)

Possible members

Lilac chlorotic leaf spot (202)
Nandina stem pitting

Derivation of Name capillo: from Latin *capillus*, a hair

REFERENCES

Ahmed, N.A.; Christie, S.R.; Zettler, F.W.: Identification and partial characterization of a closterovirus infecting *Nandina domestica*.. Phytopathology *73*:470-475 (1983).

Conti, M.; Milne, R.G.; Luisoni, E.; Boccardo, G.: A closterovirus from a stem-pitting-diseased grapevine. Phytopathology *70*:394-399 (1980).

de Sequeira, O.A.; Lister, R.M.: Purification and relationships of some filamentous viruses from apple. Phytopathology *59*:1740-1749 (1969).

Salazar, L.F.; Harrison, B.D.: Host range, purification and properties of potato virus T. Ann. appl. Biol. *89*:223-235 (1978).

Salazar, L.F.; Hutcheson, A.M.; Tollin, P.; Wilson, H.R.: Optical diffraction studies of particles of potato virus T. J. gen. Virol. *39*:333-342 (1978).

Yoshikawa, N.; Takahashi, T.: Properties of RNAs and proteins of apple stem grooving and apple chlorotic leaf spot viruses. J. gen. Virol. *69*:241-245 (1988).

Taxonomic status	English vernacular name	International name
GROUP	CARNATION LATENT VIRUS GROUP (259)	*CARLAVIRUS*

Revised by A.A. Brunt

TYPE MEMBER	CARNATION LATENT VIRUS (CLV) (61)	—

PROPERTIES OF THE VIRUS PARTICLE

Morphology Slightly flexuous filaments 610-700 nm long, 12-15 nm in diameter with helical symmetry and pitch ≈ 3.4 nm.

Physicochemical properties MW ≈ 60 x 10^6; S_{20w} = 147-176S; buoyant density in CsCl ≈ 1.3 g/cm^3.

Nucleic acid One molecule of linear positive-sense ssRNA; MW = 2.4-2.8 x 10^6; 5-6% by weight of the virus. Those of some members have been partially sequenced; the RNA molecules have 3' poly (A) tracks.

Protein One coat polypeptide, MW = 31-34 x 10^3.

Lipid None reported.

Carbohydrate None reported.

Antigenic properties Efficient immunogens. Serological relationship among some members.

REPLICATION

Particles are found occasionally scattered throughout cytoplasm but more usually occur in bundle-shaped aggregates associated with tonoplasts, cell walls or chloroplast membranes. Cytoplasm may also contain inclusions consisting of endoplasmic reticulum and some unaggregated particles.

Viral RNA has at least 6 open reading frames which have been translated *in vitro* into proteins of MW ≈ 10kDa, 33kDa, 7kDa, 12kDa, 25kDa and 41-45kDa. The 33kDa product is the coat protein and the 41-45kDa protein is possibly a viral replicase; the function of the other proteins has yet to be determined. The deduced amino acid sequences of the central regions of some coat proteins show close homology with that of potato virus X. The putative carlavirus replicase and the 7kDa, 12kDa and 25kDa proteins also show some homology with proteins of

Taxonomic status | English vernacular name | International name

similar sizes induced by potexviruses. The coat protein gene of potato virus S, like that of potexviruses, is located on a subgenomic RNA of 1.3 kb which is possibly encapsidated in filamentous particles 100-200 nm long. A single dsRNA (M ≈ 5.0-5.5 x 10^6) corresponding to the viral ssRNA has been isolated from infected plants.

BIOLOGICAL ASPECTS

Host range Individual viruses have rather narrow host ranges.

Transmission Often by aphids in a non-persistent manner; two possible members are transmitted by whiteflies (*Bemisia tabaci*). Two viruses infecting legumes are seedborne. Experimentally transmitted by mechanical inoculation.

OTHER MEMBERS

Cactus 2
Caper latent
Chrysanthemum B (110)
Dandelion latent
Elderberry carla (= Elderberry A) (263)
Helenium S (265)
Honeysuckle latent (289)
Hop (American) latent (262)
Hop latent (261)
Hop mosaic (241)
Hydrangea latent
Kalanchoe latent
Lilac mottle
Lily symptomless (96)
Mulberry latent
Muskmelon vein necrosis
Narcissus latent (= *gladiolus* ringspot) (170)
Nerine latent (= *hippeastrum* latent)
Passiflora latent
Pea streak (112) (alfalfa latent (211))
Poplar mosaic (75)
Potato M (87)
Potato S (= pepino latent) (60)
Red clover vein mosaic (22)
Shallot latent (250)
Strawberry pseudo mild yellow edge

Possible members

Aphid-borne:
- *Alstroemeria* carlavirus
- Arracacha latent

Taxonomic status	English vernacular name	International name
	Artichoke latent M	
	Artichoke latent S	
	Blueberry carlavirus	
	Butterbur mosaic	
	Cassia mild mosaic	
	Chicory yellow blotch	
	Chinese yam necrotic mosaic	
	Cynodon mosaic	
	Daphne S	
	Dulcamara A and B	
	Eggplant mild mottle (= eggplant carlavirus)	
	Evonymus mosaic	
	Fig S	
	Fuschia latent	
	Garlic mosaic	
	Gentiana carlavirus	
	Gynura latent (strain of *Chrysanthemum* B?)	
	Helleborus mosaic	
	Impatiens latent	
	Lilac ringspot	
	Nasturtium mosaic	
	Plantain 8	
	Prunus S	
	Southern potato latent	
	White bryony mosaic	
	Whitefly-borne:	
	Cassava brown streak	
	Cowpea mild mottle (= *Psophocarpus* necrotic mosaic, groundnut crinkle, tomato pale chlorosis, *Voandzeia* mosaic) (140)	

Derivation of Name carla: sigla from *car*nation *la*tent

References

Adams, A.N.; Barbara, D.J.: Host range, purification and some properties of two carlaviruses from hop (*Humulus lupulus*): hop latent and American hop latent. Ann. appl. Biol. *101*:483-494 (1982).

Allan, T.C.; McMorran, J.P.; Lawson, R.H.: Detection and identification of viruses in hydrangea. Acta Hort. *164*:85-99 (1985).

Antignus, Y.; Cohen, S.: Purification and some properties of a new strain of cowpea mild mottle virus in Israel. Ann. appl. Biol. *110*: 563-569 (1987).

Di Franco, A.; Gallitelli, D.; Vovlas, C.; Martelli, G.P.: Partial characterization of artichoke virus M. J. Phytopathology *127*: 265-273 (1989).

Fauquet, C.; Thouvenel, J.-C.: Maladies Virales des Plantes en Côte d'Ivoire. (Éditions de l'ORSTOM, Bondy, France, 1987).

Foster, G.D.; Mills, P.R.: Evidence for the role of subgenomic RNAs in the production of potato virus S coat protein during *in vitro* translation. J. gen. Virol. *71:* 1247-1249 (1990).

Gallitelli, D.; di Franco, A.: Characterization of caper latent virus. J. Phytopath. *119:*97-105 (1987).

Hammond, J.: Viruses occurring in *Plantago* species in England. Plant Path. 30:237-243 (1981).

Hampton, R.O.: Evidence suggesting identity between alfalfa latent and pea streak viruses. Phytopathology *71:*223 (1981).

Hearon, S.S.: A carlavirus from *Kalanchoe blossfeldiana.* Phytopathology *72:*838-844 (1982).

Johns, L.J.: Purification and partial characterization of a carlavirus from *Taraxacum officinale.* Phytopathology *72:*1239-1242 (1982).

Khalil, J.A.; Nelson, M.R.; Wheeler, R.E.: Host range, purification, serology, and properties of a carlavirus from eggplant. Phytopathology *72:* 1064-1068 (1982).

Koenig, R.: Recently discovered virus or virus-like diseases of ornamentals and their epidemiological significance. Acta Hort. *164:*21-31 (1985).

Mackenzie, D.J.; Tremaine, J.H.; Stace-Smith, R.: Organization and interviral homologies of the 3'-terminal portion of potato virus S RNA. J. gen. Virol. *70:* 1053-1063 (1989).

Memelink, J.; van der Vlugt, C.I.M.; Linthorst, H.J.M.; Derks, A.F.L.M.; Asjes, C.J.; Bol, J.F.: Homologies between the genomes of a carlavirus (lily symptomless virus) and a potexvirus (lily virus X) from lily plants. J. gen. Virol. *71:* 917-924 (1990).

Milne, R.G.: Taxonomy of the rod-shaped filamentous viruses. *In*, Milne, R.G. (ed.), The Plant Viruses, The filamentous plant viruses, Vol. 4, pp. 3-50. (Plenum Press, New York, 1988).

Monis, J.; de Zoeten, G.A.: Characterization and translation studies of potato virus S RNA. Phytopathology *80:* 441-445 (1990).

Monis, J.; de Zoeten, G.A.: Molecular cloning and physical mapping of potato virus S complementary DNA. Phytopathology *80:* 446-450 (1990).

Phillips, Sue; Brunt, A.A.: Four viruses of *Alstroemeria* in Britain. Acta Hort. *177.*227-233 (1986).

Plese, N.; Wrischer, M.: Filamentous virus associated with mosaic of *Euonymus japonica.* Acta Bot. Croat. *40:* 31 (1981).

Rupasov, V.V.; Morozov, S.Yu; Kanyuka, K.V.; Zavriev, S.K.: Partial nucleotide sequence of potato virus M RNA shows similarities to potexviruses in gene arrangement and the encoded amino acid sequences. J. gen. Virol. *70:* 1861-1869 (1989).

Tavantzis, S.M.: Physicochemical properties of potato virus M. Virology *133:* 427-430 (1984).

Valverde, R.A.; Dodds, J.A.; Heick, J.A.: Double-stranded ribonucleic acid from plants infected with viruses having elongated particles and undivided genomes. Phytopathology *76:* 459-465 (1986).

Wetter, C.; Milne, R.G.: Carlaviruses; *In* Kurstak, E. (ed.), Handbook of Plant Virus Infections and Comparative Diagnosis, pp. 695-730. (Elsevier/North Holland, Amsterdam, 1981).

Taxonomic status	English vernacular name	International name
GROUP	BEET YELLOWS VIRUS GROUP (260)	***CLOSTEROVIRUS***

Revised G.P. Martelli & M. Bar-Joseph

TYPE MEMBER	SUGAR BEET YELLOWS VIRUS (SBYV)(13)	—

PROPERTIES OF THE VIRUS PARTICLE

Morphology Very flexuous rods 700-2,000 nm long 12 nm wide; helical symmetry with pitch = 3.4-3.7 nm.

Physicochemical properties S_{20w} = 96-130; buoyant density in CsCl = 1.30-1.34 g/cm^3. Particles unstable in high salt concentrations.

Nucleic acid One molecule of linear positive-sense ssRNA, MW = 2.5-6.5 x 10^6; ≈ 5% by weight of virus particle.

Protein One coat polypeptide, MW = 23-43 x 10^3.

Lipid None reported.

Carbohydrate None reported.

Antigenic properties Moderately immunogenic; serological relationships between some members.

REPLICATION

Particles of most members aggregate in fibrous or cross-banded masses in phloem cells. The complete nucleotide sequence (7555 nt) and genome organization of the possible member apple chlorotic leafspot virus (ACLSV) has been determined as well as the 3' terminal half (6746 nt) of the genome of the type member SBYV. Remarkable differences exist in that: (a) the 3' end of ACLSV is polyadenilated whereas that of SBYV is not; (b) ACLSV genome has three ORFs with the coat protein cistron coterminal with the 3' end, whereas SBYV genome has as least eight ORFs with the coat protein cistron separated from the 3' end by two downstream ORFs. Multiple ds-RNAs of lower MW than those of genomic ds-RNAs have been extracted from plants infected by some members, suggesting that these ds-RNAs may be templates for transcription of subgenomic m-RNAs together with membranous vesicles containing dsRNA-like fibrils (SBYV-like vesicles). Several other members do not

Taxonomic status	English vernacular name	International name

induce formation of SBYV-like vesicles but their particles also aggregate in bundles in parenchyma cells and sieve tubes. Particles are rarely seen within nuclei.

BIOLOGICAL ASPECTS

Host range Moderately wide for individual viruses.

Transmission Few members transmissible with difficulty by mechanical inoculation. Some members transmitted by aphids, pseudococcid mealybugs (*Planococcus* and *Pseudococcus*) or whiteflies (*Bemisia, Trialeurodes*) in a semi-persistent manner.

OTHER MEMBERS

Beet yellow stunt (207)
Burdock yellows
Carnation necrotic fleck (136)
Carrot yellow leaf
Citrus tristeza (33)
Clover yellows
Festuca necrosis
Grapevine virus A
Wheat yellow leaf (157)

Possible members

Apple chlorotic leafspot (30)
Beet pseudo yellows
Cucumber yellows
Diodia yellow vein
Grapevine leafroll-associated I
Grapevine leafroll-associated II
Grapevine leafroll-associated III
Grapevine leafroll-associated IV
Grapevine leafroll-associated V
Heracleum latent (228)
Lettuce infectious yellows
Pineapple mealybug wilt-associated

Derivation of Name clostero: from Greek *kloster*, 'spindle, thread', from appearance of very long rods

---

REFERENCES

Agranovsky, A.A.; Boyke, V.F.; Karasev, A.V.; Lunina, N.A.; Koonin, E.V.; Dolja, V.V.: Nucleotide sequence of the 3' terminal half of beet yellows closterovirus RNA genome: unique arrangement of eight virus genes. J. gen. Virol. (1990) (in press).

Bar-Joseph, M.: Garnsey, S.M.; Gonsalves, D.: The closteroviruses: a distinct group of elongated plant viruses. Adv. Virus Res. *25*:93-168 (1979).

Boccardo, G.; D'Aquilio, M.: The protein and nucleic acid of a closterovirus isolated from a grapevine with stem-pitting symptoms. J. gen. Virol. *53*:179-182 (1981).

Dodds, J.A.; Bar-Joseph, M.: Double-stranded RNA from plants infected with closteroviruses. Phytopathology *73*:419-423 (1983).

Duffus, J.E.: Whitefly transmission of plant viruses; *In* Harris, K.F. (ed.), Current Topics in Vector Research, vol. 4, pp. 73-91 (Springer, Berlin,Heidelberg, New York, Tokyo, 1987).

German, S.; Candresse, T.; Lanneau, M.; Huet, J.C.; Pernollet, J.C.; Dunez, J.: Nucleotide sequence and genomic organization of apple chlorotic leafspot closterovirus. Virology *179*:104-112 (1990).

Gunasinghe, U.B.; German, T.L.: Purification and partial characterization of a virus from pineapply. Phytopathology *79*: 1337-1341 (1989).

Hu, J.S.; Gonslaves, D.; Teliz, D.: Characterization of closterovirus-like particles associated with grapevine leafroll disease. J. Phytopathol. *128*: 1-14 (1990).

Lesemann, D.E.: Cytopathology: *In* Milne, R.G. (ed.), The Plant Viruses, The filamentous plant viruses, Vol. 4, pp. 179-235 (Plenum Press, New York, 1988).

Lister, R.M.; Bar-Joseph, M.: Closteroviruses; *In* Kurstak, E. (ed.), Handbook of Plant Virus Infections and Comparative Diagnosis, pp. 809-844 (Elsevier/North Holland, Amsterdam, 1981).

Rosciglione, B.; Castellano, M.A.; Martelli, G.P.; Savino, V.; Cannizzaro, G.: Mealybug transmission of grapevine virus A. Vitis *22*: 331-347.

Taxonomic status	English vernacular name	International name
GROUP	POTATO VIRUS X GROUP (200)	*POTEXVIRUS*

Revised by R. Koenig

TYPE MEMBER	POTATO VIRUS X (PVX)(4)	–

PROPERTIES OF THE VIRUS PARTICLE

Morphology Flexuous filaments 470-580 nm long and 13 nm wide, with helical symmetry and pitch ≈ 3.4 nm.

Physicochemical properties MW ≈ 3 5 x 10^6; S_{20w} = 115-130; density in CsCl ≈ 1.31 g/cm^3; particles stable.

Nucleic acid One molecule of linear positive-sense ssRNA with 5 ORFs; MW ≈ 2.1 x 10^6, ≈ 6% by weight of the particle. 5' terminus has sequence m^{7}G$^{5'}$pppGpA. Poly(A) at 3' terminus; RNAs of PVX and white clover mosaic virus have been sequenced; RNA contains high A content (≈ 30%).

Protein One coat polypeptide, MW ≈ 18-23 x 10^3. In some viruses, protein can become partially degraded by enzymes in plant sap.

Lipid None reported.

Carbohydrate None reported.

Antigenic properties Efficient immunogens; serological relationship between some members.

REPLICATION

Fibrous cytoplasmic inclusions composed of virus particles, often banded; some members induce nuclear inclusions of different composition. Intact genomic RNA is translated into high-molecular-weight proteins, the viral coat protein from a subgenomic RNA.

BIOLOGICAL ASPECTS

Host range Narrow for individual viruses.

Transmission Readily transmissible mechanically, experimentally, and by contact between plants. No known vectors.

Taxonomic status	English vernacular name	International name

OTHER MEMBERS

Asparagus III
Cactus X (58)
Clover yellow mosaic (111)
Commelina X
Cymbidium mosaic (27)
Foxtail mosaic (264)
Hydrangea ringspot (114)
Lily X
Narcissus mosaic (45)
Nerine X
Papaya mosaic (56)
Pepino mosaic
Plantago severe mottle
Plantain X (266)
Tulip X (276)
Viola mottle (247)
White clover mosaic (41)

Possible members

Artichoke curly dwarf
Bamboo mosaic
Barley B-1
Boletus
Cassava common mosaic (90)
Centrosema mosaic
Daphne X (195)
Dioscorea latent
Lychnis potexvirus
Malva veinal necrosis
Nandina mosaic
Negro coffee mosaic
Parsley 5
Parsnip 5
Parsnip 3
Potato aucuba mosaic (98)
Rhododendron necrotic ringspot
Rhubarb 1
Smithiantha potexvirus
Wineberry latent
Zygocactus

Derivation of Name potex: sigla from *pot*ato *X*

Taxonomic status	English vernacular name	International name

Biological Aspects

Host range

Narrow for many individual viruses but other viruses infect species in up to 30 families.

Transmission

Transmitted by aphids in a non-persistent manner. Others, included as possible members of the group, are transmissible by whiteflies, mites or fungi. Transmissible experimentally by mechanical inoculation.

Similarities with Other Virus Groups or Families

Some potyviruses share significant similarities with como-, nepo- and picornaviruses, e.g. genome organization, VPg at 5'-end and poly A at 3'-end of the genomes, post-translational processing of polyproteins and similar consensus sequences among non-structural proteins.

Other Members

Alstroemeria mosaic
Amaranthus leaf mottle
Araujia mosaic
Artichoke latent
Asparagus 1
Bean common mosaic (73, 337)
Bean yellow mosaic
(= pea mosaic, crocus tomasinianus) (40)
Beet mosaic (53)
Bidens mottle (161)
Blackeye cowpea mosaic (305)
Cardamom mosaic
Carnation vein mottle (78)
Carrot thin leaf (218)
Celery mosaic (50)
Clover yellow vein (= pea necrosis) (131)
Cocksfoot streak (59)
Colombian datura
Commelina mosaic
Cowpea aphid-borne mosaic (= Azuki bean mosaic) (134)
Cowpea green vein banding
Dasheen mosaic (191)
Datura shoestring
Dendrobium mosaic
Garlic mosaic
Gloriosa stripe mosaic
Groundnut eyespot
Guinea grass mosaic (190)
Helenium virus Y

Taxonomic status	English vernacular name	International name
	Henbane mosaic (95)	
	Hippeastrum mosaic (117)	
	Iris fulva mosaic (310)	
	Iris mild mosaic (116, 324)	
	Iris severe mosaic (338) (= bearded iris mosaic) (147)	
	Johnsongrass mosaic	
	Leek yellow stripe (240)	
	Lettuce mosaic (9)	
	Maize dwarf mosaic	
	Narcissus degeneration	
	Narcissus yellow stripe (76)	
	Nothoscordum mosaic	
	Onion yellow dwarf (158)	
	Ornithogalum mosaic	
	Papaya ringspot (= watermelon mosaic 1) (63,84,292)	
	Parsnip mosaic (91)	
	Passionfruit woodiness (122)	
	Pea seed-borne mosaic (146)	
	Peanut mottle (141)	
	Peanut stripe (= peanut mild mottle, peanut chlorotic ring mottle)	
	Pepper severe mosaic	
	Pepper veinal mottle (104)	
	Plum pox (70)	
	Pokeweed mosaic (97)	
	Potato A (54)	
	Potato V (316)	
	Sorghum mosaic	
	Soybean mosaic (93)	
	Statice Y	
	Sugarcane mosaic (88)	
	Sweet potato feathery mottle (= sweet potato russett crack sweet potato A)	
	Tamarillo mosaic	
	Telfairia mosaic	
	Tobacco etch (55;258)	
	Tobacco vein mottling (325)	
	Tomato (Peru) mosaic (255)	
	Tulip chlorotic blotch	
	Tulip breaking (71)	
	Turnip mosaic (8)	
	Watermelon mosaic 2 (63;293)	
	Wisteria vein mosaic	
	Yam mosaic (314) (= *Dioscorea* green banding)	
	Zucchini yellow fleck	
	Zucchini yellow mosaic (282)	

Taxonomic status	English vernacular name	International name

Possible members

Aphid-borne (* aphid transmission not confirmed)

Anthoxanthum mosaic*
*Aquilegia**
Aracacha Y
Asystasia gangetica mottle*
Bidens mosaic
Bryonia mottle
Canavalia maritima mosaic
Carrot mosaic
Cassia yellow blotch*
Celery yellow mosaic
Chickpea busy dwarf
Chickpea filiform
Clitoria yellow mosaic
Clover (Croatian)
Cowpea Moroccan aphid-borne
Crinum mosaic*
*Cypripedium calceolus**
Daphne Y
Datura 437
Datura mosaic*
Desmodium mosaic
Dioscorea alata ring mottle
Dioscorea trifida
Dock mottling mosaic
Euphorbia ringspot
Ficus carica
Freesia mosaic
Garlic yellow streak
Guar symptomless*
Habenaria mosaic
Holcus streak*
Hungarian *Datura innoxia**
Hyacinth mosaic*
Isachne mosaic*
Kennedya Y
Lily mild mottle
Maclura mosaic (239)
(particle length and coat protein MW are atypical)
Malva vein clearing
Marigold mottle
Melilotus mosaic
Mungbean mosaic*
Mungbean mottle
Narcissus late season yellows
(= jonquil mild mosaic)
*Nerine**

Taxonomic status	English vernacular name	International name

Palm mosaic*
Papaya leaf distortion
Passionfruit ringspot
Patchouli mottle
Peanut green mosaic
Peanut mosaic
Pecteilis mosaic
Pepper mild mosaic
Pepper mottle (253) (may be synonymous with PVY)
Perilla mottle
Plantain 7
Pleioblastus mosaic
*Populus**
Primula mosaic
Reed canary mosaic
Sunflower mosaic*
Teasel mosaic
Tobacco vein banding mosaic
Tradescantia/Zebrina
Tropaeolum 1
Tropaeolum 2
Ullucus mosaic
Vallota mosaic
Vanilla mosaic
Vanilla necrosis
White Bryony mosaic
Wild potato mosaic
Zoysia mosaic

Fungal-borne

Barley yellow mosaic (143)
Oat mosaic (145)
Rice necrosis mosaic (172)
(=wheat yellow mosaic)
Wheat spindle streak mosaic (167)

Mite-borne (** mite transmission not demonstrated)

Agropyron mosaic (118)
Brome streak virus
Hordeum mosaic**
Oat necrotic mottle (169)**
Ryegrass mosaic (86)
Spartina mottle**
Wheat streak mosaic (48)

Whitefly-borne

Sweet potato mild mottle (162)

Derivation of Name poty: sigla from *pot*ato *Y*

REFERENCES

Buchen-Osmond, C.; Crabtree, K.; Gibbs, A.; McLean, G.; Viruses of Plants in Australia. (The Australian Nat. Univ., Canberra, 1988).

Demski, J.W.; Reddy, D.V.R.; Sowell, G.; Bays, D.: Peanut stripe virus - a new seed-borne potyvirus from China infecting groundnut (*Arachis hypogaea*). Ann. appl. Biol. *105*:495-501 (1984).

Domier, L.L.; Shaw, J.G.; Rhoads, R.E.: Potyviral proteins share amino acid sequence homology with picorna-, como-, and caulimoviral proteins. Virology *158*:20-27 (1987).

Dougherty, W.G.; Hiebert, E.: Translation of potyvirus RNA in a rabbit reticulocyte lysate: cell-free translation strategy and a genetic map of the potyviral genome. Virology *104*:183-194 (1984).

Dougherty, W.G.; Allison, R.F.; Parks, T.D.; Johnston, R.E.; Feild, M.J.; Armstrong, F.B.: Nucleotide sequence at the 3' terminus of pepper mottle virus genomic RNA: evidence for an alternative mode of potyvirus capsid protein gene organization. Virology *146*:282-291 (1985).

Edwardson, J.R.: Host range of viruses in the PVY group. Florida Agric. Exp. Station Monogr. Ser., vol. 5 (1974).

Edwardson, J.R.: Some properties of the potato virus Y group. Florida Agric. Exp. Station Monogr. Ser, vol. 4 (1974).

Francki, R.I.B.; Milne, R.G.; Hatta, T.: Potyvirus group, *In* Atlas of Plant Viruses, Vol. II, pp 183-218 (CRC Press, Boca Raton, Fl., 1985).

Goldbach, R.: Genome similarities between plant and animal RNA viruses. Microbiol. Sci. *4*:197-202 (1987).

Hellmann, G.M.; Thornbury, D.W.; Hiebert, E.; Shaw, J.G.; Pirone, T.P.; Rhoads, R.E.: Cell-free translation of tobacco vein mottling virus RNA. II. Immunoprecipitation of products by antisera to cylindrical inclusion, nuclear inclusion, and helper component proteins. Virology *124*:434-444 (1983).

Hiebert, E.; Thornbury, D.W.; Pirone, T.P.: Immunoprecipitation analysis of potyviral *in vitro* translation products using antisera to helper component of tobacco vein mottling virus and potato virus Y. Virology *135*:1-9 (1984).

Hollings, M.; Brunt, A.A.: Potyviruses; *In* Kurstak, E. (ed.), Handbook of Plant Virus Infections and Comparative Diagnosis, pp. 731-807 (Elsevier/North Holland, Amsterdam, 1981).

Milne, R.G.: The Plant Viruses. The Filamentous Plant Viruses, Vol. 4. (Plenum Press, New York, 1988).

Shukla, D.D.; Inglis, A.S.; McKern, N.M.; Gough, K.H.: Coat protein of potyviruses. 2. Amino acid sequence of the coat protein of potato virus Y. Virology *152*:118-125 (1986).

Shukla, D.D.; Ward, C.W.: Identification and classification of potyviruses on the basis of coat protein sequence data and serology. Arch. Virol. *106:* 171-200 (1989).

Shukla, D.D.; Ward, C.W.: Structure of potyvirus coat proteins and its application in the taxonomy of the potyvirus group. Adv. Virus Res. *36:* 273-314 (1989).

Vance, V.B.; Beachy, R.N.: Translation of soybean mosaic virus RNA *in vitro*: evidence of protein processing. Virology *132*:271-281 (1984).

Taxonomic status	English vernacular name	International name
GROUP	TOBACCO MOSAIC VIRUS GROUP (184)	***TOBAMOVIRUS***

Revised by M.H. Van Regenmortel

TYPE MEMBER	TOBACCO MOSAIC VIRUS (TMV) (COMMON OR U1 STRAIN) (151)	—

PROPERTIES OF THE VIRUS PARTICLE

Morphology Elongated rigid particles about 18 nm diameter and 300 nm long, helically symmetrical with pitch ≈ 2.3 nm.

Physicochemical properties MW ≈ 40 x 10^6; S_{20w} ≈ 194; buoyant density in CsCl ≈ 1.325 g/cm^3; particles very stable.

Nucleic acid One molecule of linear positive-sense ssRNA, MW ≈ 2 x 10^6. 5' terminus has the sequence $m^7G^{5'}ppp^{5'}Gp$. 3' terminus has a tRNA-like structure which accepts histidine.

Protein One coat polypeptide, MW = 17-18 x 10^3.

Lipid None.

Carbohydrate None.

Antigenic properties Efficient immunogens.

REPLICATION

Virus replicates in the cytoplasm, inducing characteristic viroplasms; virus particles often form large crystalline arrays, visible by light microscopy. A virus-induced polymerase is detected in infected tissues; RNA replicates via an RF or RI. Coat protein is synthesized from a small monocistronic mRNA (whose base sequence is also on the viral RNA near the 3' end); the mRNA is encapsidated in some members. Three other virus-specific proteins (MW ≈ 180, 126 and 30 x 10^3) are transcribed from full-length viral RNA. The 126 kDa non-structural protein of TMV, thought to be a component of the viral replicase, accumulates in cytoplasmic inclusions (X-bodies), whereas the 30 kDa non-structural transport protein is localised in plasmodesmata.

Taxonomic status	English vernacular name	International name

BIOLOGICAL ASPECTS

Host range Most members have moderate host range.

Transmission Readily transmitted by mechanical inoculation. Some members transmitted by seed.

SEQUENCE SIMILARITIES WITH OTHER VIRUS GROUPS OR FAMILIES

Some non-structural proteins synthesized by tobacco mosaic virus share sequence similarities with non-structural proteins of some other RNA plant viruses [e.g. tripartite viruses (alfalfa mosaic, brome mosaic and cucumber mosaic viruses), a bipartite virus (tobacco rattle virus), and a monopartite virus (carnation mottle virus)] and sindbis virus, a monopartite RNA animal virus.

OTHER MEMBERS

Cucumber green mottle mosaic (154)
(= Cucumber virus 4)
Frangipani mosaic (196)
Kyuri green mottle mosaic
Odontoglossum ringspot (155)
Pepper mild mottle (330)
Ribgrass mosaic (152)
Sammons' *Opuntia*
Sunn-hemp mosaic (153)
Tobacco mild green mosaic (351)
Tomato mosaic (156)
Ullucus mild mottle

Possible members

Chara australis
Hypochoeris mosaic (273)

Derivation of Name tobamo: sigla from *toba*cco *mo*saic

REFERENCES

Francki, R.I.B.; Hu, J.; Palukaitis, P.: Taxonomy of cucurbit infecting tobamoviruses as determined by serological and molecular hybridization analyses. Intervirology *26:* 156-163 (1986).

Goldbach, R.: Genome similarities between plant and animal RNA viruses. Microbiol. Sci. *4:*197-202 (1987).

Hills, G.J.; Plaskitt, K.A.; Young, N.D.; Dunigan, D.D.; Watts, J.W.; Wilson, T.M.A.; Zaitlin, M.: Immunogold localization of the intra-cellular sites of structural and non-structural tobacco mosaic virus proteins. Virology *161:* 488-496 (1987).

Tomenius, K.; Clapham, D.; Meshi, T.: Localization by immunogold cytochemistry of the virus-coded 30K protein in plasmodesmata of leaves infected with tobacco mosaic virus. Virology *160:* 363-371 (1987).

van Regenmortel, M.H.V.; Fraenkel-Conrat, H.: The Plant Viruses, The Rod Shaped Plant Viruses, Vol. 2. (Plenum Press, New York, 1986).

Taxonomic status	English vernacular name	International name
GROUP	COWPEA MOSAIC VIRUS GROUP (199)	*COMOVIRUS*

Revised by R. Goldbach

TYPE MEMBER	COWPEA MOSAIC VIRUS (CPMV) (SB ISOLATE) (47;197)	—

PROPERTIES OF THE VIRUS PARTICLE

Morphology

All three sedimenting components, T, M and B respectively, possess isometric particles ≈ 28 nm in diameter. The shell consists of 60 subunits of each of the two structural proteins assembled in a T = 1 icosahedral structure. There are 12 pentamers of the larger protein at the 5-fold vertices and 20 trimers of the smaller protein at the positions of 3-fold symmetry. The two structural proteins are folded into 3 antiparallel β-barrel structures; the smaller protein forms one barrel and the large forms two barrels. The 60 copies of each protein type in the virus generate 180 β-barrel domains that are arranged in a manner very similar to a T = 3 capsid. M particles contain a single molecule of RNA-2, and B particles a single molecule of RNA-1.

Physicochemical properties

Particles are usually very stable and sediment as three components, T, M, and B, respectively containing ≈ 0, 25 and 37% RNA by weight with S_{20w} ≈ 58, 98 and 118 and MWs ≈ 3.8, 5.2, and 6.2 x 10^6; buoyant densities in CsCl ≈ 1.29 (T), 1.41 (M) and 1.44 - 1.46 (B) g/cm^3. Partial proteolytic degradation of the smaller coat protein results in particles with increased electrophoretic mobility.

Nucleic acid

Two species of linear positive-sense ssRNA of 5889 nucleotides (RNA-1) and 3481 nucleotides (RNA-2). Complete nucleotide sequences determined. The two RNA molecules each have a high content of A + U but have little base sequence homology. Each molecule has a poly (A) tract of variable length (RNA-1: 50-150 residues; RNA-2: 50-300 residues) at their 3' end and a VPg (MW ≈ 4K) covalently linked by a serine residue to its 5' end. Enzymatic degradation of this protein does not diminish the infectivity of the RNA. Both RNA-1 and RNA-2 contain an ORF specifying a "polyprotein".

Protein

Two coat polypeptides, MWs ≈ 22 and 42 x 10^3. The smaller, and in some members both, polypeptide(s) may become partially degraded by proteolytic cleavage *in vivo* and *in vitro*.

Lipid

None reported.

Carbohydrate

The coat proteins may be glycosylated. The amino acid sequence of both proteins indicates putative N-type glycosylation sites.

Antigenic properties

Good immunogens. All members are serologically interrelated, often distantly. The coat proteins of red clover mottle virus and bean pod mottle virus show approximately 50% identity in amino acid sequence to the CPMV coat proteins.

REPLICATION

Unfractionated RNA is highly infective but neither RNA species alone can infect plants. RNA-1 can replicate in protoplasts but in the absence of RNA-2 (which carries the coat protein cistrons), no virus particles are produced. RNA-1 carries all information for viral RNA replication, including the core polymerase. Both RNA species are translated into polyproteins that are cleaved to form the functional proteins. Final translation products of RNA-1 are proteins, MWs ≈ 87 x 10^3 (viral core polymerase), 4 x 10^3 (VPg), 58 x 10^3 (membrane protein), 24 x 10^3 (viral proteinase), 32 x 10^3 (proteinase co-factor). The RNA-2 molecules of all comoviruses tested are translated into two overlapping polyproteins. The final products of RNA-2 translation are two overlapping proteins, MW ≈ 58 and 48 x 10^3 (putative virus transport proteins), and the two coat proteins (VP37 and VP23). Membranous vesicles and electron-dense amorphous masses are the characteristic cytopathological structures found in the cytoplasm of infected cells. They contain all viral non-structural proteins, the (membrane-bound) viral polymerase activity, two dsRNA species corresponding to each of the particle RNA, and complementary RNA. Newly formed virus particles accumulate in the cytoplasm, sometimes in crystalline arrays but not in association with any cell organelle. Cell-to-cell transport probably occurs as particles, through tubular structures that penetrate through cell walls.

BIOLOGICAL ASPECTS

Host range

Individual members have narrow host ranges, 9 out of 13 members being restricted to a few *Leguminosae* species. Mosaic and mottle symptoms are characteristic.

Taxonomic status	English vernacular name	International name
GROUP	CARNATION RINGSPOT VIRUS GROUP	***DIANTHOVIRUS***

Revised by R.I. Hamilton

TYPE MEMBER	CARNATION RINGSPOT VIRUS (CRSV) (21)	—

PROPERTIES OF THE VIRUS PARTICLE

Morphology Polyhedral particles 31-34 nm in diameter. The arrangement of the two RNA species within particles has not been established.

Physicochemical properties MW $\approx 7.1 \times 10^6$; $S_{20w} \approx 133$; buoyant density in CsCl $\approx$ 1.37 g/cm^3; alkaline pH (7-8) induces swelling of virus particles.

Nucleic acid Two molecules of positive-sense ssRNA, MW $\approx$ 1.5 and 0.5×10^6. The larger RNA contains the coat protein cistron. RNA-1 (3889 nt) and RNA-2 (1448 nt) of red clover necrotic mosaic virus have been sequenced. Both have a 5' m^{7}GpppA; neither is polyadenylated or contains a VpG. RNA-1 contains three ORFs for proteins, MW $\approx$ 27, 37 (capsid) and 57×10^3; RNA-2 contains a single ORF for protein of MW = 35×10^3.

Protein One coat polypeptide, MW $\approx 40 \times 10^3$.

Lipid None reported.

Carbohydrate None reported.

Antigenic properties Efficient immunogens. Single precipitin line in gel diffusion tests.

REPLICATION

Particles located in the cytoplasm, scattered and clustered; patches of densely stained, amorphous material also seen in cytoplasm of some cells. RNA-1 of red clover necrotic mosaic virus can replicate alone in cowpea and tobacco protoplasts. Two dsRNAs corresponding to the genomic ssRNAs and a third corresponding to a subgenomic RNA derived from RNA-1 have been detected in infected plants. Evidence suggests that coat protein may be translated from subgenomic RNA derived from RNA-1. Three proteins of

Taxonomic status	English vernacular name	International name

MW ≈ 39 (capsid protein) 36 and 34 x 10^3 are translated *in vitro*.

BIOLOGICAL ASPECTS

Host range Each virus has a wide host range.

Transmission Transmitted through soil. Readily transmissible experimentally by mechanical inoculation.

OTHER MEMBERS

Red clover necrotic mosaic (181)
Sweet clover necrotic mosaic (321)

Derivation of Name diantho: from *Dianthus*, the generic name of carnation.

REFERENCES

Dodds, J.A.; Tremaine, J.H.; Ronald, W.P.: Some properties of carnation ringspot virus single- and double-stranded ribonucleic acid. Virology *83:*322-328 (1977).

Hiruki, C.; Rao, D.V.; Chen, M.H.; Okuno, T.; Figueiredo, G.: Characterization of sweet clover necrotic mosaic virus. Phytopathology *74:*482-486 (1984).

Hiruki, C.: The dianthoviruses: a distinct group of isometric plant viruses with bipartite genome. Adv. Virus Res. *33:*257-300 (1987).

Lommel, S.A.; Weston-Fina, M.; Xiong, Z.; Lomonossoff, G.P.: The nucleotide sequence and gene organisation of red clover necrotic mosaic virus RNA-2. Nuc. Acids Res. *16:* 8587-8602 (1988).

Morris-Krsinich, B.A.M.; Forster, R.L.S.; Mossop, D.W.: Translation of red clover necrotic mosaic virus RNA in rabbit reticulocyte lysate: identification of the virus coat protein cistron on the larger RNA strand of the bipartite genome. Virology *124:*349-356 (1983).

Okuno, T.; Hiruki, C.; Rao, D.V.; Figueiredo, G.C.: Genetic determinants distributed in two genomic RNAs of sweet clover necrotic mosaic, red clover necrotic mosaic and clover primary leaf necrosis viruses. J. gen. Virol. *64:*1907-1914 (1983).

Osman, T.A.M.; Buck, K.W.: Replication of red clover necrotic mosaic virus RNA in cowpea protoplasts: RNA-1 replicates independently of RNA 2. J. gen. Virol. *68:* 289-296 (1987).

Osman, T.A.M.; Buck, K.W.: Double-stranded RNAs isolated from plant tissue infected with red clover necrotic mosaic virus correspond to genomic and subgenomic single-stranded RNAs. J. gen. Virol. *71:* 945-948 (1990).

Xiong, Z.; Lommel, S.A.: The complete nucleotide sequence and genome organization of red clover necrotic mosaic virus RNA-1. Virology *171:* 543-554 (1989).

Taxonomic status	English vernacular name	International name
GROUP	BROAD BEAN WILT VIRUS GROUP	***FABAVIRUS***

Revised by R. Milne

TYPE MEMBER	BROAD BEAN WILT VIRUS (BBWV), SEROTYPE I (81)	—

PROPERTIES OF THE VIRUS PARTICLE

Morphology All three sedimenting components consist of isometric particles ≈ 30 nm in diameter with hexagonal outlines. M particles contain a single molecule of RNA-2, B particles a single molecule of RNA-1.

Physicochemical properties Particles very stable, sedimenting as three components, T, M and B, respectively containing ≈ 0, 25 and 35% RNA by weight with S_{20w} of 56-63, 93-100 and 113-126.

Nucleic acid Two species of linear positive-sense ssRNA with MW ≈ 2.1 x 10^6 (RNA-1) and 1.5 x 10^6 (RNA-2).

Protein Two coat polypeptides of MW ≈ 43 x 10^3 and 27 x 10^3.

Lipid None reported.

Carbohydrate None reported.

Antigenic properties Efficient immunogens. Distant serological relationships between members.

REPLICATION

Both RNA species are necessary for infectivity. Inclusion bodies consisting of large masses of convoluted membranes and vesicles occur in the cytoplasm. Virus particles occur scattered in the cytoplasm, or in single files within cytoplasmic tubules and in the plasmodesmata. The virus particles also tend to aggregate to form crystals or striking tubular or rectangular arrays.

BIOLOGICAL ASPECTS

Host range Wide. Symptoms range from ringspots, mottles, mosaic, distortion, wilting and apical necrosis to symptomless infection.

Transmission Transmitted by several species of aphid in the non-persistent manner. Readily transmissible experimentally

Taxonomic status	English vernacular name	International name
	by mechanical inoculation. Not known to be seed-transmitted.	

OTHER MEMBERS

Broad bean wilt virus, serotype II
Lamium mild mosaic virus

Derivation of Name faba: L. *Faba*, bean; also *Vicia faba*, broad bean.

REFERENCES

Doel, T.R.: Comparative properties of type, *nasturtium* ringspot and *petunia* ringspot strains of broad bean wilt virus. J. gen. Virol. *26*:95-108 (1975).

Francki, R.I.B.; Milne, R.G.; Hatta, T.: Comovirus group, *In* Atlas of Plant Viruses, Vol. II, pp. 1-22 (CRC Press, Boca Raton, Fl., 1985).

Lisa, V.; Luisoni, E.; Boccardo, G.; Milne, R.G.; Lovisolo, O.: *Lamium* mild mosaic virus: a virus distantly related to broad bean wilt. Ann. appl. Biol. *100*:467-476 (1982).

Uyemoto, J.K.; Provvidenti, R.: Isolation and identification of two serotypes of broad bean wilt virus. Phytopathology *64*:1547-1548 (1974).

Taxonomic status	English vernacular name	International name
GROUP	TOBACCO RINGSPOT VIRUS GROUP (185)	***NEPOVIRUS***

Revised by G.P. Martelli

TYPE MEMBER	TOBACCO RINGSPOT VIRUS (TOBRV) (17;309)	—

PROPERTIES OF THE VIRUS PARTICLE

Morphology

All three sedimenting components possess isometric particles ≈ 28 nm in diameter, often with hexagonal outlines. M particles contain a single molecule of RNA-2, B particles a single molecule of RNA-1; some members have a second type of B particle containing two molecules of RNA-2.

Physicochemical properties

Particles of most members are stable and sediment as three components, T, M and B, respectively containing ≈ 0, 27-40 and 42-46% RNA by weight, with S_{20w} = 49-56, 86-128 and 115-134, and MWs (x 10^6) = 3.2-3.4, 4.6-5.8, and 6.0-6.2; buoyant densities in CsCl ≈ 1.28 (T), 1.43-1.48 (M), and 1.51-1.53 (B) g/cm^3. Satellite RNAs become packaged in helper virus capsids to form additional sedimenting and buoyant density components.

Nucleic acid

Two species of linear positive-sense ssRNA with MW = 2.4-2.8 x 10^6 (RNA-1) (7365 nt and 7212 nt in tomato black ring and grapevine chrome mosaic, respectively) and 1.3-2.4 x 10^6 (RNA-2) (4662 and 4441 nt in tomato black ring and grapevine chrome mosaic, respectively). The two RNA molecules have little base sequence homology. Each RNA molecule has a poly (A) tract at its 3' end and a Vpg (MW = 3-6 x 10^3) covalently linked to its 5' end; enzymatic degradation of this polypeptide (genome-linked protein) decreases or abolishes the infectivity of the RNA. 'Satellite' RNA molecules of two sizes and types are associated with some members. 'Large' satellite RNAs are linear messenger molecules of MW 0.37-0.47 x 10^6 (1114-1375 nt) encoding a polypeptide of MW = 38-48 x 10^3. 'Small' satellite RNAs are non-messenger molecules of MW 0.10-0.16 x 10^6 (300-457 nt) which have a strong modulating effect on symptom expression. *In vitro* transcripts of cDNA clones of TobRV and tomato black ring satellite RNAs are biologically active.

Protein One coat polypeptide, MW = 55-60 x 10^3; probably 60 copies per species of particle. Most of the possible members are so listed because they have two or three polypeptides of lower MW.

Lipid None reported.

Carbohydrate None reported.

Antigenic properties Efficient immunogens. Few instances of serological cross-reactivity between members.

REPLICATION

Unfractionated RNA induces many local lesions in assay hosts, but separated RNA species induce few or none. RNA-1 can replicate in protoplasts but, in the absence of RNA-2 (which carries the coat protein cistron), no virus particles are produced. RNA-1 carries information for the polymerase function and for the Vpg. Virus-induced RNA-dependent RNA polymerase is present in TobRV-infected tissue along with short dsRNA molecules of unknown function. Inhibitor studies indicate that nepovirus proteins are synthesized on cytoplasmic ribosomes. Both RNA species are translated *in vitro* into large polypeptides approaching in size of their theoretical coding capacity; these 'polyproteins' must be cleaved *in vivo* to form the functional proteins. The RNA-2 polyprotein has the coat polypeptide at its C-terminal end. 'Homology' comparisons suggest a 'transport protein' gene in RNA-2 and polymerase and protease domains in the RNA-1 polyprotein. Characteristic vesiculated inclusion bodies occur in the cytoplasm, usually adjacent to the nucleus. Virus particle antigen accumulates in these structures, which may be the sites of synthesis or assembly of virus components. Newly formed virus particles accumulate in the cytoplasm. They are also commonly found in the plasmodesmata and in single files within tubules in the cytoplasm.

BIOLOGICAL ASPECTS

Host range Wide. Ringspot symptoms are characteristic, but spotting or mottling symptoms are probably more frequent. Leaves produced later are often symptomless though infected ('recovery'). Symptomless infection is common.

Transmission Seed transmission (via either gamete) is very common. There is circumstantial evidence for transmission of one member to plants pollinated with pollen from infected

Taxonomic status	English vernacular name	International name

plants. Most members are transmitted by soil-inhibiting longidorid nematodes, but one is reported also to be transmitted aerially and the vectors of others are unknown. Nematodes retain ability to transmit virus for weeks or months but cease to transmit after moulting. The viruses do not multiply in the vector. Readily transmissible experimentally by mechanical inoculation.

Similarities with other Virus Groups or Families

Tomato black ring virus shares significant similarities with como-, poty- and picorna viruses, e.g. genome organization, VPg at 5'-end and poly A at 3'-end of the genomes, post-translational processing of polyproteins and sequence similarities among non-structural proteins.

Other Members

Arabis mosaic (16)
Arracacha A (216)
Artichoke Italian latent (176)
Artichoke yellow ringspot (271)
Blueberry leaf mottle (267)
Cassava American latent
Cassava green mottle
Cherry leaf roll (80; 306)
Chicory yellow mottle (132)
Cocoa necrosis (173)
Crimson clover latent
Cycas necrotic stunt
Grapevine Bulgarian latent (186)
Grapevine chrome mosaic (103)
Grapevine fanleaf (28)
Grapevine Tunisian ringspot
Hibiscus latent ringspot (233)
Lucerne Australian latent (225)
Mulberry ringspot (142)
Myrobalan latent ringspot (160)
Olive latent ringspot (301)
Peach rosette mosaic (150)
Potato black ringspot (206)
Potato U
Raspberry ringspot (6; 198)
Tomato black ring (38)
Tomato ringspot (18;290)

Possible members

Arracacha B (270)
Artichoke vein banding (285)

Taxonomic status	English vernacular name	International name
	Cherry rasp leaf (159) Lucerne Australian symptomless *Rubus* Chinese seed-borne Satsuma dwarf (208) Strawberry latent ringspot (126) Tomato top necrosis	

Derivation of Name nepo: sigla from *ne*matode, *po*lyhedral to distinguish these viruses from the tobravirus group.

REFERENCES

Brault, V.; Hibrand, L.; Candresse, T.; Le Gall, O.; Dunez, J.: Nucleotide sequence and genetic organization of Hungarian grapevine chrome mosaic nepovirus RNA 2. Nuc. Acids Res. *17:* 7809-7819 (1989).

Fritsch, C.; Mayo, M.A.; Murant, A.F.: Translation products of genome and satellite RNAs of tomato black ring virus. J. gen. Virol. *46:*381-389 (1980).

Gerlach, W.L.; Buzayan, J.M.; Schneider, I.R.; Bruening, G.: Satellite tobacco ringspot virus RNA: biological activity of DNA clones and their *in vitro* transcripts. Virology *151:* 172-185 (1986).

Goldbach, R.: Genome similarities between plant and animal RNA viruses. Microbiol. Sci. *4:*197-202 (1987).

Greif, C.; Hemmer, O.; Demangeat, G.; Fritsch, C.: *In vitro* synthesis of biologically active transcripts of tomato black ring virus satellite RNA. J. gen. Virol. *71:* 907-915 (1990).

Greif, C.; Hemmer, O.; Fritsch, C.: Nucleotide sequence of tomato black ring virus RNA-1. J. gen. Virol. *69:* 1517-1529 (1988).

Le Gall, O.; Candresse, T.; Brault, V.; Dunez, J.: Nucleotide sequence of Hungarian grapevine chrome mosaic nepovirus RNA 1. Nuc. Acids Res. *17:* 7795-7807 (1989).

Martelli, G.P.; Taylor, C.E.: Distribution of viruses and their nematode vectors; *In* Harris, K.F. (ed.), Advances in Disease Vector Research, Vol. 6, pp. 151-189. (Springer, Wien, New York, 1989).

Mayo, M.A.; Barker, H.; Harrison, B.D.: Polyadenylate in the RNA of five nepoviruses. J. gen. Virol. *43:*603-610 (1979).

Mayo, M.A.; Barker, H.; Harrison, B.D.: Specificity and properties of the genome-linked proteins of nepoviruses. J. gen. Virol. *59:*149-162 (1982).

Meyer, M.; Hemmer, O.; Mayo, M.A.; Fritsch, C.: The nucleotide sequence of tomato black ring virus RNA-2. J. gen. Virol. *67:*1257-1271 (1986).

Mircetich, S.M.; Sanborn, R.R.; Ramos, D.E.: Natural spread, graft-transmission, and possible etiology of walnut blackline disease. Phytopathology *70:*962-968 (1980).

Murant, A.F.: Nepoviruses; *In* Kurstak, E. (ed.), Handbook of Plant Virus Infections and Comparative Diagnosis, pp. 197-238. (Elsevier/North Holland, Amsterdam, 1981).

Robinson, D.J.; Barker, H.; Harrison, B.D.; Mayo, M.A.: Replication of RNA-1 of tomato black ring virus independently of RNA-2. J. gen. Virol. *51:*317-326 (1980).

Rubino, L.; Tousignant, M.E.; Steger, G.; Kaper, J.M.: Nucleotide sequence and structural analysis of two satellite RNAs associated with chicory yellow mottle virus. J. gen. Virol. *71:* 1897-1903 (1990).

Stace-Smith, R.; Ramsdell, D.C.: Nepoviruses of the Americas; *In* Harris, K.F. (ed.), Current Topics in Vector Research, Vol. 5, pp. 131-166. (Springer, Wien, New York, 1987).

Walter, B.; Ladeveze, I.; Etienne, L.; Fuchs, M.: Some properties of a previously undescribed virus from cassava: Cassava American latent virus. Ann. appl. Biol. *115:* 279-289 (1989).

Taxonomic status	English vernacular name	International name
FAMILY	NODAMURA VIRUS GROUP	*NODAVIRIDAE*

Revised by R.R. Rueckert

GENUS	—	*NODAVIRUS*
TYPE SPECIES	NODAMURA VIRUS	—

PROPERTIES OF THE VIRUS PARTICLE

Morphology Virus particles are unenveloped, roughly spherical, 29-30 nm in diameter. Icosahedral shell symmetry (T = 3). Structure of the protein shell of BBV has been solved to atomic dimensions.

Physicochemical properties MW 8×10^6; S_{20w} = 135-142; buoyant density in CsCl = 1.30-1.35 g/cm^3 (varies with species). Stable in 1% sodium dodecyl sulfate except Boolara virus; black beetle and flock house viruses are stable at pH 3 and form stable crystals; resistant to organic solvents.

Nucleic acid Two ssRNA molecules, one each of 1.1 and 0.48×10^6 in the same particle, 16% RNA by weight; both molecules of the isolated RNA are required for infection.

Protein One major polypeptide species (β) of MW 39×10^3 and one minor species (γ) of MW= 4.5×10^3; derived by proteolytic cleavage of a precursor protein (α) of MW= 44×10^3. Mature virions often contain some uncleaved precursor protein.

Lipid Probably none.

Carbohydrate Not determined.

Antigenic properties All are cross-reactive by double-diffusion precipitin line test but members are distinguishable by other properties such as neutralization, electric charge and host range.

REPLICATION

The virus replicates in the cytoplasm. RNA synthesis is resistant to actinomycin D. Infected cells contain three ssRNAs: RNA 1 (MW = 1.1×10^6), RNA 2 (MW = 0.48×10^6) and RNA 3 (MW = 0.15×10^6). RNA 3 is not packaged into virions. RNA 1 codes for protein A (MW = 105×10^3); the latter is probably a component of the viral RNA polymerase. RNA 2 codes for coat protein (MW = 44×10^3); and RNA 3, for protein B (MW = 10×10^3) of

Taxonomic status	English vernacular name	International name
	unknown function. Cells infected with isolated RNA 1 synthesize RNA 1 and RNA 3 but no RNA 2. Both RNA 1 and RNA 2 are required for production of virions. RNA 2 strongly inhibits synthesis of RNA 3. Messenger activity of the RNAs in infected cells is RNA 3 > RNA 2 > RNA 1. Cultured virus forms defective-interfering particles readily if not passaged at low multiplicity of infection. Readily generates persistently infected cells, resistant to infection by wild type virus.	

BIOLOGICAL ASPECTS

Host range

Natural - All species were isolated from insects, Diptera, Coleoptera or Lepidoptera. Viruses are not notably host-specific. Experimental - Most, if not all, can be propagated in the common wax moth, *Galleria mellonella*. Nodamura virus, unlike other members, grows in suckling mice but not in cultured *Drosophila* cells. All except Nodamura virus form plaques on *Drosphila* cell monolayers.

Transmission

Nodamura virus is transmissible to suckling mice by *Aedes aegypti*.

OTHER MEMBERS

Black beetle virus
Flock house virus
Gypsy moth virus
Boolarra virus
Manawatu virus

Derivation of Name

Nodamura: village in Japan where type species was isolated.

REFERENCES

Dasgupta, R.; Sgro, J.-Y.: Nucleotide sequences of three Nodavirus RNA2's: the messengers for their coat protein precursors. Nuc. Acids Res. *17*: 7525-7526 (1989).

Dasmahapatra, B.; Dasgupta, R.; Ghosh, A.; Kaesberg, P.: Structure of the Black Beetle Virus genome and its functional implications. J. Mol. Biol. *182*:183-189 (1985).

Dasmahapatra, B.; Dasgupta, R.; Saunders, K.; Selling, B.H.; Gallagher, T.M.; Kaesberg, P.: Infectious RNA derived by transcription from cloned cDNA copies of the genomic RNA of an insect virus. Proc. Natl. Acad. Sci. USA *83*: 63-66 (1986).

Friesen, P.D.; Rueckert, R.R.: Synthesis of Black Beetle Virus proteins in cultured *Drosophila* cells: Differential expression of RNAs 1 and 2. J. Virol. *37*: 876-886 (1981).

Gallagher, T.M.; Friesen, P.D.; Rueckert, R.R.: Autonomous replication and expression of RNA 1 from Black Beetle Virus. J. Virol. *46*:481-489 (1983).

Gallagher, T.M.; Rueckert, R.R.: Assembly-dependent maturation cleavage in provirions of a small icosahedral insect ribovirus. J. Virol. *62*: 3399-3406 (1988).

Greenwood, L.K.; Moore, N.F.: Purification and partial characterization of a small RNA-virus from Lymantria - Identification of a Nodamura-like virus. Microbiologica *5*:49-52 (1982).

Guarino, L.A.; Ghosh, A.; Dasmahapatra, B.; Dasgupta, R.; Kaesberg, P.: Sequence of the Black Beetle Virus subgenomic RNA and its location in the viral genome. Virology *139*: 199-203 (1984).

Hosur, M.V.; Schmidt, T.; Tucker, R.C.; Johnson, J.E.; Gallagher, T.M.; Selling, B.H.; Rueckert, R.R.: Structure of an insect virus at 3.0 Å resolution. Proteins: structure, function, and genetics *2*: 167-176 (1987).

Hosur, M.V.; Schmidt, T.; Tucker, R.C.; Johnson, J.E.; Selling, B.H.; Rueckert, R.R.: Black Beetle Virus - Crystallization and particle symmetry. Virology *133*:119-127 (1984).

Kaesberg, P.; Dasgupta, R.; Sgro, J.-Y.; Wery, J.-P.; Selling, B.H.; Hosur, M.V.; Johnson, J.E.: Structural homology among four nodaviruses as deduced by sequencing and x-ray crystallography. J. Mol. Biol. *214*: 423-435 (1990).

Moore, N.F.; Tinsley, T.W.: The small RNA viruses of insects. Arch. Virol. *72*:229-245 (1982).

Reavy, B.; Moore, N.F.: The replication of small RNA-containing viruses of insects. Microbiologica *5*:63-84 (1982).

Reinganum, C.; Bashiruddin, J.B.; Cross, G.F.: Boolarra virus: a member of the *Nodaviridae* isolated from *Oncopera intricoides* (Lepidoptera: *Hepialidae*). Intervirology *24*: 10-17 (1985).

Scotti, P.D.; Dearing, S.; Mossop, D.W.: Flock House Virus: A Nodavirus isolated from *Costelytra zealandica* (White) (Coleoptera: *Scarabaediae*). Arch. Virol. *75*:181-189 (1983).

Selling, B.H.; Allison, R.F.; Kaesberg, P.: Genomic RNA of an insect virus directs synthesis of infectious virions in plants. Proc. Natl. Acad. Sci. USA *87*: 434-438 (1990).

Selling, B.H.; Rueckert, R.R.: Plaque assay for Black Beetle Virus. J. Virology *51*: 251-253 (1984).

Wery, J.-P.; Johnson, J.E.: Molecular biology at atomic resolution: the structure of an insect virus and its functional implications. Analyt. Chem. *61*: 1341A-1350A (1989).

Taxonomic status	English vernacular name	International name
GROUP	PEA ENATION MOSAIC VIRUS GROUP	—

Revised by R. Hull & S. Salquero

TYPE MEMBER	PEA ENATION MOSAIC VIRUS (PEMV) (25;257)	—

PROPERTIES OF THE VIRUS PARTICLE

Morphology Polyhedral particles, ≈ 28 nm in diameter.

Physicochemical properties Particles of two types (B and T) with MWs ≈ 5.7 x 10^6 (B) and ≈ 4.6 x 10^6 (T); S_{20w} ≈ 112 (B) and ≈ 99 (T); buoyant density in CsCl ≈ 1.42 g/cm^3 for B component; T component is disrupted; in Cs_2SO_4 ≈ 1.38 g/cm^3 for both components. Particles readily disrupted in neutral chloride salts.

Nucleic acid Two molecules of linear positive-sense ssRNA, MWs ≈ 1.7 and 1.3 x 10^6. Some strains also contain a third RNA component with MW ≈ 0.3 x 10^6 which is considered to be a satellite.

Protein Major coat polypeptide, MW ≈ 22 x 10^3, minor polypeptide (MW ≈ 28 x 10^3) associated with aphid transmissibility.

Lipid None reported.

Carbohydrate None reported.

Antigenic properties Weakly to moderately immunogenic. One or two precipitin lines are formed in gel diffusion tests.

REPLICATION

Virus particles found in the nucleus. Vesicular cytopathological structures originating from nuclear membranes develop in infected cells. *In vitro* translation of RNA-1 yields two major proteins, vp2 (MW ≈ 88 x 10^3) and vp4 (MW ≈ 30 x 10^3); a minor protein, vp1 (MW ≈ 147 x 10^3) is also obtained. RNA-2 is translated into vp3 (MW ≈ 45 x 10^3). PEMV antiserum precipitated vp2, suggesting that this protein contains sequences related to those of the coat protein. A VPg-like protein (MW ≈ 17.5 x 10^3) is linked to the 5'-ends of both genomic RNAs; neither are polyadenylated.

Taxonomic status	English vernacular name	International name
	BIOLOGICAL ASPECTS	
Host range	Infects many legumes but few species in other families.	
Transmission	Transmitted by aphids in a persistent manner. Readily transmissible experimentally by mechanical inoculation, often with loss of aphid transmissibility.	

REFERENCES

Demler, S.A.; de Zoeten, G.A.: Characterization of a satellite RNA associated with pea enation mosaic virus. J. gen. Virol. *70:* 1075-1084 (1989).

de Zoeten, G.A.; Gaard, G.; Diez, F.B.: Nuclear vesiculation associated with pea enation mosaic virus-infected plant tissue. Virology *48:*638-647 (1972).

Gabriel, C.J.; de Zoeten, G.A.: The *in vitro* translation of pea enation mosaic virus RNA. Virology *139:*223-230 (1984).

Hull, R.: Particle differences related to aphid-transmissibility of a plant virus. J. gen. Virol. *34:*183-187 (1977).

Hull, R.: Pea enation mosaic virus; *In* Kurstak, E. (ed.), Handbook of Plant Virus Infections and Comparative Diagnosis, pp. 239-256. (Elsevier/North Holland, Amsterdam, 1981).

Powell, C.A.; de Zoeten, G.A.: Replication of pea enation mosaic virus RNA in isolated pea nuclei. Proc. Natl. Acad. Sci. USA *74:* 2919-2922 (1977).

Powell, C.A.; de Zoeten, G.A.; Gaard, G.: The localization of pea enation mosaic virus-induced RNA-dependent RNA polymerase in infected peas. Virology *78:*135-143 (1977).

Reisman, D.; de Zoeten, G.A.: A covalently linked protein at the 5'-ends of the genomic RNAs of pea enation mosaic virus. J. gen. Virol. *62:*187-190 (1982).

Taxonomic status	English vernacular name	International name
GROUP	SOIL-BORNE WHEAT MOSAIC VIRUS GROUP	***FUROVIRUS***

Compiled by A.A. Brunt

TYPE MEMBER	SOIL-BORNE WHEAT MOSAIC VIRUS (SBWMV) (77)	—

PROPERTIES OF THE VIRUS PARTICLE

Morphology
Fragile rod-shaped particles ≈ 20 nm in diamter with predominant lengths of 92-160 nm and 250-300 nm; two possible members also have particles 380-390 nm long. Particle helically symmetrical; the protein helix of beet necrotic yellow vein virus particles is right-handed with a pitch of 2.6 nm and 12 1/4 subunits per turn.

Physicochemical properties
Two or more sedimenting components, number depending on member. S_{20w} = 220-230 (long particles), 170-225 (shorter particles), and 126-177 (deletion mutants); buoyant density in CsCl ≈ 1.32 g/cm^3.

Nucleic acid
Two molecules of linear ssRNA, RNA-1 = 5.9-6.9 kb (MW = 1.83-2.42 x 10^6), RNA-2 = 3.5-4.3 kb (MW = 1.23-1.83 x 10^6) and deleted molecules = 2.1-2.4 kb (MW = 0.74-0.84 x 10^6). The two RNAs are not polyadenylated at 3'-end and do not have a 5'-cap structure of Vpg.

Beet necrotic yellow vein virus is unusual in having four ssRNAs (6.75, 4.61, 1.77 and 1.47 kb, excluding poly A tails): all four have been sequenced and shown to be 3'-polyadenylated (65-140 residues) and to have 5'-terminal caps (m^7 GpppA); RNAs 3 and 4 also have unusually long (445 and 379 nucleotides, respectively) 5'-non-coding regions.

Protein
Single polypeptide MW = 19.7-23.0 x 10^3 (but mostly ≈ 20 x 10^3).

Lipid
None reported.

Carbohydrate
None reported.

Antigenic properties
Most members are fairly good immunogens. Type member is serologically fairly distantly related to potato mop top, broadbean necrosis, oak golden stripe and sorghum chlorotic spot viruses.

Taxonomic status	English vernacular name	International name
GROUP	TOBACCO RATTLE VIRUS GROUP	***TOBRAVIRUS***

Revised by D.J. Robinson

TYPE MEMBER	TOBACCO RATTLE VIRUS (TRV) (PRN ISOLATE) (346)	—

Properties of the Virus Particle

Morphology Tubular particles with helical symmetry and pitch of 2.5 nm; 20.3-23.1 nm in diameter (electron microscopy) or 20.5-22.5 nm (X-ray). RNA-1 and RNA-2 contained in tubular particles of 180-215 nm length (L) and 46-114 nm length (S), the latter length depending on the isolate.

Physicochemical properties MWs = 48-50 x 10^6 (L) and 11-29 x 10^6 (S); S_{20w} = 286-306 (L) and 155-245 (S); buoyant density in CsCl = 1.306-1.324 g/cm^3. Particles stable.

Nucleic acid Two strands of linear positive-sense ssRNA with MWs ≈ 2.4 x 10^6 (RNA-1) and 0.6-1.4 x 10^6 (RNA-2), the size of the latter depending on the isolate; 5' terminus has the sequence $m^7G^{5'}ppp^{5'}Ap$... RNA-1 is infective; RNA-2 is not infective but it contains the cistron for the capsid protein; both RNAs are required for production of progeny long (L) and short (S) particles. RNA-2 sequences differ considerably between isolates of each member virus and also between the three member viruses. In contrast, RNA-1 sequences of different isolates of each member virus are substantially similar, though entirely different from those of other member viruses.

Protein One coat polypeptide; MW ≈ 22 x 10^3.

Lipid None reported.

Carbohydrate None reported.

Antigenic properties Moderately immunogenic; considerable antigenic heterogeneity between isolates. Little or no serological relationship between members.

Replication

Accumulation of virus particles sensitive to cycloheximide but not to chloramphenicol, suggesting cytoplasmic ribosomes are involved in viral protein synthesis; L

Taxonomic status	English vernacular name	International name

particles accumulate in early part of infection cycle, whereas S particles tend to accumulate in the later stages; isolates unable to produce nucleoprotein particles (NM isolates) are obtained from inocula containing only L particles; such isolates are also found in naturally infected plants.

BIOLOGICAL ASPECTS

Host range Wide, including monocotyledonous and dicotyledonous families.

Transmission Primarily by nematodes (*Paratrichodorus* and *Trichodorus* spp.) in which the virus may persist, but is not retained through the moult; there is no evidence of replication within the vector. Also transmitted by seed and experimentally by mechanical inoculation, but with difficulty for NM isolates.

SEQUENCE SIMILARITIES WITH OTHER VIRUS GROUPS OR FAMILIES

Some non-structural proteins synthesized by tobacco rattle virus share sequence similarities with non-structural proteins of some other RNA plant viruses [e.g. tripartite virus (alfalfa mosaic, brome mosaic and cucumber mosaic viruses) and a monopartite plant (carnation mottle virus)] and animal viruses [e.g. Sindbis virus].

OTHER MEMBERS

Pea early-browning (120)
Pepper ringspot

Derivation of Name tobra: sigla from *tob*acco *ra*ttle.

REFERENCES

Goldbach, R.: Genome similarities between plant and animal RNA viruses. Microbiol. Sci. *4:*197-202 (1987).

Harrison, B.D.; Robinson, D.J.: The tobraviruses. Adv. Virus Res. *23:*25-77 (1978).

Harrison, B.D.; Robinson, D.J.: Tobraviruses; *In* Kurstak, E. (ed.), Handbook of Plant Virus Infections and Comparative Diagnosis, pp. 515-540. (Elsevier/North Holland, Amsterdam, 1981).

Harrison, B.D.; Robinson, D.J.: Tobraviruses; *In* van Regenmortel, M.H.V.; Fraenkel-Conrat, H. (eds.), The Plant Viruses, The rod-shaped plant viruses, Vol. 2, pp. 339-369 (Plenum Press, New York, 1986).

Robinson, D.J.; Harrison, B.D.: Unequal variation in the two genome parts of tobraviruses and evidence for the existence of three separate viruses. J. gen. Virol. *66:*171-176 (1985).

Taxonomic status	English vernacular name	International name
GROUP	BROME MOSAIC VIRUS GROUP (215)	***BROMOVIRUS***

Revised by E.P. Rybicki

TYPE MEMBER	BROME MOSAIC VIRUS (BMV) (3;180)	—

PROPERTIES OF THE VIRUS PARTICLE

Morphology Polyhedral particles ≈ 26 nm in diameter, with icosahedral T = 3 surface lattice symmetry. Although all particles have approximately the same S_{20w} (85), three different particles exist, one containing one molecule of RNA-1, one containing one molecule of RNA 2 and one containing one molecule each of RNA 3 and RNA 4.

Physicochemical properties MW ≈ 4.6 x 10^6; S_{20w} ≈ 85; buoyant density in CsCl ≈ 1.35 g/cm^3. Particles swell reversibly in presence of Ca^{2+} or Mg^{2+} or pH increase above 7.0, with concomitant changes in capsid conformation; pronounced loss of stability; salt-, detergent-, protease- and ribonuclease-susceptible in swollen form.

Nucleic acid Three genomic molecules of linear positive-sense ssRNA of 3.2 kb (RNA 1), 2.8 kb (RNA 2) and 2.1 kb (RNA 3); 0.8 kb coat protein mRNA (RNA 4) is also encapsidated. 5'-termini capped ($m^7G^{5'}ppp^{5'}Gp$...); 3'-termini have a tRNA-like structure which accepts tyrosine *in vivo* and *in vitro*; however, encapsidated RNA is not aminoacylated.

Protein Single coat polypeptide, MW ≈ 20 x 10^3, 189 amino acids. Highly basic NH_2-terminus (≈ 25 residues); often partially degraded *in vivo* and *in vitro*.

Lipid None reported.

Carbohydrate None reported.

Antigenic properties Moderately poor immunogens unless stabilised by glutaraldehyde or formaldehyde cross-linking. Serological reactions of virions best performed below pH 7.0; serological differences between compact and swollen forms, and artificial empty capsids and intact virions. Moderate to distant serological relationships between all members.

REPLICATION

RNAs of BMV, cowpea chlorotic mottle and broad bean mottle viruses can be *in vitro* translated into 4 major proteins: RNAs 1 and 2 are monocistronic and encode proteins of MW ≈ 110 and ≈ 95 x 10^3 respectively; dicistronic RNA 3 encodes proteins of MW ≈ 35 (3a) and ≈ 20 x 10^3 (coat protein), 3a protein only produced *in vitro*; RNA 4 is a subgenomic monocistronic mRNA for coat protein. Genomic RNAs replicate via full-length complementary (-) sense RNA, in membrane-associated replicase complex containing RNA 1 and RNA 2 encoded proteins. Replicase recognition of genomic RNA depends upon integrity of tRNA-like structure; 3'-adenylate residue is added autocatalytically; 3'-terminal sequence is apparently telomeric. *In vitro* transcripts of cloned cDNA copies of BMV are infectious. Recombination can occur during replication to restore native sequences. BMV RNA 3 contains an intercistronic variable-length oligo(A) tract which is restored in a template-independent manner if deleted. RNA 4 arises by internal initiation of replicase on (-)strand of RNA 3 using a specific promoter sequence; this RNA is not replicated. An intact coat protein gene, especially the NH_2-terminal 25 amino acids, is necessary for RNA encapsidation, and 3a and coat proteins are necessary for cell-to-cell movement. Virions assemble in cytoplasm, and granular inclusions are found, sometimes in crystalline arrays. Particles are found in both cytoplasm and nuclei of old infected cells. The viruses do not appear to be tissue-specific.

BIOLOGICAL ASPECTS

Host range Narrow.

Transmission Some members transmitted by beetles. Readily transmissible experimentally by mechanical inoculation.

SEQUENCE SIMILARITIES WITH OTHER VIRUS GROUPS OR FAMILIES

Predicted amino acid sequences from single ORFs of RNA 1 and RNA 2 share significant sequence homology with analogous sequences from cucumoviruses and alfalfa mosaic virus, and with replication-associated proteins produced by tobamoviruses, tobraviruses and togaviruses: this would put the bromoviruses in the Sindbis-like virus "superfamily". The RNA 1 ORF contains a putative nucleotide binding domain; the RNA 2 ORF contains the putative (+)strand RNA virus polymerase domain. Distant

Taxonomic status	English vernacular name	International name

similarities can be seen between the 3a protein (movement protein) sequences of bromoviruses and cucumoviruses, and more distantly between these and ilarviruses and alfalfa mosaic virus.

OTHER MEMBERS

Broad bean mottle (101)
Cassia yellow blotch
Cowpea chlorotic mottle (49)
Melandrium yellow fleck (236)
Spring beauty latent

Derivation of Name bromo: sigla from *bro*me *mo*saic; also, from plant generic name *Bromus*, brome.

REFERENCES

Ahlquist, P.; Dasgupta, R.; Kaesberg, P.: Nucleotide sequence of the brome mosaic virus genome and its implications for viral replication. J. Mol. Biol. *172:*369-383 (1984).

Ahlquist, P.; French, R.; Janda, M.; Loesch-Fries, L.S.: Multicomponent RNA plant virus infection derived from cloned viral cDNA. Proc. Natl. Acad. Sci. USA *81:*7066-7070 (1984).

Allison, R.F.; Janda, M.; Ahlquist, P.: Sequence of cowpea chlorotic mottle virus RNAs 2 and 3 and evidence of a recombination event during bromovirus evolution. Virology *172:* 321-330 (1989).

Allison, R.; Thompson, C.; Ahlquist, P.: Regeneration of a functional RNA virus genome by recombination between deletion mutants and requirement for cowpea chlorotic mottle virus 3a and coat genes for systemic infection. Proc. Natl. Acad. Sci. USA *87:* 1820-1824 (1990).

Bujarski, J.J.; Kaesberg, P.: Genetic recombination between RNA components of a multipartite plant virus. Nature *321:* 528-531 (1986).

Dale, J.L.; Gibbs, A.J.; Behncken, G.M.: *Cassia* yellow blotch virus: a new bromovirus from an Australian native legume, *Cassia pleurocarpa*. J. gen. Virol. *65:*281-288 (1984).

Davies, J.W.; Verduin, B.J.M.: *In vitro* synthesis of cowpea chlorotic mottle virus polypeptides. J. gen. Virol. *44:*545-549 (1979).

Francki, R.I.B.: The Plant Viruses, Polyhedral Virions with Tripartite Genomes. vol. 1 (Plenum Press, New York, London, 1985).

Francki, R.I.B.; Milne, R.G.; Hatta, T.: Bromovirus group; *In* Atlas of Plant Viruses, Vol. II, pp. 69-80. (CRC Press, Boca Raton, Fl., 1985).

Goldbach, R.: Genome similarities between plant and animal RNA viruses. Microbiol. Sci. *4:*197-202 (1987).

Karpova, O.V.; Tyulkina, L.G.; Atabekov, K.J.; Rodionova, N.P.; Atabekov, J.G.: Deletion of the intercistronic poly (A) tract from brome mosaic virus RNA 3 by ribonuclease H and its restoration in progeny of the religated RNA 3. J. gen. Virol. *70:* 2287-2297 (1989).

Lane, L.C.: Bromoviruses; *In* Kurstak, E. (ed.), Handbook of Plant Virus Infections and Comparative Diagnosis, pp. 333-376 (Elsevier/North Holland, Amsterdam, 1981).

Loesch-Fries, L.S.; Hall, T.C.: Synthesis, accumulation and encapsidation of individual brome mosaic virus RNA components in barley protoplasts. J. gen. Virol. *47:*323-332 (1980).

Miller, W.A.; Bujarski, J.J.; Dreher, T.W.; Hall, T.C.: Minus-strand initiation by brome mosaic virus replicase within the 3' tRNA-like structure of native and modified RNA templates. J. Mol. Biol. *187:* 537-546 (1986).

Miller, W.A.; Dreher, T.W.; Hall, T.C.: Synthesis of brome mosaic virus subgenomic RNA *in vitro* by internal initiation on (-) sense genomic RNA. Nature *313:*68-70 (1985).

Pfeiffer, P.: Changes in the organization of bromegrass mosaic virus in response to cation binding as probed by changes in susceptibility to degradative enzymes. Virology *102:*54-61 (1980).

Rao, A.L.N.; Dreher, T.W.; Marsh, L.E.; Hall, T.C.: Telomeric function of the transfer RNA-like structure of brome mosaic virus RNA. Proc. Natl. Acad. Sci. USA *86:* 5335-5339 (1989).

Rao, A.L.N.; Hall, T.C.: Requirement for a viral trans-acting factor encoded by brome mosaic virus RNA 2 provides strong selection *in vivo* for functional recombinants. J. Virol. *64:* 2437-2441 (1990).

Roenhorst, J.W.; van Lent, J.W.M.; Verduin, B.J.M.: Binding of cowpea chlorotic mottle virus to cowpea protoplast and relation of binding to virus entry and infection. Virology *164:* 91-98 (1988).

Rybicki, E.P.; von Wechmar, M.B.: The serology of the bromoviruses. I. Serological interrelationships of the bromoviruses. Virology *109:*391-402 (1981).

Sacher, R.; Ahlquist, P.: Effects of deletions in the N-terminal basic arm of brome mosaic virus coat protein on RNA packaging and systemic infection. J. Virol. *63:* 4545-4552 (1989).

Valverde, R.A.: Properties of the RNA and coat protein of spring beauty latent virus and comparison with those of four bromoviruses. Phytopathology *79:* 1214-1215 (1989).

Valverde, R.A.: Spring beauty latent virus: a new member of the bromovirus group. Phytopathology *75:* 395-398 (1985).

Taxonomic status	English vernacular name	International name
GROUP	CUCUMBER MOSAIC VIRUS GROUP	*CUCUMOVIRUS*

Revised by H. Lot

TYPE MEMBER	CUCUMBER MOSAIC VIRUS (CMV) (S ISOLATE) (1;213)	—

PROPERTIES OF THE VIRUS PARTICLE

Morphology

Polyhedral particles ≈ 29 nm in diameter, with T = 3 surface lattice symmetry. Although all particles have approximately the same S_{20w}, three particles exist, one containing one molecule of RNA-1, one containing one molecule of RNA-2 and one containing one molecule each of RNA-3 and RNA-4.

Physicochemical properties

MW ≈ 6 x 10^6; S_{20w} ≈ 99; buoyant density in CsCl ≈ 1.37 g/cm^3; particles readily disrupted in neutral chloride salts and by SDS; particles sensitive to RNAse.

Nucleic acid

Three genomic molecules of the linear positive-sense ssRNA; RNA-1 (3357-3389 nt), RNA-2 (3035-3050 nt), RNA-3 (2197-2216 nt); a sub-genomic coat protein mRNA (RNA-4, 1027 nt) is also encapsidated. Satellite RNA (333-393 according to isolates) which depends on genomic RNA for replication and encapsidation, occurs in some CMV and peanut stunt virus isolates. There is little sequence similarity between the satellite and genomic RNAs. 5' termini of all four RNAs have the sequence: m^{7}Gppp... The 3' termini of all RNAs contain long (200 nt) regions of sequence similarity characteristic of each member; the termini are not poly-adenylated but they can be aminoacylated by tyrosine.

Protein

Single coat polypeptide, MW ≈ 26.2 x 10^3.

Lipid

None reported.

Carbohydrate

None reported.

Antigenic properties

Poor immunogens. Serological reactions complicated by sensitivity of virus particles to salts. Distant serological relationships among members.

REPLICATION

The RNAs of CMV can each be translated *in vitro* to yield 4 major proteins; RNAs 1, 2, 3 and 4 code for proteins of

MW = 111, 94, and 30 x 10^3 and coat protein, respectively. RNAs replicate via corresponding negative-sense strands. Coat protein readily detected in infected cells and protoplasts but other translation products have not been found. Virus particles assemble in the cytoplasm and accumulate there as scattered particles. Sometimes, virus particles also occur in nuclei and vacuoles, rarely forming crystals. Chloroplasts with extensively modified internal structure are characteristic of cells infected by some virus strains. Small vesicles associated with the tonoplast may be the sites of RNA replication.

Biological Aspects

Host range

Type member has wide host range (≈ 1000 species); other members have more restricted host ranges.

Transmission

Seed transmission in several host plants. Transmitted by aphids in non-persistent manner. Readily transmissible experimentally by mechanical inoculation.

Sequence Similarities with Other Virus Groups or Families

Some putative non-structural proteins coded for by RNAs 1 and 2 share sequence similarities with similar proteins of other plant viruses [e.g. tripartite genome viruses (alfalfa mosaic, brome mosaic), a bipartite genome virus (tobacco rattle), monopartite genome viruses (carnation mottle and tobacco mosaic)], and Sindbis virus, a RNA animal virus with a monopartite genome.

Other Members

Peanut stunt (91) (= *Robinia* mosaic) (65)
Tomato aspermy (79)

Probable member

Cowpea ringspot

Derivation of Name

cucumo: sigla from *cucu*mber *mo*saic.

REFERENCES

Davies, C.; Symons, R.H.: Further implications for the evolutionary relationships between tripartite plant viruses based on cucumber mosaic virus RNA 3. Virology *165:* 216-224 (1988).

Francki, R.I.B.; Milne, R.G.; Hatta, T.: Cucumovirus group; *In* Atlas of Plant Viruses, vol. II, pp. 53-68 (CRC Press, Boca Raton, Fl., 1985).

Goldbach, R.: Genome similarities between plant and animal RNA viruses. Microbiol. Sci. *4:*197-202 (1987).

Hull, R.; Maule, A.J.: Virus Multiplication; *In* Francki, R.I.B. (ed.), The Plant Viruses, Polyhedral Virions with Tripartite Genomes, Vol. 1, pp. 83-115 (Plenum Press, New York, 1985).

Kaper, J.M.; Waterworth, H.E.: Cucumoviruses; *In* Kurstak, E. (ed.), Handbook of Plant Virus Infections and Comparative Diagnosis, pp. 257-332 (Elsevier/North Holland, Amsterdam, 1981).

Owen, J.; Shintaku, M.; Aeschleman, P.; Ben Tahar, S.; Palukaitis, P.: Nucleotide sequence and evolutionary relationships of cucumber mosaic virus (CMV) strains: CMV RNA 3. (In Press) (1990).

Phatak, H.C.; Diaz-Ruiz, J.R.; Hull, R.: Cowpea ringspot virus: a seed transmitted cucumovirus. Phytopath. Z. *87:*132-142 (1976).

Rezaian, M.A.; Williams, R.H.V.; Gordon, K.H.J.; Gould, A.R.; Symons, R.H.: Nucleotide sequence of cucumber mosaic virus RNA 2 reveals a translation product significantly homologous to corresponding proteins of other viruses. Eur. J. Biochem. *143:* 277-284 (1984).

Rezaian, M.A.; Williams, R.H.V.; Symons, R.H.: Nucleotide sequence of cucumber mosaic virus RNA 1: Presence of a sequence complementary to part of the viral satellite RNA and homologies with other viral RNAs. Eur. J. Biochem. *150:*331-339 (1985).

Rizzo, T.M.; Palukaitis, P.: Nucleotide sequence and evolutionary relationships of cucumber mosaic virus (CMV) strains: CMV RNA 1. J. gen. Virol. *70:* 1-11 (1989).

Rizzo, T.M.; Palukaitis, P.: Nucleotide sequences and evolutionary relationships of cucumber mosaic virus (CMV) strains: CMV RNA 2. J. gen. Virol. *69:* 1777-1787 (1988).

Symons, R.H.: Viral Genome Structure; *In* Francki, R.I.B. (ed.), The Plant Viruses, Polyhedral Virions with Tripartite Genomes, Vol. 1, pp. 57-81 (Plenum Press, New York , 1985).

Taxonomic status	English vernacular name	International name
GROUP	TOBACCO STREAK VIRUS GROUP (275)	*ILARVIRUS*

Revised by R.I.Hamilton

TYPE MEMBER	TOBACCO STREAK VIRUS (TSV) (44)	—

PROPERTIES OF THE VIRUS PARTICLE

Morphology Particles are quasi-isometric or occasionally bacilliform. Particles of different components, although differing in size, are mostly 26-35 nm in diameter.

Physicochemical properties Several particle types, S_{20w} = 80-120; buoyant density of all particle types ≈ 1.36 g/cm^3 in CsCl; particles readily disrupted in neutral chloride salts and by SDS.

Nucleic acid Three molecules of linear positive-sense ssRNA; MW ≈ 1.1 (RNA-1), 0.9 (RNA-2) and 0.7 (RNA-3) x 10^6; coat protein mRNA of MW ≈ 0.3 x 10^6 (RNA-4), a subgenomic fragment of RNA-3 is also encapsidated.

Protein Single coat polypeptide, MW ≈ 25 x 10^3.

Lipid None reported.

Carbohydrate None reported.

Antigenic properties Weakly to moderately immunogenic. Serological relationship among some members. Several sub-groups (I-X) with type member and closely related strains as the only members of subgroup I.

REPLICATION

Besides RNAs 1-3, coat protein or RNA-4 is required for infectivity. Coat protein of most ilarviruses (and also of alfalfa mosaic virus) are interchangeable in this respect. For some members it has been shown that RNAs 1 and 2 can be translated *in vitro* into proteins of MW corresponding to the total genetic information present in these RNAs. RNA-3-directs the synthesis of protein, MW ≈ 34 x 10^3, and RNA-4 directs the synthesis of coat protein.

BIOLOGICAL ASPECTS

Host range Wide.

Transmission

Some viruses transmitted by seeds and by pollen to flower-bearing plants. Thrips may be involved in pollen transmission. Readily transmissible experimentally by mechanical inoculation.

Sequence Similarities with other Virus Groups or Families

The MW ≈ 34 x 10^3 protein directed by tobacco streak virus RNA-3 shares some sequence similarity with the MW ≈ 35 x 10^3 protein directed by alfalfa mosaic virus RNA-3.

Other Members

Subgroup II
- *Asparagus* virus II (288)
- *Citrus* leaf rugose (164)
- *Citrus* variegation (164)
- Elm mottle (139)
- Tulare apple mosaic (42)

Subgroup III
- *Prunus* necrotic ringspot (5) (= some isolates of rose mosaic)
- Blueberry scorch
- Cherry rugose
- Hop C
- Apple mosaic (83) (= some isolates of rose mosaic)
- Danish plum line pattern
- Hop A

Subgroup IV
- Prune dwarf (19)

Subgroup V
- American plum line pattern (280)

Subgroup VI
- Spinach latent (281)

Subgroup VII
- Lilac ring mottle (201)

Subgroup VIII
- *Hydrangea* mosaic

Subgroup IX
- *Humulus japonicus*

Subgroup X
- *Parietaria* mottle

Derivation of Name

ilar: sigla from *i*sometric *la*bile *r*ingspot.

REFERENCES

Adams, A.N.; Clark, M.F.; Barbara, D.J..: Host range, purification and some properties of a new ilarvirus from *Humulus japonicus*. Ann. appl. Biol. *114:* 497-508 (1989).

Caciagli, P.; Boccardo, G.; Lovisolo, O.: Parietaria mottle virus, a possible new ilarvirus from *Parietaria officinalis* (*Urticaceae*). Plant Pathol. *38:* 577-584 (1989).

Cornelissen, B.J.C.; Janssen, H.; Zuidema, D.; Bol, J.F.: Complete nucleotide sequence of tobacco streak virus RNA 3. Nuc. Acids Res. *12:* 2427-2437 (1984).

Francki, R.I.B.: The Plant Viruses, Polyhedral virions with tripartite genomes, Vol. 1 (Plenum Press, New York, 1985).

Fulton, R.W.: Ilar-like characteristics of American plum line pattern virus and its serological detection in *Prunus*. Phytopathology *72:*1345-1348 (1982).

Fulton, R.W.: Ilarviruses; *In* Kurstak, E. (ed.), Handbook of Plant Virus Infections and Comparative Diagnosis, pp. 377-413 (Elsevier/North Holland, Amsterdam, 1981).

Huttinga, H.; Mosch, W.H.M.: Lilac ring mottle virus: a coat protein-dependent virus with a tripartite genome. Acta Hort. *59:*113-118 (1976).

MacDonald, S.G.; Martin, R.R.; Bristow, P.R.: Characterization of an ilarvirus associated with blueberry scorch disease. (Submitted).

Sdoodee, R.; Teakle, D.S.: Transmission of tobacco streak virus by *Thrips tabaci*: a new method of plant virus transmission. Plant Pathology *36:* 377-380 (1987).

Thomas, B.J.; Barton, R.J.; Tuszynski, A.: *Hydrangea* mosaic virus, a new ilarvirus from *Hydrangea macrophylla* (Saxifragaceae). Ann. appl. Biol. *103:*261-270 (1983).

Taxonomic status	English vernacular name	International name
GROUP	ALFALFA MOSAIC VIRUS GROUP	—

Revised by R. Goldbach

TYPE MEMBER	ALFALFA MOSAIC VIRUS (AMV)(46;229)	—

PROPERTIES OF THE VIRUS PARTICLE

Morphology

Bacilliform particles, 56 x 18 nm (B), 43 x 18 nm (M), 35 x 18 nm (Tb) and a particle (Ta) that occurs in both bacilliform (Ta-b; 30 x 18 nm) and ellipsoidal shape (Ta-t). The three largest particles contain a single RNA molecule each: RNA-1 (B), RNA-2 (M), RNA-3 (Tb); Ta contains two molecules of RNA-4.

Physicochemical properties

MW of particles (B,M,Tb and Ta) range from 6.9 to 3.5 x 10^6; S_{20w} ≈ 94 (B), 82 (M), 73 (Tb) and 66 (Ta); buoyant density in Cs_2SO_4 ≈ 1.28 g/cm^3, in CsCl (after fixation) ≈ 1.37 g/cm^3 (components differ slightly in banding densities). RNA content of all particle species between 15-17%. Particles disrupted in neutral chloride salts; sensitive to ribonuclease at pH 6-7, but do not appear to swell.

Nucleic acid

Three molecules of linear positive-sense ssRNA of 3644 nucleotides (RNA-1), 2593 nucleotides (RNA-2) and 2142 nucleotides (RNA-3); a 881 nucleotide mRNA (RNA-4), encoding the coat protein, is also encapsidated. 5' termini of the four RNAs have the sequence of $m^7G^{5'}ppp^{5'}Gp$. The last 145 nucleotides at the 3'-termini of all four RNA species are similar.

Protein

One coat polypeptide, MW ≈ 24 x 10^3. Some degradation of the N terminus may occur *in vitro*.

Lipid

None reported.

Carbohydrate

None reported.

Antigenic properties

Poor immunogens. Biologically distinct strains are antigenically similar.

REPLICATION

Besides RNAs-1, -2 and -3, coat protein or RNA-4 is required for infectivity. Coat proteins from ilarviruses are also able to activate the AMV genome. RNAs-1, -2, and

Taxonomic status	English vernacular name	International name

-3 encode non-structural proteins of MWs = 126 (P1), 90 (P2) and 32 x 10^3 (P3), respectively. Coat protein is translated from a subgenomic mRNA (RNA-4) derived from RNA-3. All four proteins are produced by *in vitro* translation of these RNAs and have been detected in infected leaves and inoculated protoplasts. P1 and P2 are involved in viral RNA synthesis. P3 is the putative transport protein, involved in cell-to-cell movement of virus. Virus particles accumulate in the cytoplasm and sometimes in vacuoles, either scattered or as whorled aggregates.

Biological Aspects

Host range Wide host range, including many leguminous plants.

Transmission Seed transmission in some plants. Transmitted by aphids in a nonpersistent manner. Readily transmissible experimentally by mechanical inoculation.

Sequence Similarities with other Virus Groups or Families

The proteins directed by RNA-1 and RNA-2 show significant sequence similarity with the proteins directed by RNA-1 and RNA-2 of ilar-, bromo-, cucumo-, tobamo- and tobraviruses as well as with the proteins directed by the genomic RNA of alphaviruses. AMV, bromo- and cucumoviruses are, moreover, very similar in tripartite genome organization.

References

Berna, A.; Briand, J.-P.; Stussi-Garaud, C.; Godefroy-Colburn, T.: Kinetics of accumulation of the three non-structural proteins of alfalfa mosaic virus in tobacco plants. J. gen. Virol. *67:*1135-1147 (1986).

Cornelissen, B.J.C.; Bol, J.F.: Homology between the proteins encoded by tobacco mosaic virus and two tricornaviruses. Plant Mol. Biol. *3:*379-384 (1984).

Francki, R.I.B.: The Plant Viruses, Polyhedral virions with tripartite genomes, Vol. 1 (Plenum Press, London, New York, 1985).

Goldbach, R.: Genome similarities between plant and animal RNA viruses. Microbiol. Sci. *4:*197-202 (1987).

Goldbach, R.; Wellink, J.: Evolution of plus-strand RNA viruses. Intervirology *29:* 260-267 (1988).

Haseloff, J.; Goelet, P.; Zimmern, D.; Ahlquist, P.; Dasgupta, R.; Kaesberg, P.: Striking similarities in amino acid sequence among nonstructural proteins encoded by RNA viruses that have dissimilar genomic organization. Proc. Natl. Acad. Sci. USA *81:*4358-4362 (1984).

Jaspars, E.M.J.: Interactions of alfalfa mosaic virus nucleic acid and protein. *In* Davies, J.W. (ed.), Molecular Plant Virology, Vol. 1, pp. 155-221 (CRC Press, Boca Raton, Fl., 1985).

Stussi-Garaud, C.; Garaud, J.-C.; Berna, A.; Godefroy-Colburn, T.: In situ location of an alfalfa mosaic virus non-structural protein in plant cell walls: Correlation with virus transport. J. gen. Virol. *68:* 1779-1784 (1987).

van Pelt-Heerschap, H.; Verbeek, H.; Huisman, M.J.; Loesch-Fries, L.S.; van Vloten-Doting, L.: Non-structural proteins and RNAs of alfalfa mosaic virus synthesized in tobacco and cowpea protoplasts. Virology *161:*190-197 (1987).

van Vloten-Doting, L.; Francki, R.I.B.; Fulton, R.W.; Kaper, J.M.; Lane, L.C.: Tricornaviridae - a proposed family of plant viruses with tripartite, single-stranded RNA genomes. Intervirology *15:*198-203 (1981).

Taxonomic status	English vernacular name	International name
GROUP	BARLEY STRIPE MOSAIC VIRUS GROUP	***HORDEIVIRUS***

Revised by R.I. Hamilton

TYPE MEMBER	BARLEY STRIPE MOSAIC VIRUS (BSMV) (68; 344)	—

PROPERTIES OF THE VIRUS PARTICLE

Morphology Elongated rigid particles about 20 x 110-150 nm; helically symmetrical with pitch ≈ 2.5 nm.

Physicochemical properties Major sedimenting species S_{20w} = 182-193 S; other species S_{20w} = 165-200S, depending on the strain.

Nucleic acid Three molecules of positive sense ssRNA of 3768 nt (RNA α), 3289 nt (RNA β) and 3164 (RNA γ) are present in the type strain; other strains contain similar RNAs. In the Argentine mild strain, a fourth RNA arises from a deletion in RNA. Other RNAs of 800-2900 nt are found, depending on the strain and may represent subgenomic RNAs. There is no appreciable sequence similarity between RNA α and the other genomic RNAs of BSMV, and none between those of BSMV and poa semilatent virus. Each RNA has m^7GpppGUA at its 5'-end and a poly A tract of 8-40 nt followed by a 236-238 nt tRNA-like structure at its 3'-end which accepts tyrosine.

Protein Single polypeptide, MW = 22.15 x 10^6.

Lipid None reported.

Carbohydrate Capsid protein is reported to be glycosylated.

Antigenic properties Efficient immunogens. Members are distantly related serologically.

REPLICATION

Virus particles accumulate in both cytoplasm and nuclei, most being in the cytoplasm. RF RNAs corresponding to all viral ssRNAs can be isolated from infected plants. RNA α of the type strain has an ORF which is translated *in vitro* to produce a protein, MW ≈ 129.6 x 10^3, possibly the virus replicase. RNA β is translated *in vivo* into capsid protein (βa), MW ≈ 22.15 x 10^3 and a second one (βb), MW ≈ 58.1 x 10^3; two other ORFs code for proteins, MW

Taxonomic status	English vernacular name	International name

≈ 17.4 and 14.1 x 10^3, the functions of which are unknown. RNA γ contains ORFs for two proteins, MW ≈ 87.3 and 17.2 x 10^3 whose functions are also unknown.

Biological Aspects

Host range Narrow host range, mostly among *Gramineae.*

Transmission By mechanical inoculation and through seed. No vector is known.

Sequence similarities with other Virus Groups or Families

There are similarities in genome organization with that of beet necrotic yellows virus (furovirus group), and potato virus X and white clover mosaic virus (potexvirus group).

Other Members

Anthoxanthum latent blanching
Lychnis ringspot
Poa semilatent

Derivation of Name hordei: from Latin *hordeum*, 'barley'.

References

Carroll, T.W.: Hordeiviruses: biology and pathology. *In* van Regenmortel, M.H.V.; Fraenkel-Conrat, H. (eds.), The Plant Viruses, The Rod-shaped Plant Viruses, Vol. 2, pp. 373-395. (Plenum Press, New York, 1986).

Edwards, M.L.; Kelley, S.E.; Arnold, M.K.; Cooper, J.I.: Properties of a hordeivirus from *Anthoxanthum odoratum.* Plant Pathology *38:* 209-218 (1989).

Gustafson, G.; Armour, S.L.: The complete nucleotide sequence of RNAß from the type strain of barley stripe mosaic virus. Nuc. Acids Res. *14:* 3895-3909 (1986).

Gustafson, G.; Armour, S.L.; Gamboa, G.C.; Burgett, S.G.; Shepherd, J.W.: Nucleotide sequence of barley stripe mosaic virus RNAα: RNAα encodes a single polypeptide with homology to corresponding proteins from other viruses. Virology *170:* 370-377 (1989).

Gustafson, G.; Hunter, B.; Hanau, R.; Armour, S.L.; Jackson, A.O.: Nucleotide sequence and genetic organization of barley stripe mosaic virus RNAγ. Virology *158:*394-406 (1987).

Hunter, B.G.; Heaton, L.A.; Bracker, C.E.; Jackson, A.O.: Structural comparison of poa semilatent virus and barley stripe mosaic virus. Phytopathology *76:* 322-326 (1986).

Hunter, B.G.; Smith, J.; Fattouh, F.; Jackson, A.O.: Relationship of *lychnis* ringspot virus to barley stripe mosaic virus and poa semilatent virus. Intervirology *30:* 18-26 (1989).

Jackson, A.O.; Hunter, B.G.; Gustafson, G.D.: Hordeivirus relationships and genome organization. Ann. Rev. Phytopathol. *27:* 95-121 (1989).

Kozlov, Y.V.; Rupasov, V.V.; Adyshev, D.M.; Belgelarskaya, S.N.; Agranovsky, A.A.; Mankin, A.S.; Morozov, S.Yu.; Dolja, V.V.; Atabekov, J.G.: Nucleotide sequence of the 3'-terminal tRNA-like structure in barley stripe mosaic virus genome. Nuc. Acids Res. *12:* 4001-4009 (1984).

Partridge, J.E.; Shannon, L.M.; Gumpf, D.J.; Colbaugh, P.: Glycoprotein in the capsid of plant viruses as a possible determinant of seed transmissibility. Nature *247:* 391-392 (1974).

Taxonomic status	English vernacular name	International name
GROUP	RICE STRIPE VIRUS	***TENUIVIRUS***

Revised by K. Tomaru

TYPE MEMBER	RICE STRIPE VIRUS (RSV) (269)	—

PROPERTIES OF THE VIRUS PARTICLE

Morphology Filamentous particles ≈ 8 nm in diameter and varying lengths, occasionally branched and composed of a supercoiled ribonucleoprotein ≈ 3 nm diameter.

Physicochemical properties Several (2-5) nucleoprotein components each probably containing a single species of RNA distinguished by rate zonal centrifugation in sucrose gradients; buoyant density in CsCl ≈ 1.28 g/cm^6. Nucleoproteins contain ≈ 5% RNA.

Nucleic acid Four molecules of linear ssRNA (possibly minus sense), MW ≈ 3.0, 1.6, 1.1 and 0.9 x 10^6 (five ssRNAs of MW ≈ 3.01, 1.18, 0.8, 0.78 and 0.52 x 10^6 have been reported for another member). Four or five species of dsRNA are also detected in purified virus preparations.

Protein Single coat polypeptide, MW ≈ 32 x 10^3. Virion-associated RNA-dependent RNA polymerase has been found in purified preparations of type member and the serologically related member, rice grassy stunt virus.

Lipid None reported.

Carbohydrate None reported.

Antigenic properties Serological relationships between some members.

REPLICATION

Virus particles occur in the cytoplasm and occasionally in the nuclei of leaf cells. Large amounts of a non-structural protein are found in infected cells.

BIOLOGICAL ASPECTS

Host range Individual members may have broad host ranges; hosts are restricted to the *Gramineae*.

Taxonomic status	English vernacular name	International name
Transmission	Transmitted by leafhoppers in a persistent manner; transovarial transmission by viruliferous females to progeny. Experimental sap transmission difficult.	

OTHER MEMBERS

Maize stripe virus (300)
Rice grassy stunt virus (320)

Possible members

Echinochloa hoja blanca virus
European wheat striate mosaic virus
Rice hoja blanca virus (299)
Winter wheat mosaic virus

Derivation of Name tenui: from Latin *tenuis*, 'thin, fine weak'.

REFERENCES

Falk, B.W.; Tsai, J.H.: Identification of single- and double-stranded RNAs associated with maize stripe virus. Phytopathology *74:*909-915 (1984).

Falk, B.W.; Tsai, J.H.; Lommel, S.A.: Differences in levels of detection for the maize stripe virus capsid and major non-capsid proteins in plant and insect hosts. J. gen. Virol. *68:*1801-1811 (1987).

Falk, B.W.; Morales, F.J.; Tsai, J.H.; Niessen, A.I.: Serological and biochemical properties of the capsid and major noncapsid proteins of maize stripe, rice hoja blanca, and *Echinochloa* hoja blanca viruses. Phytopathology *77:*196-201 (1987).

Gingery, R.E.; Nault, L.R.; Yamashita, S.: Relationship between maize stripe virus and rice stripe virus. J. gen. Virol. *64:*1765-1770 (1983).

Toriyama, S.: Ribonucleic acid polymerase activity in filamentous nucleoproteins of rice grassy stunt virus. J. gen. Virol. *68:*925-929 (1987).

Toriyama, S.: An RNA-dependent RNA polymerase associated with the filamentous nucleoproteins of rice stripe virus. J. gen. Virol. *67:*1247-1255 (1986).

Toriyama, S.: Characterization of rice stripe virus: A heavy component carrying infectivity. J. gen. Virol. *61:* 187-195 (1982).

Toriyama, S.; Watanabe, Y.: Characterization of single- and double-stranded RNAs in particles of rice stripe virus. J. gen. Virol. *70:* 505-511 (1989).

Taxonomic status	English vernacular name	International name

SATELLITES

Compiled by M. Mayo

TYPE MEMBER	CUCUMBER MOSAIC VIRUS RNA5 (CARNA5) (269)	—

DEFINITION

Satellites are nucleic acid molecules that depend for their multiplication on co-infection of a host cell with a helper virus. Satellite nucleic acids have no appreciable sequence homology with their helper virus genome and are not a part of its genome.

DISTINCTIVE FEATURES

Satellite nucleic acids are distinct from other types of dependent nucleic acid such as sub-genomic nucleic acids (e.g. defective interfering and messenger molecules), genome parts, and transmission-defective but independently replicating viruses. Some satellites may contribute advantageous characters to their helper virus; the distinction between these and genome parts is sometimes not clear-cut.

CLASSIFICATION

Most reported satellites are associated with plant viruses and these have been arbitrarily classified into four types according to physical and messenger properties of the satellite RNA. These are,

Type A - RNA is large (> 0.7 kb) and encodes a capsid protein that forms satellite-specific particles.

Type B - RNA is large (> 0.7 kb) and encodes a non-structural protein.

Type C - RNA is small (< 0.7 kb), lacks significant mRNA properties and does not form circular RNA.

Type D - RNA is small (<0.7 kb), lacks mRNA activity and forms circular molecules during replication.

KNOWN EXAMPLES OF SATELLITES

Most records of satellites are of those associated with plant viruses. Table 1 lists these, together with some of their properties. Satellites have also been found associated with viruses of other taxonomic groups. Examples are bacteriophage P4, which is a dsDNA satellite virus dependent on bacteriophage P2, adeno-associated viruses (*Dependovirus*: ***Parvoviridae***) which are ssDNA satellite viruses dependent on adenoviruses or herpesviruses, hepatitis delta virus which is a large, but circular, satellite RNA dependent on hepatitis B virus and a ssRNA satellite virus which is associated with chronic bee-paralysis virus.

Table 1: Plant virus satellites and their associated satellites

Helper Virus	Virus Group	RNA size	Type
Tobacco necrosis	Necrovirus	1.2 kb	A
Tobacco necrosis	Necrovirus	0.52 kb	C
Tobacco mosaic	Tobamovirus	1.1 kb	A
Panicum mosaic	(unclassified)	0.3 kb	A
Panicum mosaic	(unclassified)	0.4 kb	C
Maize white line mosaic	(unclassified)	c.1.3kb	A
Tomato black ring	Nepovirus	1.4 kb	B
Strawberry latent ringspot	Nepovirus	c.1.2 kb	B
Arabis mosaic	Nepovirus	1.1 kb	B
Arabis mosaic	Nepovirus	0.3 kb	D
Myrobalan latent ringspot	Nepovirus	c.1.2 kb	B
Chicory yellow mottle	Nepovirus	1.1 kb	B
Chicory yellow mottle	Nepovirus	0.46 kb	D
Grapevine fanleaf	Nepovirus	1.1 kb	B
Grapevine Bulgarian latent	Nepovirus	c.1.7 kb	B
Tobacco ringspot	Nepovirus	0.3 kb	D
Beet western yellows	Luteovirus	3.1 kb	B
Groundnut rosette	(unclassified)	0.9 kb	?B
Pea enation mosaic	(monotypic)	c. 0.8 kb	?B
Cucumber mosaic	Cucumovirus	0.3 kb	C
Peanut stunt	Cucumovirus	0.4 kb	C
Turnip crinkle	Carmovirus	0.2-0.3 kb	C
Cymbidium ringspot	Tombusvirus	0.7 kb	D
Tomato bushy stunt	Tombusvirus	0.7 kb	?C
Artichoke mottled crinkle	Tombusvirus	0.7 kb	?C
Carnation Italian ringspot	Tombusvirus	0.7 kb	?C
Petunia asteroid mosaic	Tombusvirus	0.7 kb	?C
Pelargonium leaf curl	Tombusvirus	0.7 kb	?C
Lucerne transient streak	Sobemovirus	0.32 kb	D
Velvet tobacco mottle	Sobemovirus	0.37 kb	D
Solanum nodiflorum mottle	Sobemovirus	0.38 kb	D
Subterranean clover mottle	Sobemovirus	0.33+0.39 kb	D
Barley yellow dwarf	Luteovirus	0.32 kb	D

REFERENCES

Ball, B.V.; Overton, H.A.; Buck, K.W.; Bailey, L.; Perry, J.N.: Relationships between the multiplication of chronic bee-paralysis virus and its associate particle. J. gen. Virol. *66:* 1423 - 1429 (1985).

Demler, S.A.; de Zoeten, G.A.: Characterization of a satellite RNA associated with pea enation mosaic virus. J. gen. Virol. *70:* 1075-1084 (1989).

Falk, B.W.; Chin, L.-S.; Duffus, J.E.: Complementary DNA cloning and hybridization analysis of beet western yellows luteovirus RNAs. J. gen. Virol. *70:* 1301-1309 (1989).

Francki, R.I.B.: Plant virus satellites. Ann. Rev. Microbiol. *39:* 151-174 (1985).

Fritsch, C.; Mayo, M.A.: Satellites of plant viruses. *In* Mandahar, C.L. (ed.), Plant Viruses, Structure and Replication, Vol. I, pp. 289-321 (CRC Press, Boca Raton, Fl., 1989).

Miller, W.A.; Hercus, T.; Waterhouse, P.M.; Gerlach, W.L.: Characterization of a satellite RNA associated with a luteovirus. Abst. VII Int. Congress of Virology, Edmonton, p299 (1987).

Murant, A.F.; Mayo, M.A.: Satellites of plant viruses. Ann. Rev. Phytopathol. *20:* 49-70 (1982).

Murant, A.F.; Rajeshwari, R.; Robinson, D.J.; Raschke, J.H.: A satellite RNA of groundnut rosette virus that is largely responsible for symptoms of groundnut rosette disease. J. gen. Virol. *69:* 1479-1486 (1988).

Piazolla, P.; Rubino, L.; Tousignant, M.E.; Kaper, J.M.: Two different types of satellite RNA associated with chicory yellow mottle virus. J. gen. Virol. *70:* 949-954 (1989).

Six, E.W.; Klug, C.A.C.: Bacteriophage P4: a satellite virus depending on a helper such as prophage P2. Virology. *51:* 327-344 (1973).

Wang, K.-S.; Choo, Q.-L.; Weiner, A.J.; Ou, J.-H.; Najarian, R.C.; Thayer, R.M.; Mullenbach, G.T.; Denniston, K.J.; Gerin, J.L.; Houghton, M.: Structure, sequence and expression of the hepatitis delta (δ) viral genome. Nature *323:* 508 - 514 (1986).

Taxonomic status	English vernacular name	International name

VIROIDS

Compiled by J.W. Randles and M.A. Rezaian

TYPE MEMBER	POTATO SPINDLE TUBER VIROID (PSTV)	—

DEFINITION

Viroids are unencapsidated, low molecular weight, circular, single-stranded infectious RNAs pathogenic to plants.

PROPERTIES OF VIROIDS

Physical properties

Non-denatured viroid molecules adopt extensive internal base pairing to give rod-like structures ≈ 50 nm long. These denature by cooperative melting to single-stranded circles of ≈ 100 nm contour length. MW = 80-122 x 10^3; S_{20w} = 8-10; Tm in 10 mM Na^+ ≈ 50°C; density in Cs_2SO_4 ≈ 1.6 g/cm^3.

Chemical properties

Comprise 246 to over 370 nucleotides; all except ASBVd are GC rich with central conserved regions. Oligomers have potential to form palindromic structures involving the upper part of the central conserved region. CCCVd, CLVd, AGVd, CbVd, show sequence rearrangements indicative of probable RNA recombination in viroids. No evidence for encoding protein.

Antigenic properties

No antigenicity demonstrated.

REPLICATION

Differ fundamentally from viruses which parasitise host translation; viroids parasitise host transcription possibly using RNA polymerase II. Multimers isolated *in vivo* may be replicative intermediates produced by a rolling circle mechanism. ASBVd multimers self-cleave *in vitro* to produce unit length viroid but others do not, and may rely on host factors for cleavage. PSTVd accumulates in nucleoli.

Biological Aspects

Host range Some with wide, others narrow host range in the angiosperms. CCCVd and CTiVd infect monocotyledons, remainder dicotyledons.

Transmission Most distributed by vegetative propagation but some transmissible by seed, aphids, or mechanical damage.

Classification

Sequences are the primary basis for comparison. The sequence of the central conserved region allows all characterized viroids to be classed into four groups. Variation occurs within each viroid "species" and an arbitrary level of 90% sequence similarity currently separates variants from species.

Other Members

Table 1: Grouping viroids using core sequence affinities

Viroid	Acronym	Size (Nuc.)	Group
Apple scar skin	ASSVd	330	ASSVd
Australian grapevine	AGVd	369	ASSVd
Avocado sunblotch (254)	ASBVd	247	ASBVd
Burdock stunt	BSVd	n.a.	
Chrysanthemum stunt	CSVd	356	PSTVd
Citrus exocortis (226)	CEVd	370–375	PSTVd
Coconut cadang-cadang (287)	CCCVd	246	PSTVd
Coconut tinangaja	CTiVd	254	PSTVd
Coleus blumei	CbVd1	n.a.	
	CbVd2	n.a.	
	CbVd3	248	
	CbVd	n.a.	
Columnea latent	CLVd	370	PSTVd
Grapevine yellow speckle 1	GYSVd 1	367	ASSVd
Grapevine yellow speckle 2	GYSVd 2+	363	ASSVd
Hop latent	HLVd	256	PSTVd
Hop stunt (326) *	HSVd	297–303	PSTVd
Peach latent mosaic	PLMVd	n.a.	
Potato spindle tuber (66)	PSTVd	359	PSTVd
Tomato apical stunt	TASVd	360	PSTVd
Tomato bunchy top	TBTVd	n.a.	
Tomato planto macho	TPMVd	360	PSTVd

* Agent also of cucumber pale fruit, dapple fruit of plum and peach, and isolated from citrus and grapevine.
n.a. not available; + synonymous with GVd1B (Koltunow et al., 1989)

Taxonomic status	English vernacular name	International name
	Possible members	
	Brazilian coleus viroid Carnation stunt viroid *Chrysanthemum* chlorotic mottle viroid *Citrus* viroids	

Derivation of Name viroid: from the name given to the sub-viral RNA agent of potato spindle tuber disease.

References

Diener, T.O.: Potato spindle tuber "virus" IV. A replicating, low molecular weight RNA. Virology *45:* 411-428 (1971).

Diener, T.O.: Viroids. Adv. Virus Res. *28:* 241-283 (1983).

Duran-Vila, N.; Roistacher, C.N.; Rivera-Bustamante, R.; Semancik, J.S.: A definition of citrus viroid groups and their relationship to the exocortis disease. J. gen. Virol. *69:* 3069-3080 (1988).

Hammond, R.; Smith, D.R.; Diener, T.O.: Nucleotide sequence and proposed secondary structure of *Columnea* latent viroid: a natural mosaic of viroid sequences. Nuc. Acids Res. *17:* 10083-10094 (1989).

Hashimoto, J.; Koganezawa, H.: Nucleotide sequence and secondary structure of apple scar skin viroid. Nuc. Acids Res. *15:*: 7045–7052 (1987).

Horst, R.K.: *Chrysanthemum* chlorotic mottle; *In* Diener, T.O. (ed.), The Viroids (The Viruses), pp. 291-295 (Plenum Press, New York, 1987).

Keese, P.; Osorio-Keese, M.E.; Symons, R.H.: Coconut tinangaja viroid: sequence homology with coconut cadang-cadang viroid and other potato spindle tuber viroid related RNAs. Virology *162:* 508-510 (1988).

Keese, P.; Symons, R.H.: Physical-chemical properties: molecular structure (primary and secondary); *In* Diener, T.O. (ed.), The Viroids (The Viruses). pp. 37-62 (Plenum Press, New York, 1987).

Koltunow, A.M.; Krake, L.R.; Johnson, S.D.; Rezaian, M.A.: Two related viroids cause grapevine yellow speckle disease independently. J. gen. Virol. *70:* 3411-3419 (1989).

Koltunow, A.M.; Rezaian, M.A.: A scheme for viroid classification. Intervirology *30:* 194-201 (1989).

Puchta, H.; Ramm, K.; Sänger, H.L.: The molecular structure of hop latent viroid (HLV), a new viroid occurring worldwide in hops. Nuc. Acids Res. *16.* 4197-4216 (1988).

Rezaian, M.A.: Australian grapevine viroid - evidence for extensive recombination between viroids. Nuc. Acids Res. *18:* 1813-1818 (1990).

Sano, T.; Hataya, T.; Terai, Y.; Shikata, E.: Hop stunt viroid strains from dapple fruit disease of plum and peach in Japan. J. gen. Virol. *70:* 1311-1319 (1989).

Schumacher, J.; Randles, J.W.; Riesner, D.: A two-dimensional electrophoretic technique for the detection of circular viroids and virusoids. Analytical Biochemistry *135:* 288-295 (1983).

Spieker, R.L.; Haas, B.; Charng, Y-C.; Freimüller, K.; Sänger, H.L.: Primary and secondary structure of a new viroid "species" (CbVd1) present in the *Coleus blumei* cultivar "Bienvenue". Nuc. Acids Res. *18*: 3998 (1990).

Author Index

Abad-Zapatero, C. 329
Abdel-Meguid, S.S. 329
Abe, H. 379
Abo El-Nil, M.M. 350
Abou-Elnasr, M.A. 215
Accotto, G.P. 215
Acharya, R. 325
Ackermann, H.-W. 125, 156, 158, 162, 164, 166, 180, 183, 185, 308
Adair, B.M. 325
Adams, A.N. 343, 391
Adams, M.J. 379
Adang, M.J. 122
Adler, J.J. 204
Adrian, T. 143, 144
Adyshev, D.M. 397
Aeschleman, P. 388
Agol, V.I. 325
Agranovsky, A.A. 347, 397
Ahern, K.G. 262
Ahl, R. 302
Ahlquist, P. 363, 384, 385, 393
Ahmed, N.A. 340
Ahne, W. 261, 262
Akashi, H. 326
Akusjarvi, G. 143
Alestrom, P. 143
Alexander, D.J. 246
Allan, T.C. 343
Allison, A.C. 149
Allison, R. 384, 356, 374
Almond, J.W. 246
Almond, M.R. 210
Anders, K. 136
Anderson, D.K. 232
Anderson, L.J. 143
Ando, T. 302
Antignus, Y. 343
Antoniw, J.F. 215
Archard, L.C. 102
Arif, B.A. 121
Arikawa, J. 283
Arita, I. 101, 102
Armour, S.L. 396
Armstrong, F.B. 356
Arnold, M.K. 396
Arnott, H.J. 121
Ashby, J.W. 311
Asjes, C.J. 344
Atabekov, J.G. 384, 397
Atabekov, K.J. 384
Auperin, D.D. 288
Bachrach, H.L. 52, 302, 325
Baer, L. 262
Baer, R. 110
Bailey, L. 402
Balazs, E. 305
Ball, B.V. 402
Baltimore, D. 52, 299
Bamford, D.H. 156
Bandy, B.P. 211
Bankier, A.T. 110
Banner, L.R. 236
Bannert, H. 299
Banttari, E.E. 315
Bar-Joseph, M. 347
Barash, I. 211
Barbara, D.J. 343, 391
Barber, T.L. 198
Barker, H. 371
Barnett, I.T.R. 325
Barnett, O.W. 52
Barrell, B.G. 110
Barsoum, N.J. 101
Bartha, A. 143, 144
Barton, R.J. 204, 210, 391
Bashiruddin, J.B. 325, 374
Bates, R.C. 172
Batterson, W. 110
Baxby, D. 101
Bays, D. 356
Beachy, R.N. 356
Beaudreau, G.S.
Becerra, V.M. 102
Becht, H. 202
Beck, D. 379
Becker, B. 139
Beczner, L. 350
Behncken, G.M. 363, 384
Belgelarskaya, S.N. 397
Belloncik, S. 198
Ben Tahar, S. 388
Ben-Zvi, B. 211
Benko, M. 143
Bergold, G.H. 289
Berna, A. 393, 394
Bernards, R. 144
Berns, K.I. 172
Berry, E.S. 136
Berzofsky, J.A. 299
Bevan, M.W. 350
Bibb, J.A. 205

Biggin, M.D. 110
Biggs, P.M. 110
Binns, M.M. 236
Bishop, D.H.L. 52, 246, 249, 261, 283, 288, 289
Bishop, J.M. 299
Bitton, G. 308
Black, D.N. 101
Blinov, V.M. 232, 329
Blissard, G.W. 131
Bobek, L.A. 205
Boccardo, G. 198, 215, 340, 347, 367, 391
Bohm, P. 122
Bol, J.F. 215, 344, 350, 391, 393
Bomu, W. 363
Bonami, J.-R. 172
Borden, E.C. 101
Border, D.J. 205
Bornkamm, G.W. 110
Bos, L. 319
Bostian, K.A. 205
Bouissou, C. 288
Boulanger, P. 143
Bouley, J.P. 338
Bouloy, M. 233
Boursnell, M.E.G. 236
Bouzoubaa, S. 379
Bowen, E.T.A. 241, 249
Bowyer, J.W. 363
Boyke, V.F. 347
Boyt, P.M. 302
Bozarth, R.F. 204, 205, 210, 211
Bracker, C.E. 396
Bradley, D.W. 232
Brahic, M. 325
Brakke, M.K. 379
Brandt, W.E. 232, 233
Brauer, D.H.K. 110
Brault, V. 371
Bredenbeek, P.J. 221, 222, 236
Brennan, V.E. 205
Brian, D.A. 236
Briand, J.-P. 393
Bricogne, G. 335
Bricout, F. 144
Bridger, J.C. 198
Brierley, I. 236
Brinton, M.A. 222, 233
Brisco, M.J. 215
Bristow, P.R. 391
Brown, D.R. 102
Brown, F. 52, 122, 198, 261, 302, 325, 326
Brown, J.F. 198
Brown, P. 299
Brown, T.D.K. 236
Brown-Luedi, M. 152
Browne, D.. 122
Bruening, G. 363, 371
Bruenn, J.A. 205
Bruns, M. 288
Brunt, A.A. 344, 350, 356, 379
Buchen-Osmond, C. 356
Buchman, T.G. 110
Buchmeier, M.J. 249, 288, 289
Buck, K.W. 205, 210, 211, 365, 379, 402
Buckley, S.M. 288
Bugert, J. 101, 102
Bujarski, J.J. 384, 385
Bulla, L.A.Jr. 122
Burand, J.P. 122, 123
Burbank, D.E. 139
Burgett, S.G. 396
Burgyan, J. 334, 335
Burn, V. 205
Burroughs, J.N. 52, 198, 302
Bussey, H. 205
Butel, J.S. 149
Buzayan, J.M. 371
Caciagli, P. 391
Cahour, A. 233
Calendar, R. 308
Calisher, C.H. 52, 221, 232, 261, 283
Candresse, T. 347, 371
Cannizzaro, G. 347
Carmichael, L.E. 110
Carrington, J.C. 305, 335
Carroll, T.W. 396
Carstens, E.B. 121
Carter, B.J. 172
Carter, M.J. 302
Casals, J. 52, 232, 288, 289
Castellano, M.A. 317, 347
Cavanagh, D. 236
Cedeno, R. 288
Chakerian, R. 121
Chamberlain, R.W. 232
Chambers, T.J. 232
Chamorro, M. 325
Chang, S.-F. 183
Chanock, R.M. 110, 198
Charnay, P. 116
Charng, Y-C. 405
Chen, M.H. 365
Chen, Z. 363
Chin, L.-S. 402
Chirnside, E.D. 221
Choo, Q.L. 232, 402
Choppin, P.W. 246
Choudhury, M.M. 313
Christie, R.G. 350
Christie, S.R. 340
Chu, P.W.G. 329

Chumakov, M.P. 52
Cihak, J. 288
Cizdziel, P. 205
Clapham, D. 359
Clare, J.J. 205
Clark, J.M. 311
Clark, M.F. 391
Cleveland, P.H. 249
Coffin, J. 299
Cohen, G.N. 288, 289
Cohen, J. 198
Cohen, S. 343
Coia, G. 233
Colbaugh, P. 397
Collett, M.S. 232
Compans, R.W. 246, 249
Condit, C. 317
Consigli, R.A. 122, 123
Conti, M. 262, 340
Cook, D. 131
Cook, E.H. 232
Cooley, A.J. 101
Cooper, J.I. 396
Cooper, P.D. 52, 325
Cornelissen, B.J.C. 350, 391, 393
Corteyn, A.H. 198
Couch, J.A. 122
Covey, S.N. 152
Cowley, J. 198
Cox, N.J. 249
Crabtree, K. 356
Crandall, R.A. 102, 302
Cream, J.J. 102
Crestani, O.A. 338
Crick, J. 261
Crook, N.E. 122, 198
Cropp, C.B. 283
Cross, G.F. 374
Crowther, R. 325, 331
Cubitt, W.D. 302
Cuppels, D.A. 139
D'Aquilio, M. 347
D'Arcy, C.J. 311
Daheli, M. 288
Dai, H. 183
Dale, J.L. 363, 384
Dales, S. 101
Dall, D.J. 198
Dalrymple, J.M. 232, 283
Dalrymple, M.A. 110
Dalton, R.E. 205
Daniel, M.D. 110
Darai, G. 101, 102, 136, 299
Dasgupta, R. 329, 373, 374, 384, 393
Dasmahapatra, B. 373, 374
Daubert, S.D. 363
Davies, C. 329, 383
Davies, J.W. 177, 363, 384, 394
Davis, S.W. 101
Davison, A.J. 110
Dawe, V.H. 145
Day, P.R. 205
de Haan, P. 283
de Sequeira, O.A. 340
de Vries, A.A.F. 221
de Zoeten, G.A. 344, 376, 402
de-The, G. 110
Dearing, S. 374
DeFries, R.U. 249
Deinhardt, F. 110
Deininger, P.L. 110
Dejean, A. 116
Delius, H. 110, 136
Demangeat, G. 371
Demler, S.A. 376, 402
Demski, J.W. 356
den Boon, J.A. 222
Denhardt, D.T. 180
Denniston, K.J. 402
Derks, A.F.L.M. 344
Descoteaux, J-P. 101
Despres, P. 233
Di Franco, A. 317, 335, 344, 379
Diaco, R. 311
Diamond, M.E. 205
Diaz-Ruiz, J.R. 388
Diener, T.O. 405
Dietzgen, R.G. 261
Diez, F.B. 376
Digard, P. 236
Digoutte, J.-P. 261, 283
Dijkstra, J. 262
Dimock, K. 288
Dobos, P. 202
Dodds, J.A. 344, 347, 365
Doel, T.R. 367
Doerfler, W. 121, 122
Doi, Y. 215
Dolan, A. 110
Dolja, V.V. 347, 397
Domier, L.L. 356, 363
Dominguez, G. 221
Donald, S. 110
Donchenko, A.P. 232, 329
Doolittle, R.F. 299
Dougherty, W.G. 356
Dowhanick, J.J. 205
Downs, W.G. 222
Dreher, T.W. 385
Dreizin, R.S. 143, 144
DuBow, M.S. 125, 156, 158, 162, 164, 166, 180, 183, 185, 308

Hattman, S. 139
Havens, W.M. 205
Hayashi, Y. 302
Hearne, P.Q. 334
Hearon, S.S. 344
Heaton, L.A. 261, 305, 396
Hedrick, R.P. 199
Heick, J.A. 344
Heinz, F.X. 222, 232
Hellmann, U. 222, 283
Hellmann, G.M. 356
Hemida, S.K. 313, 319
Hemmer, O. 371
Henderson, D.A. 101
Hendry, D.A. 331
Hercus, T. 311, 402
Hermodson, M.A. 329
Hess, R.T. 331
Hibino, H. 311
Hibrand, L. 371
Hiebert, E. 350, 356
Hierholzer, J.C. 143, 144
Hill, B.J. 202
Hill, J.H. 311
Hillman, B.I. 261, 305, 334, 335
Hills, G.J. 359
Hirth, L. 152, 338
Hiruki, C. 365
Hogle, J.M. 305
Hohmann, A.W. 122
Hohn, T. 152
Holland, J.J. 261
Hollings, M. 210, 356
Holmes, K.V. 236
Holzmann, H. 232
Hoops, P. 122
Hopper, J.E. 205
Horiuchi, K. 183
Horner, G.W. 143
Horst, R.K. 405
Horwitz, M.S. 143
Horzinek, M.C. 221, 222, 232, 233, 236, 238
Hosur, M.V. 374
Hotchin, J. 289
Houghton, M. 232, 402
Howard, C.R. 116, 288, 289
Howarth, A.J. 152, 177
Howley, P.M. 149
Hsieh, C.H. 123
Hsu, H.-K. 311
Hsu, Y.H. 379
Hu, J.S. 347, 359
Huang, C.-M. 183
Huang, E.-S. 110
Huang, H.V. 221
Hudson, G.S. 110
Huet, J.C. 347
Huijberts, N. 319
Huisman, M.J. 350, 394
Hull, R. 122, 152, 198, 215, 261, 329, 334, 350, 376, 388
Hung, S.L. 236
Hunt, R.E. 313
Hunter, B.G. 261, 396
Hunter, K. 143
Hutcheson, A.M. 319, 340
Huttinga, H. 319, 391
Hyman, L. 311
Iacono-Connors, L. 283
Iapalucci, S. 288, 289
Icho, T. 205
Igarashi, A. 222, 232, 233
Inaba, Y. 326
Inglis, A.S. 356
Inglis, S.C. 236
Isaacson, M. 241, 249
Iwaki, M. 311, 317
Jackson, A.O. 261, 396
Jacoby, D.R. 289
Jamil, N. 205
Janda, M. 384
Janssen, H. 391
Jarvis, A.W. 164, 166
Jaspars, E.M.J. 394
Jeggo, M.H. 198
Jennings, D.M. 198
Jezek, Z. 101, 102
Jochim, M.M. 198
Johns, L.J. 344
Johnson, G.P. 101
Johnson, J.E. 329, J.E. 363, 374
Johnson, K. 241, 249, 261, 289
Johnson, M.S. 299
Johnson, S.D. 405
Johnston, R.E. 356
Johnstone, G.R. 311
Joklik, W.K. 198
Jolly, C.A. 311
Jonard, G. 152, 305, 379
Jones, A.T. 215, 329
Jones, P. 379
Josephs, S.F. 299
Jupin, I. 379
Kaariainen, L. 221, 222, 233, 283
Kaesberg, P. 329, 373, 374, 384, 393
Kalkkinen, N. 221, 222
Kallender, H. 379
Kallio, A. 221
Kalmakoff, J. 122
Kalter, S.S. 143, 144
Kamer, G. 363
Kaminjolo, J.S. 102

Kanyuka, K.V. 344
Kaper, J.M. 371, 388, 394, 402
Kapikian, A.Z. 198
Karabatsos, N. 221, 232, 261, 283
Karasev, A.V. 347
Karpova, O.V. 384
Kasdorf, G.G.F. 326
Kassanis, B. 215
Kasza, L. 325
Kawamura, H. 143
Keck, J.G. 236
Keefer, M.A. 180
Keene, J.D. 249
Keese, P. 338, 405
Keith, G. 262
Kelley, S.E. 396
Kells, D.T.C. 202
Kelly, D.C. 136, 172
Kempson-Jones, G.F. 211
Kendall, T.L. 379
Kennedy, S. 325
Khaless, N. 315
Khalil, J.A. 344
Kiley, M.P. 241, 249
Kilpatrick, B.A. 110
Kilpatrick, D. 102
Kim, J.W. 211
Kimura, T. 262
King, L.A. 325, 331
Kingsbury, D.W. 241, 246, 272
Kisary, J. 144
Kit, S. 149
Kitajima, E.W. 317, 338
Klug, C.A.C. 402
Knight, J.C. 199, 102, 246, 283, 299
Knipe, D.M. 110, 143
Knorr, D.A. 334
Knowles, N.J. 325
Knudson, D.L. 198
Koch, H.G. 136
Koenig, R. 305, 325, 329, 334, 335, 338, 344, 350
Koganezawa, H. 405
Kolakofsky, D. 246
Koltin, Y. 205, 211
Koltunow, A.M. 405
Kombe, A.H. 102
Koonin, E.V. 232, 329, 347
Kozlov, Y.V. 397
Krake, L.R. 405
Krell, J.D. 122
Krell, P.J. 131, 202
Krug, R.M. 272
Kruse, J. 152
Kuczmarski, D. 139
Kuhn, C.W. 145
Kuismanen, E. 283
Kunz, C. 232
Kuo, G. 232
Kuo, T.-T. 183
Kurath, G. 262
Kurstak, E. 172, 261, 262, 283, 334, 344, 347, 350, 356, 371, 376, 381, 384, 388, 391
Kuzio, J. 121
La Torre, J.L. 325
Ladeveze, I. 371
Ladnyi, I.D. 101
Lai, M.M.C. 236
Lana, A.F. 338
Lane, L.C. 139, 384, 394
Lang, D. 211
Langenberg, W.G. 379
Lannan, C.N. 199
Lanneau, M. 347
Lapenotiere, H.F. 283
LaPlaca, M. 325
Larson, R. 232
Laude, H. 236
Lawler, K.A. 122
Lawson, R.H. 343
Lazarowitz, S.G. 177
Le Gall, O. 371
Lee, P.E. 136
Lehmann-Grube, F. 288, 289
Leider, J. 236
Leis, J. 299
Leisy, D.J. 121, 122
Lemaire, O. 379
Lemius, J. 317
Lemke, L. 210
Lenches, E.M. 232
Lennon, A.M. 315
Leon, P. 315
Leong, J.C. 262
Lesemann, D.-E. 334, 338, 347
Leslie, A.G.W. 329
Lesnaw, J.A. 205, 317
Leunissen, J. 363
Levine, A.J. 149
Levis, R. 221
Levy, M. 101
Lhoas, P. 205
Li, X.H. 305
Li, Y. 363
Lightner, D.V. 172
Lin, M.T. 317, 338
Lin, Y.-H. 183
Linthorst, H.J.M. 215, 344, 350
Lipton, H. 325
Lisa, V. 215,367
Lister, R.M. 311, 340, 347

Lockhart, B.E.L. 154, 315
Loesch-Fries, L.S. 384, 394
Lohuis, D. 262
Lommel, S.A. 305, 365, 379, 399
Lomonossoff, G. 363, 365
Lonberg-Holm, K. 325
Lopez, N. 289
Lopez, R. 288, 289
Lorenzen, N. 262
Lovisolo, O. 367, 391
Lucero, M. 289, 288
Luisoni, E. 215, 340, 367
Lunina, N.A. 347
Luytjes, W. 236
Lvov, D.K. 52, 222, 233
Lyttle, D.J. 102
Maat, D.Z. 319
MacCallum, F.O. 326
MacDonald, S.G. 391
Mackenzie, D.J. 344
Mackey, J.K. 143
Madeley, C.R. 52, 302
Madin, S.H. 52, 302
Maeda, A. 305
Mahy, B. 246, 325
Makkouk, K.M. 334
Mandahar, C.L. 402
Mandart, E. 116
Mandel, B. 325
Mandelbrot, A. 210
Mandl, C.W. 232
Maniloff, J. 125, 180, 183
Mankin, A.S. 397
Marani, F. 379
Marciel-Zambolin, E. 350
Marcoli, R. 158
Mari, J. 172
Maria, E.R.A. 317
Marinho, V.L.A. 338
Marion, P.L. 116
Marr, L.D. 221
Marriott, A.C. 283
Marsh, L.E. 385
Marshall, I.D. 52,
Martelli, G.P. 261, 334, 335, 343, 347, 371, 379
Martin, J.L. 325
Martin, R.R. 311, 391
Martini, G. 249
Maruniak, J.E. 122
Marzachi, C. 215
Masenga, V. 262
Mason, W.S. 116
Massalski, P.R. 177
Mata, M. 164, 166
Matsuura, Y. 289
Matthews, R.E.F. 52, 210
Maule, A.J. 152, 388
Maurer, B. 299
Mautner, V. 143
Maynard, J.E. 232
Mayo, M.A. 215, 311, 325, 329, 371, 379, 402
McCaustland, K.A. 232
McClure, M.A. 299
McCormick, J.B. 241, 249, 288
McCrae, M.A. 198
McFadden, G. 102
McFadden, J.J.P. 210, 211
McFerran, J.B. 144, 325
McGeoch, D.J. 110
McGinty, R.M. 211
McGregor, S. 325
McKenzie, R.A. 102
McKern, N.M. 356
McKillop, E.R. 325
McLean, G.D. 329, 356
McMorran, J.P. 343
McNab, D. 110
McNulty, M.S. 246, 325
Meegan, J.M. 283
Meints, R.H. 139
Meints, S.M. 139
Meister, B. 139
Melnick, J.L. 110, 149, 325
Memelink, J. 344
Mengeling, W.L. 302
Mercenier, A. 164, 166
Merdinoglu, D. 379
Merryweather, A.T. 331
Mertens, P.P.C. 198
Meshi, T. 359
Messing, J. 152
Meyer, M. 371
Meyers, G. 232
Meyers, T.R. 199
Mihok, S. 101
Miki, T. 302
Miles, J.A.R. 149
Miller, L.K. 122
Miller, R.L. 207
Miller, W.A. 311, 385, 402
Mills, P.R. 344
Milne, B. 110
Milne, R.G. 152, 177, 198, 215, 261, 262, 283, 311, 317, 329, 338, 340, 344, 347, 356, 363, 367, 384, 388
Mims, C.A. 289
Mindich, L. 158, 185
Mircetich, S.M. 371
Mislivec, P.B. 211
Mitchel, L. 101

Moennig, V. 232
Moens, Y. 102
Mohanty, S.B. 325
Monath, T.P. 198, 221
Monis, J. 344
Monsarrat, A. 338
Montagnier, L. 299
Moore, M.B:
Moore, N.F. 325, 331, 374
Morales, F.J. 399
Morch, M.-D. 338
Morozov, S.Y. 344, 397
Morris, T.J. 305, 331, 334, 335
Morris-Krsinich, B.A.M. 365
Morrison, T.G. 246
Mosch, W.H.M. 391
Mosmann, T.R. 110
Moss, B. 101, 102
Mossop, D.W. 365, 374
Mowat, W.P. 350
Muhlemann, M.F. 102
Mulder, C. 110
Mullenbach, G.T. 402
Muller, G. 288
Murant, A.F. 313, 319, 325, 311, 371, 402
Murphy, F.A. 101, 232, 241, 249, 261, 283, 289
Murti, K.G. 136
Muscio, O.A. 325
Myers, G. 299
Nagel, J. 350
Nagy, E. 202
Nagy, P.D. 334
Nahmias, A.J. 110
Najarian, R.C. 402
Nakano, J.H. 101, 102
Nathanson, N. 283
Natsuaki, K.T. 215
Natsuaki, T. 215
Nault, L.R. 313, 315, 399
Neill, J.D. 302
Nelson, M.R. 344
Nemeroff, M.E. 205
Nesson, M.H. 121
Neumann, H. 126, 128
Neurath, A.R. 143, 236,.246
Neve, H. 164, 166
Niessen, A.I. 399
Nishizawa, T. 262
Norrby, E. 143, 246
Noten, A.F.H. 236
Novembre, F.J. 102
Nowak, T. 232
Nunmaker, C.E. 102
Nuss, D.L. 198
Nuttall, P.A. 283
Nutter, R.C. 305
Obijeski, J.F. 101, 261
Ochoa, A. 289
Odink, K.G.-
Ohashi, M. 302
Ohlinger, V.F. 302
Okada, S. 302
Oker-Blom, C. 52, 221
Okuda, S. 215
Okuno, T. 365
Oldstone, M.B.A. 238, 289
Olesen, N.J. 262
Oliver, S.G. 205
Olson, A.J. 335
Onogi, S. 317
Oroszlan, S. 299
Orvell, C. 246
Oseko, N. 262
Osman, T.A.M. 365
Osorio-Keese, M.E. 338, 405
Ou, J.-H. 402
Overby, L.R. 232
Overton, H.A. 402
Owen, J. 388
Owens, M.J. 102
Ozden, S. 325
Pachuk, C.J. 236
Pagano, J.S. 149
Palese, P. 236, 272
Palm, P. 126, 128
Palmenberg, A. 325
Palmer, E.L. 101
Palukaitis, P. 359, 388
Paluku, K.M. 102
Panganiban, L.C. 305
Paoletti, E. 101
Parekh, B.S. 288, 289
Parker, M.D. 233
Parker, R.L. 205
Parks, T.D. 356
Partanen, P. 221
Partridge, J.E. 397
Patterson, J.L. 283
Pattison, J.R. 172
Pattyn, S.R. 241, 249
Payne, C.C. 122, 198
Pearson, G.D. 262
Pedley, S. 198
Pei, M.-Y. 311
Perkus, M.E. 101
Pernollet, J.C. 347
Perry, J.N. 402
Perry, L.J. 110
Peters, C.J. 283
Peters, D. 241, 249, 261, 262, 283
Peterson, J.F. 350

Pettersson, R.F. 52, 221,. 222, 283
Pettersson, U. 143, 144
Pfeiffer, P. 152, 385
Phatak, H.C. 388
Phillips, S. 344, 350
Piazolla, P. 402
Pietras, D.F. 205
Piggott, J.D.A. 350
Pilcher, K.S. 202
Pinnock, D.E. 331
Pirone, T.P. 356
Plaskitt, K.A. 359
Plese, N. 344
Plowright, W. 110
Plumb, J.A. 199
Plummer, G. 325
Poch, O. 262
Pogo, B.G.T. 101
Porter, C.D. 102
Porterfield, J.S. 52, 222, 232, 233
Postlethwaite, R. 102
Pourcel, C. 116
Povey, R.C. 52, 302, 325
Powell, C.A. 376
Powell, I.B. 164, 166
Pozniak, D. 211
Price, K.H. 205
Pring, D.R. 315
Pringle, C.R. 241,246, 261
Prior, H. C. 102
Provvidenti, R. 367
Prozesky, O.W. 241, 249
Puchta, H. 405
Pullin, J.S.K. 325
Purchio, A.F. 232
Purcifull, D.E. 350
Putz, C. 379
Quillet, L. 379
Rabson, A.B. 299
Rajeshwari, R. 402
Raju, R. 221
Ramirez, P. 315
Ramm, K. 405
Ramos, D.E. 371
Ramsdell, D.C. 371
Rana, G.I.L. 317
Randles, J.W. 329, 379, 405
Ranu, R.S. 180
Rao, A.L.N. 385
Rao, D.V. 365
Rapp, F. 110
Raschke, J.H. 402
Rasschaert, D. 236
Ratti, G. 205
Rawlinson, C.J. 210, 211
Ray, D.S. 183
Ray, P. 198
Rayment, I. 329
Reavy, B. 331, 374
Reddy, D.V.R. 356, 379
Redman, R.M. 172
Reed, W. 102
Regnery, R.L. 241, 249
Reichmann, M.E. 261, 317
Reilly, J.D. 205
Reinganum, C. 325, 331, 374
Reinhard, M.K. 101
Reisman, D. 376
Reisner, H. 102
Reisser, W. 139
Reiter, W.-D 126, 128
Rettenberger, M. 126, 128
Rey, O. 288, 289
Rezaian, M.A. 388, 405
Rezelman, G. 363
Rhoads, R.E. 356, 363
Rice, C.M. 221, 222, 232
Richards, K. 152, 305, 379
Richins, R.D. 152
Richman, D.D. 249
Riesner, D. 405
Rigden, P. 143
Rinehart, C.A. 329
Rivera, C. 315
Rivera-Bustamante, R. 405
Riviere, Y. 289
Rixon, F.J. 110
Rizzo, T.M. 388
Roberts, I.M. 177, 319, 379
Roberts, R.C. 180
Roberts, R.J. 143
Roberts, T.E. 202
Robertson, J.S. 136, 326, 331
Robinson, A.J. 102
Robinson, D.J. 177, 311, 371, 379, 381, 402
Robinson, H. 299
Robinson, W.S. 116
Rochon, D. 334, 335
Rodionova, N.P. 384
Roechan, M. 311
Roehrig, J.T. 233
Roenhorst, J.W. 385
Rohde, W. 379
Rohrmann, G.F. 121
Roistacher, C.N. 405
Roizman, B. 110
Romanos, M.A. 210
Romanowski, V. 288, 289
Ronald, W.P. 365
Ronda, C. 164, 166
Ronnholm, R. 283
Rosciglione, B. 347

Rose, J. 262
Rosen-Wolf, A. 101, 102
Rosenkranz, E. 313
Roslaykov, A.A. 102
Ross, L.J.N. 110
Rossman, M.G. 329
Rottier, P.J.M. 222
Rougeon, F. 262
Rouhandeh, H. 102
Rouhiainen, L. 156
Rowe, L.B. 205
Rowe, W.P. 143, 289
Rowhani, A. 350
Rowlands, D. 325
Rubino, L. 335, 371, 402
Rubinstein, R. 198
Rueckert, R.R. 325, 373, 374
Ruiz-Linares, A. 233
Rumenapf, T. 232
Rupasov, V.V. 344, 397
Russell, D.L. 122
Russell, P.K. 52, 222, 233
Russell, W.C. 143, 144
Russo, M. 334, 335, 379
Rutgers, T. 329
Ruti, K. 102
Rybicki, E.P. 326, 385
Sacher, R. 385
Sachs, L. 149
Saif, L.J. 238
Salazar, L.F. 340
Salerno-Rife, T. 329
Salzman, N.P. 149
Samalecos, C. 136
Sanborn, R.R. 371
Sanchez, A. 249
Sanderlin, R.S. 211
Sänger, H.L. 405
Sano, T. 405
Saraste, J. 283
Sasso, D.R. 288
Satake, H. 158
Satchwell, S.C. 110
Satir, B. 283
Saunders, K. 373
Savino, V. 347
Sawicki, D.L. 236
Sawicki, S.G. 236
Saxelin, M. 164, 166
Schable, C.A. 232
Schäfer, R. 158
Schaffer, F.L. 52, 302, 325
Scheets, K. 305
Schell, J. 177
Schiff, L.A. 199
Schlesinger, M.J. 222, 233
Schlesinger, R.W. 143, 232
Schlesinger, S. 222, 233
Schmaljohn, C.S. 233
Schmidt, T. 363, 374
Schneider, D. 158
Schneider, I.R. 371
Schneider, L.G. 261
Schneider, R. 116
Schödel, F. 116
Scholz, J. 102
Schubert, M. 262
Schumacher, J. 405
Schuster, A.M. 139
Sciaky, D. 143
Scodeller, E.A. 325
Scott, F. 52, 302
Scott, H.A. 363
Scott, J.E. 110
Scotti, P.D. 374
Sdoodee, R. 391
Sears, A.E. 110
Seeley, N.D. 317
Seguin, C. 110
Sekine, S. 302
Seligy, V.L. 136
Selling, B. 373, 374
Semancik, J.S. 405
Semler, B.R. 325
Sethi, P. 325
Sethna, P.B. 236
Sgro, J.-Y. 373, 374
Shanks, M. 363
Shannon, L.M. 397
Shapiro, D. 233
Shaw, J.G. 356, 363
Shea, T.B. 136
Sheets, R.L. 232
Shelbourn, S.L. 205
Sheldrick, P. 110
Shelokov, A. 233
Shenk, T. 144
Shepherd, J.W. 396
Shepherd, R.J. 152
Shikata, E. 405
Shin, S.J. 232
Shintaku, M. 388
Shirako, Y. 379
Shope, R.E. 52, 110, 232, 261, 283
Shukla, D.D. 356
Siddell, S.G. 236
Siegert, R. 249
Siegl, G. 172
Simon, A.E. 305
Simons, J.F. 283
Simpson, D.I.H. 241, 249, 261
Singh, M.K. 289

Six, E.W. 402
Skalka, A.M. 299
Skrdla, M.P. 139
Skrzeczkowski, L.J. 350
Slenczka, W. 241, 249
Smith, A. 205
Smith, A.W. 52, 302
Smith, D.R. 405
Smith, G.E. 122, 123
Smith, I.R.L. 122
Smith, J. 396
Smith, K.M. 121
Smith, O.P. 131
Smith, T.F. 299
Sneh, B. 211
Snijder, E.J. 238
Söderlund, H. 156
Southern, P.J. 288, 289
Sowell, G. 356
Spaan, W.J.M. 221, 222, 236, 238
Speight, G. 233
Spencer, R.A. 122
Spieker, R.L. 405
Sprengel, R. 116
St. George, T.D. 261
Stace-Smith, R. 344, 371
Stanley, J. 177, 363
Stauffacher, C. 363
Steck, F. 238
Steger, G. 371
Steinlauf, R. 205
Stollar, V. 232
Stoltz, D.B. 131
Stone, O.M. 350
Stott, E.J. 246
Strauss, E.G. 221, 222, 232
Street, B.K. 205
Strick, D. 232
Struthers, J.K. 331
Stuart, D. 325
Studdert, M.J. 52, 302
Sturgess, J.M. 101
Stussi-Garaud, C. 317, 393, 394
Su, T.M. 172
Suck, D. 329
Summers, D.F. 261
Summers, J. 116, 122, 123, 131
Sundell, G. 144
Sureau, P. 241, 249
Swaby, A.G. 379
Swinton, D. 139
Symons, R.H. 388, 405
Szczeniowski, M. 102
Szybiak, U. 338
Tagaya, I. 325
Taguchi, F. 236
Tajbakhsh, S. 136
Takahashi, M. 110
Takahashi, T. 340
Takkinen, K. 156, 222
Talbot, P.J. 236
Tamada, T. 379
Tangy, F. 325
Tantera, D.M. 311
Tattersall, P. 172
Tavantzis, S.M. 211, 344
Taylor, C.E. 371
Taylor, G. 246
Taylor, J. 198
Teakle, D.S. 391
Teich, N. 299
Teliz, D. 347
Temin, H.M. 299
Teninges, D. 202
Terai, Y. 405
Teranaka, M. 215
Tesh, R.B. 261, 262, 283
Teuber, M. 164, 166
Thayer, R.M. 402
Theiler, M. 222
Theilmann, D.A. 131
Thiel, H.-J. 232
Thomas, B.J. 391
Thomas, J.E. 177, 311
Thompson, C. 384
Thornbury, D.W. 356
Thorne, E.T. 102
Thottapilly, G. 311
Thouvenel, J.-C. 338, 343
Tijssen, P. 172
Timmins, P. 152
Tinsley, T.W. 326, 331, 374
Tiollais, P. 116
Tipper, D.J. 205
Tjia, S.T. 121
Tochihara, H. 311
Tollin, P. 340
Tomenius, K. 359
Tomley, F.M. 236
Tomlinson, J.A. 317
Tordo, N. 262
Toriyama, S. 399
Tousignant, M.E. 371, 402
Townsend, R. 177
Traub, E. 289
Travassos da Rosa, A.P.A. 261, 262
Travassos da Rosa, J.S. 262
Tremaine, J.H. 335, 344, 365
Trent, D.W. 222, 233
Tripathy, D.N. 102
Trousdale, M.D. 233
Tsai, J.H. 315, 399

Tsai, K.S. 102
Tsukihara, T. 329
Tu, C.C. 205
Tucker, R.C. 374
Tuffnell, P.S. 110
Tuszynski, A. 391
Tweeten, K.A. 122
Tyler, K.L. 199
Tyrrell, D.A.J. 325
Tyulkina, L.G. 384
Ulmanen, I. 221
Utagawa, E. 302
Uyemoto, J.K. 317, 367
Vafai, A. 102
Vaheri, A. 221
Valentin, P. 379
Valverde, R.A. 344, 385
van Beek, N.A.M. 262
van Berlo, M.F. 222
van der Eb, A.J. 144
van der Groen, G. 241, 249
van der Vlugt, C.I.M. 344
van Duin, J. 308
van Etten, J.L. 139
van Kammen, A. 363
van Lent, J.W.M. 363, 385
van Pelt-Heerschap, H. 394
van Regenmortel, M.H.V. 236, 246, 359, 381, 396
van Vloten-Doting, L. 394
Vance, V.B. 356
Varmus, H.E. 116, 299
Varsanyi, T.M. 144
Verbeek, H. 394
Verduin, B.J.M. 384, 385
Vestergard Jorhensen, P.E. 262
Vidgren, G. 222
Vinson, S.B. 131
Vlak, J.M. 122, 123
Vlasak, R. 236
Vogt, V. 299
Volkman, L.E. 123
von Wechmar, M.B. 326, 385
Vonka, V. 149
Voroshilova, M. 325
Vovlas, C. 343
Wadell, G. 143, 144
Wagemakers, L. 283
Wagner, R.R. 143, 158, 180, 183, 241, 246, 261, 262, 272, 283, 302
Walker, P.J. 198
Walter, B. 371
Wang, A.L. 207
Wang, C.-Y. 221
Wang, K.-S. 402
Wang, M. 283
Ward, C.W; 356
Ward, D. 172
Ward, V.K. 283
Watanabe, Y. 399
Waterhouse, P.M. 311, 402
Waterworth, H.E. 388
Watts, J.W. 359
Webb, B.A. 131
Webb, M.J.W. 317
Webb, P.A. 241, 249, 289
Weiland, F. 302
Weimer, T. 116
Weiner, A.J. 232, 402
Weiss, M. 238
Weiss, R. 299
Weiss, S.R. 236
Weissman, M.B. 205
Wellink, J. 363, 39?
Wengler, G. 232
Wenner, H.A. 325
Werner, F.-J. 110
Wernstedt, C. 222
Wery, J.-P. 374
Westaway, E.G. 52, 222, 232, 233
Weston-Fina, M. 365
Wetter, C. 344
Wheeler, R.E. 344
White, M.I. 102
White, R.F. 215
White, T.C. 207
Wickner, R.B. 205
Wiegers, F.P. 123
Wigand, R. 143, 144
Wildy, P. 52
Will, H. 116
Williams, E.S. 102
Williams, J. 144
Williams, R.H.V. 388
Williamson, C. 326
Willis, D.B. 136
Wilson, H.R. 340
Wilson, M.E. 122, 123
Wilson, T.M. 102, 359
Wilusz, J. 249
Winberg, G. 144
Winqvist, G. 102
Winslow, J.P. 101
Winterfield, R.W. 102
Winton, J.R. 199
Wittek, R. 101
Witz, J. 152
Wodnar-Filipowicz, A. 350
Wokatsch, R. 102
Wold, W.S.M. 143
Wolf, K. 110
Wong-Staal, F. 299

Wood, H.A. 123, 204, 210,. 211
Wooddell, M.K. 288
Woode, G.N. 238
Woods, R.D. 215
Wrischer, M. 344
Wu, S. 329
Wulff, H. 241, 249
Wunner, W.H. 262
Xie, W. 215
Xiong, Z. 365
Yabuuchi, K. 302
Yamamoto, T. 199
Yamashita, H. 326
Yamashita, S. 215, 399
Yoshida, M. 299
Yoshikawa, N. 340
Yoshimizu, M. 262
Young, C.S.H. 143
Young, N.D. 359
Young, P.R. 289
Yu, T.F. 311
Zagorski, W. 338
Zaitlin, M. 359
Zakin, M.M. 288, 289
Zavriev, S.K. 344
Zeller, H. 261, 283
Zetina, C. 288
Zettler, F.W. 340, 350
Zeyen, R.J. 315
Ziegler, V. 379
Ziegler-Graff, V. 379
Zillig, W. 126, 128
Zimmern, D. 363, 393
Zinder, N.D. 183, 308
Zoltick, P.W. 236
Zsak, L. 144
Zuidema, D. 261, 305, 391

Virus Index

0c1r, ***Inoviridae***, 183
1A, ***Siphoviridae***, 164
1β6, ***Siphoviridae***, 164
1ϕ1, ***Microviridae***, 179
1ϕ3, ***Microviridae***, 179
1ϕ7, ***Microviridae***, 179
1ϕ9, ***Microviridae***, 179
2BV, ***Podoviridae***, 166
2D/13, ***Microviridae***, 179
3, ***Myoviridae***, 162
3A, ***Siphoviridae***, 164
3T+, ***Myoviridae***, 162
5, ***Siphoviridae***, 164
06N-22PII, ***Myoviridae***, 162
06N-58P, ***Corticoviridae***, 157
7-7-7, ***Siphoviridae***, 164
7-11, ***Podoviridae***, 166
7/26, ***Podoviridae***, 166
7s, ***Leviviridae***, 308
9/0, ***Myoviridae***, 162
10tur, ***Inoviridae***, 183
11F, ***Myoviridae***, 162
12S, ***Myoviridae***, 162
13p2 reovirus, ***Reoviridae***, 192
13p2, ***Reoviridae***, 192
16-2-12, ***Siphoviridae***, 164
16-19, ***Myoviridae***, 162
22, ***Podoviridae***, 166
29, ***Myoviridae***, 162
32, ***Siphoviridae***, 164
37, ***Myoviridae***, 162
43, ***Myoviridae***, 162
44AHJD, ***Podoviridae***, 166
44RR2.8t, ***Myoviridae***, 162
50, ***Myoviridae***, 162
51, ***Myoviridae***, 162
59.1, ***Myoviridae***, 162
63U11, ***Bunyaviridae***, 275
66F, ***Myoviridae***, 162
75V-2374, ***Bunyaviridae***, 276
75V-2621, ***Bunyaviridae***, 276
77, ***Siphoviridae***, 164
78V-2441, ***Bunyaviridae***, 276
107, ***Siphoviridae***, 164
108/106, ***Myoviridae***, 162
114, ***Podoviridae***, 165
119, ***Siphoviridae***, 164
121, ***Myoviridae***, 162
182, ***Podoviridae***, 166
187, ***Siphoviridae***, 164
222a, ***Myoviridae***, 162
223, ***Siphoviridae***, 164
317, ***Siphoviridae***, 164
936, ***Siphoviridae***, 164
949, ***Siphoviridae***, 164
1307, ***Plasmaviridae***, 124
1358, ***Siphoviridae***, 164
1483, ***Siphoviridae***, 164
2389, ***Siphoviridae***, 164
2671, ***Siphoviridae***, 164
2685, ***Siphoviridae***, 164
2848A, ***Siphoviridae***, 164
4211, ***Myoviridae*** 162
4996, ***Podoviridae***, 166
5845, ***Myoviridae*** 162
7480b, ***Podoviridae***, 166
8764, ***Siphoviridae***, 164
8893 , ***Myoviridae***, 162
9266, ***Myoviridae*** 162
A strain, ***Necrovirus***, 317
A, ***Siphoviridae***, 164
α, ***Siphoviridae***, 164
A-4(L), ***Podoviridae***, 166
α1, ***Myoviridae***, 162
A1-Dat, , ***Siphoviridae***, 164
α3, ***Microviridae***, 179
α3a, ***Siphoviridae***, 164
A5/A6, ***Siphoviridae***, 164
A6, ***Myoviridae***, 162
α10, ***Microviridae***, 179
α15, ***Leviviridae***, 308
A19, ***Myoviridae***, 162
A23, ***Picornaviridae***, 322
A25, ***Siphoviridae***, 164
AA-1, ***Podoviridae***, 165
AAV type 2, ***Parvoviridae***, 170
AAV type 3, ***Parvoviridae***, 170
AAV type 4, ***Parvoviridae***, 170
AAV type 5, ***Parvoviridae***, 170
Abras, ***Bunyaviridae***, 276
Abu Hammad, ***Bunyaviridae***, 279
Abu Mina, ***Bunyaviridae***, 279
Abutilon mosaic, ***Geminivirus***, 176
AbV-4, mushroom 4, *Partitiviridae*, 209
AC-1, ***Podoviridae***, 166
Acara, ***Bunyaviridae***, 276
Acherontia atropas, ***Tetraviridae***, 331
Acheta Densovirus, ***Parvoviridae***, 171
Acholeplasma Phage L2, ***Plasmaviridae***, 124
Acholeplasma phage L51, ***Inoviridae***, 182
AcMNPV, ***Baculoviridae***, 119
Acrobasis zelleri, ***Poxviridae***, 99
Adelaide River, ***Rhabdoviridae***, 255

Adeno-associated type 1, ***Parvoviridae,*** 170
AE2, ***Inoviridae,*** 182
Aedes aegypti, ***Poxviridae***, 100
Aedes albopictus cell fusing agent, ***Flaviviridae,*** 231
Aedes Densovirus, ***Parvoviridae,*** 171
Aeh2, ***Myoviridae,*** 162
African cassava mosaic, ***Geminivirus,*** 176
African horse sickness, ***Reoviridae,*** 189
AfV-S, ***Totiviridae,*** 204
AG83-497, ***Bunyaviridae,*** 276
Ag83-1746, ***Bunyaviridae,*** 275
Agaricus bisporus 4, ***Partitiviridae***, 209
Agraulis Densovirus, ***Parvoviridae,*** 171
Agraulis vanillae, ***Tetraviridae,*** 331
Agropyron mosaic, ***Potyvirus,*** 355
Aguacate, ***Bunyaviridae,*** 278
AGVd, Viroid, 404
Aino, ***Bunyaviridae,*** 276
Akabane, ***Bunyaviridae,*** 276
Alajuela, ***Bunyaviridae,*** 276
Albatrosspox, ***Poxviridae***, 100
Alenquer, ***Bunyaviridae,*** 278
Alentian mink disease parvovirus, ***Parvoviridae,*** 168
Alfalfa cryptic 1, ***Cryptovirus,*** 213
Alfalfa cryptic 2, ***Cryptovirus,*** 214
Alfalfa latent, ***Carlavirus,*** 342
Alfalfa mosaic, Alfalfa mosaic virus, 392
Alfuy, ***Flaviviridae,*** 226
Almpiwar, ***Rhabdoviridae,*** 256
Alstroemeria carlavirus, ***Carlavirus,*** 342
Alstroemeria mosaic, ***Potyvirus,*** 352
Alteromonas Phage PM2, ***Corticoviridae,*** 157
Amapari, ***Arenaviridae,*** 288
Amaranthus leaf mottle, ***Potyvirus,*** 352
American plum line pattern, ***Ilarvirus,*** 390
Amsacta moorei, ***Poxviridae***, 99
AMV, Alfalfa mosaic virus, 392
AN25S-1, ***Podoviridae,*** 166
ana1-ana2, ***Adenoviridae,*** 143
Ananindeua, ***Bunyaviridae,*** 276
Andean potato mottle, ***Comovirus,*** 362
Anhanga, ***Bunyaviridae,*** 278
Anhembi, ***Bunyaviridae,*** 275
Anomala cuprea, ***Poxviridae***, 99
Anopheles A, ***Bunyaviridae,*** 275
Anopheles B, ***Bunyaviridae,*** 275
ans1-ans3, ***Adenoviridae,*** 143
Antequera, ***Bunyaviridae,*** 282
Antheraea eucalypti, ***Tetraviridae,*** 331
Anthoxanthum latent blanching, ***Hordeivirus,*** 396
Anthoxanthum mosaic, ***Potyvirus,*** 354
Anthriscus strain, Parsnip yellow fleck virus , 319
Anthriscus yellows, Maize chlorotic dwarf virus, 312
AnV-S, ***Totiviridae,*** 204
AoV, ***Partitiviridae,*** 209
AP50, ***Tectiviridae,*** 155
Apeu, ***Bunyaviridae,*** 275
Aphid lethal paralysis, ***Picornaviridae,*** 324
Aphodius tasmaniae, ***Poxviridae***, 99
Aphthovirus A, ***Picornaviridae,*** 324
Aphthovirus Asia 1, ***Picornaviridae,*** 324
Aphthovirus C, ***Picornaviridae,*** 324
Aphthovirus O, ***Picornaviridae,*** 324
Aphthovirus SAT1, ***Picornaviridae,*** 324
Aphthovirus SAT2, ***Picornaviridae,*** 324
Aphthovirus SAT3, ***Picornaviridae,*** 324
Apoi, ***Flaviviridae,*** 226
Apple chlorotic leafspot, ***Closterovirus,*** 346
Apple mosaic, ***Ilarvirus,*** 390
Apple scar skin, Viroid, 404
Apple stem grooving, ***Capillovirus,*** 339
Aquilegia necrotic mosaic, ***Caulimovirus,*** 151
Aquilegia, ***Potyvirus,*** 354
Ar 577, ***Reoviridae,*** 190
Ar 578 , ***Reoviridae,*** 190
Arabis mosaic, ***Nepovirus,*** 370
Arabis mosaic, Satellite, 401
Aracacha Y, ***Potyvirus,*** 354
Aransas Bay, ***Bunyaviridae,*** 282
Araujia mosaic, ***Potyvirus,*** 352
Arbia, ***Bunyaviridae,*** 278
Arboledas, ***Bunyaviridae,*** 278
ArGV, ***Baculoviridae***, 120
Aroa, ***Flaviviridae,*** 227
Arp, ***Siphoviridae,*** 164
Arphia conspersa, ***Poxviridae***, 100
Arracacha A, ***Nepovirus,*** 370
Arracacha B, ***Nepovirus,*** 371
Arracacha latent, ***Carlavirus,*** 342
Arrhenatherum blue dwarf, ***Reoviridae,*** 196
Artichoke curly dwarf, ***Potexvirus,*** 349
Artichoke Italian latent, ***Nepovirus,*** 370
Artichoke latent M, ***Carlavirus,*** 343
Artichoke latent S, ***Carlavirus,*** 343
Artichoke latent, ***Potyvirus,*** 352
Artichoke mottle crinkle, ***Tombusvirus,*** 334
Artichoke mottled crinkle, Satellite, 401
Artichoke vein banding, ***Nepovirus,*** 371
Artichoke yellow ringspot, ***Nepovirus,*** 370
Artogeia rapae granulosis, ***Baculoviridae***, 120
Aruac, ***Rhabdoviridae,*** 256

Arumowot, ***Bunyaviridae,*** 278
AS-1, ***Myoviridae,*** 162
ASBVd, Viroid, 404
ASGV, ***Capillovirus,*** 339
Asparagus 1, ***Potyvirus,*** 352
Asparagus III, ***Potexvirus,*** 349
Asparagus virus II, ***Ilarvirus,*** 390
Aspergillus foetidus S, ***Totiviridae,*** 204
Aspergillus niger S, ***Totiviridae,*** 204
Aspergillus ochraceous, ***Partitiviridae,*** 209
ASSVd, Viroid, 404
Asystasia gangetica mottle, ***Potyvirus,*** 354
Ateline herpesvirus 2, ***Herpesviridae***, 109
Atlantic cod ulcus syndrome, ***Rhabdoviridae,*** 256
Atropa belladonna, ***Rhabdoviridae,*** *260*
AU, ***Myoviridae,*** 162
Aucuba ringspot, Commelina yellow mottle, 154
Aura, ***Togaviridae,*** 218
Ausduk disease, ***Poxviridae***, 94
Australian grapevine, Viroid, 404
Autographa californica nuclear polyhedrosis, ***Baculoviridae***, 119
AV-1, ***Podoviridae,*** 165
Avalon, ***Bunyaviridae,*** 280
Avian AAV, ***Parvoviridae,*** 170
Avian infectious bronchitis, ***Coronaviridae,*** 234
Avian leukosis, ***Reoviridae,*** 295
Avian orthoreoviruses, ***Reoviridae,*** 187
Avian paramyxovirus 2 (Yucaipa), ***Paramyxoviridae,*** 244
Avian paramyxovirus 3, ***Paramyxoviridae,*** 244
Avian paramyxovirus 4, ***Paramyxoviridae,*** 244
Avian paramyxovirus 5 (Kunitachi), ***Paramyxoviridae,*** 244
Avian paramyxovirus 6, ***Paramyxoviridae,*** 244
Avian paramyxovirus 7, ***Paramyxoviridae,*** 244
Avian paramyxovirus 8, ***Paramyxoviridae,*** 244
Avian paramyxovirus 9, ***Paramyxoviridae,*** 244
Avian paramyxovirus type 1, ***Paramyxoviridae,*** 244
Avian reticuloendotheliosis, ***Reoviridae,*** 295
Avian rotavirus, ***Reoviridae,*** 191
Avian sarcoma and leukemia, ***Reoviridae,*** 296
Avocado sunblotch, Viroid, 404
Azuki bean mosaic, ***Potyvirus,*** 352
B, ***Herpesviridae***, 106
β, ***Leviviridae,*** 308
β, ***Siphoviridae,*** 164
B/LEE/40, ***Orthomyxoviridae,*** 271
β4, ***Siphoviridae,*** 164
B6, ***Leviviridae,*** 308
B7, ***Leviviridae,*** 308
B11-M15, ***Siphoviridae,*** 164
B19, ***Parvoviridae,*** 168
Babahoyo, ***Bunyaviridae,*** 276
Babanki, ***Togaviridae,*** 218
Baboon herpesvirus, ***Herpesviridae***, 109
Bagaza, ***Flaviviridae,*** 227
Bahia Grande, ***Rhabdoviridae,*** 256
Bahig, ***Bunyaviridae,*** 276
Bajra streak, ***Geminivirus,*** 174
Bakau, ***Bunyaviridae,*** 275
Bam35, ***Tectiviridae,*** 155
Bamboo mosaic, ***Potexvirus,*** 349
Banana streak, Commelina yellow mottle, 154
Bandia, ***Bunyaviridae,*** 279
Bangoran, ***Rhabdoviridae,*** 256
Bangui, ***Bunyaviridae,*** 282
Banzi, ***Flaviviridae,*** 227
Barley B-1, ***Potexvirus,*** 349
Barley stripe mosaic, ***Hordeivirus,*** 395
Barley yellow dwarf, ***Luteovirus,*** 309
Barley yellow dwarf, Satellite, 401
Barley yellow mosaic, ***Potyvirus,*** 355
Barley yellow striate mosaic, ***Rhabdoviridae,*** 258
Barmah Forest, ***Togaviridae,*** 218
Barranqueras, ***Bunyaviridae,*** 282
Barur, ***Rhabdoviridae,*** 256
Batai, ***Bunyaviridae,*** 275
Batama, ***Bunyaviridae,*** 276
Batken, ***Bunyaviridae,*** 282
BBWV, ***Fabavirus,*** 366
BCTV, ***Geminivirus,*** 175
Bdellovibrio Phage, ***Microviridae,*** 180
BE/1, ***Microviridae,*** 179
BeAn 157575, ***Rhabdoviridae,*** 253
Bean common mosaic, ***Potyvirus,*** 352
Bean golden mosaic, ***Geminivirus,*** 175
Bean leaf roll, ***Luteovirus,*** 310
Bean mild mosaic, ***Carmovirus,*** 304
Bean pod mottle, ***Comovirus,*** 362
Bean rugose mosaic, ***Comovirus,*** 362
Bean summer death, ***Geminivirus,*** 175
Bean yellow mosaic, ***Potyvirus,*** 352
BeAr328208, ***Bunyaviridae,*** 275
bearded iris mosaic, ***Potyvirus,*** 353
Bebaru, ***Togaviridae,*** 218
Beccles, ***Myoviridae,*** 162
Bee acute paralysis, ***Picornaviridae,*** 324
Bee slow paralysis, ***Picornaviridae,*** 324

Bee X, ***Picornaviridae,*** 324
Beet cryptic 1, ***Cryptovirus,*** 213
Beet cryptic 2, ***Cryptovirus,*** 213
Beet cryptic 3, ***Cryptovirus,*** 213
Beet curly top, ***Geminivirus,*** 175
Beet leaf curl, ***Rhabdoviridae,*** 259
Beet mild yellowing, ***Luteovirus,*** 310
Beet mosaic, ***Potyvirus,*** 352
Beet necrotic yellow vein, ***Furovirus,*** 378
Beet pseudo yellows, ***Closterovirus,*** 346
Beet soil-borne, ***Furovirus,*** 378
Beet western yellows, ***Luteovirus,*** 310
Beet western yellows, Satellite, 401
Beet yellow net, ***Luteovirus,*** 310
Beet yellow stunt, ***Closterovirus,*** 346
Belem, ***Bunyaviridae,*** 282
Belladonna mottle, ***Tymovirus,*** 337
Belmont, ***Bunyaviridae,*** 282
Belterra, ***Bunyaviridae,*** 278
Benevides, ***Bunyaviridae,*** 276
Benfica, ***Bunyaviridae,*** 276
Bermuda grass etched-line, ***Marafivirus,*** 315
Berna, ***Torovirus,*** 237
Berrimah, ***Rhabdoviridae,*** 255
Bertioga, ***Bunyaviridae,*** 276
BGMV, ***Geminivirus,*** 175
Bhanja, ***Bunyaviridae,*** 282
Bidens mosaic, ***Potyvirus,*** 354
Bidens mottle, ***Potyvirus,*** 352
Bimbo, ***Rhabdoviridae,*** 256
Bimiti, ***Bunyaviridae,*** 276
Bir, ***Siphoviridae,*** 164
Birao, ***Bunyaviridae,*** 275
Bivens Arm, ***Rhabdoviridae,*** 255
BJ5-T, ***Siphoviridae,*** 164
BK, ***Papovaviridae,*** 148
BL3, ***Siphoviridae,*** 164
Black beetle, ***Nodaviridae,*** 373
Blackeye cowpea mosaic, ***Potyvirus,*** 352
Blackgram mottle, ***Carmovirus,*** 304
BLE, ***Siphoviridae,*** 164
Blueberry carlavirus, ***Carlavirus,*** 343
Blueberry leaf mottle, ***Nepovirus,*** 370
Blueberry red ringspot, ***Caulimovirus,*** 151
Blueberry scorch, ***Ilarvirus,*** 390
Blueberry shoestring, ***Sobemovirus,*** 328
Bluetongue, ***Reoviridae,*** 188, 189
BmSNPV, ***Baculoviridae***, 119
BMV, ***Bromovirus,*** 382
Bobaya, ***Bunyaviridae,*** 282
Bobia, ***Bunyaviridae,*** 276
Boletus, ***Potexvirus,*** 349
Bombyx Densovirus, ***Parvoviridae,*** 171
Bombyx mori S Nuclear Polyhedrosis, ***Baculoviridae***, 119
Boolarra, ***Nodaviridae,*** 373
Boraceia, ***Bunyaviridae,*** 275
Border disease (of sheep), ***Flaviviridae,*** 229
bos1-bos10, ***Adenoviridae,*** 142
Botambi, ***Bunyaviridae,*** 276
Boteke, ***Rhabdoviridae,*** 253
Bouboui, ***Flaviviridae,*** 227
Bovine (alpha) herpesvirus 1, ***Herpesviridae***, 106
Bovine AAV, ***Parvoviridae,*** 170
Bovine coronavirus , ***Coronaviridae,*** 235
Bovine enteroviruses 1-2, ***Picornaviridae,*** 322
Bovine ephemeral fever, ***Rhabdoviridae,*** 255
Bovine imunodeficiency, ***Reoviridae***, 298
Bovine leukemia, ***Reoviridae,*** 297
Bovine papular stomatitis, ***Poxviridae***, 94
Bovine parvovirus, ***Parvoviridae,*** 168
Bovine respiratory syncytial, ***Paramyxoviridae,*** 245
Bovine rhinoviruses 1 and 2, ***Picornaviridae,*** 323
Bovine syncytial, ***Reoviridae,*** 296
Bovine viral diarrhea, ***Flaviviridae,*** 228
Bozo, ***Bunyaviridae,*** 275
Br1, ***Myoviridae,*** 162
Brazilian coleus, Viroid, 405
Breda, ***Torovirus,*** 237
Broad bean mottle, ***Bromovirus,*** 384
Broad bean stain, ***Comovirus,*** 362
Broad bean true mosaic, ***Comovirus,*** 362
Broad bean wilt, serotype I, ***Fabavirus,*** 366
Broad bean wilt, serotype II, ***Fabavirus,*** 367
Broadbean necrosis, ***Furovirus,*** 378
Broccoli necrotic yellows, ***Rhabdoviridae,*** 258
Brome mosaic, ***Bromovirus,*** 382
Brome streak, ***Potyvirus,*** 355
Bromus striate mosaic, ***Geminivirus,*** 174
Bruconha, ***Bunyaviridae,*** 275
Brus Laguna, ***Bunyaviridae,*** 276
Bryonia mottle, ***Potyvirus,*** 354
BSMV, ***Hordeivirus,*** 395
BSVd, Viroid, 404
Buenaventura, ***Bunyaviridae,*** 278
Buffalopox, ***Poxviridae***, 93
Buggy Creek, ***Togaviridae,*** 218
Bujaru, ***Bunyaviridae,*** 278
Bukalasa bat, ***Flaviviridae,*** 226
Bunyamwera, ***Bunyaviridae,*** 275
Bunyamwera, ***Bunyaviridae,*** 275
Burdock stunt, Viroid, 404
Burdock yellows, ***Closterovirus,*** 346
Bushbush, ***Bunyaviridae,*** 276
Bussuquara, ***Flaviviridae,*** 227
Butterbur mosaic, ***Carlavirus,*** 343

Buttonwillow, ***Bunyaviridae,*** 276
BVDV, ***Flaviviridae,*** 228
Bwamba, ***Bunyaviridae,*** 275
BYDV, ***Luteovirus,*** 309
χ, ***Siphoviridae,*** 164
C-1, ***Leviviridae,*** 308
C-2, ***Inoviridae,*** 182
C-2, ***Siphoviridae,*** 164
C2, ***Leviviridae,*** 308
c2, ***Siphoviridae,*** 164
C3, L3, ***Podoviridae,*** 166
C16, ***Myoviridae,*** 162
Cabbage b, Davis isolate, ***Caulimovirus,*** 150
Cacao swollen shoot, Commelina yellow mottle, 154
Cacao yellow mosaic, ***Tymovirus,*** 337
Cacao, ***Bunyaviridae,*** 278
Cache Valley, ***Bunyaviridae,*** 275
Cacipacore, ***Flaviviridae,*** 227
Cactus 2, ***Carlavirus,*** 342
Cactus X, ***Potexvirus,*** 349
Caddo Canyon, ***Bunyaviridae,*** 282
Caimito, ***Bunyaviridae,*** 278
Calchaqui, ***Rhabdoviridae,*** 253
California encephalitis, ***Bunyaviridae,*** 276
Callistephus chinensis chlorosis, ***Rhabdoviridae,*** 260
Camel contagious ecthyma, ***Poxviridae***, 94
Camelpox, ***Poxviridae***, 93
*Campoletis sonorensis, **Polydnaviridae,*** 130
*Camptochironomus tentans, **Poxviridae***, 100
CaMV, ***Caulimovirus,*** 150
can1-can2, ***Adenoviridae,*** 142
Cananeia, ***Bunyaviridae,*** 276
Canarypox, ***Poxviridae***, 95
Canavalia maritima mosaic, ***Potyvirus,*** 354
Candiru, ***Bunyaviridae,*** 278
Canid herpesvirus, ***Herpesviridae***, 106
Canine AAV, ***Parvoviridae,*** 170
Canine calicivirus., ***Caliciviridae,*** 301
Canine coronavirus, ***Coronaviridae,*** 235
Canine distemper, ***Paramyxoviridae,*** 244
Canine herpesvirus, ***Herpesviridae***, 106
Canine parvovirus, ***Parvoviridae,*** 168
Canna yellow mottle, Commelina yellow mottle, 154
cap1, ***Adenoviridae,*** 142
Caper latent, ***Carlavirus,*** 342
Caper vein yellowing, ***Rhabdoviridae,*** 260
Capim, ***Bunyaviridae,*** 276
Caprine arthritis/encephalitis, ***Reoviridae,*** 298
Carajas, ***Rhabdoviridae,*** 253
Caraparu, ***Bunyaviridae,*** 275
Cardamom mosaic, ***Potyvirus,*** 352
Carey Island, ***Flaviviridae,*** 226
CarMV, ***Carmovirus,*** 303
Carnation bacilliform, ***Rhabdoviridae,*** 260
Carnation cryptic 1, ***Cryptovirus,*** 213
Carnation cryptic 2, ***Cryptovirus,*** 213
Carnation etched ring, ***Caulimovirus,*** 151
Carnation Italian ringspot, Satellite, 401
Carnation Italian ringspot, ***Tombusvirus,*** 334
Carnation latent , ***Carlavirus,*** 341
Carnation mottle, ***Carmovirus,*** 303
Carnation necrotic fleck, ***Closterovirus,*** 346
Carnation ringspot , ***Dianthovirus,*** 364
Carnation stunt, Viroid, 405
Carnation vein mottle, ***Potyvirus,*** 352
Carnation yellow stripe, ***Necrovirus,*** 317
Carrot latent, ***Rhabdoviridae,*** 259
Carrot mosaic, ***Potyvirus,*** 354
Carrot mottle, ***Togaviridae,*** 221
Carrot red leaf, ***Luteovirus,*** 310
Carrot temperate 1, ***Cryptovirus,*** 213
Carrot temperate 2, ***Cryptovirus,*** 214
Carrot temperate 3, ***Cryptovirus,*** 213
Carrot temperate 4, ***Cryptovirus,*** 213
Carrot thin leaf, ***Potyvirus,*** 352
Carrot yellow leaf, ***Closterovirus,*** 346
Cassava American latent, ***Nepovirus,*** 370
Cassava brown streak, ***Carlavirus,*** 343
Cassava common mosaic, ***Potexvirus,*** 349
Cassava green mottle, ***Nepovirus,*** 370
Cassava symptomless, ***Rhabdoviridae,*** 260
Cassava vein mosaic, ***Caulimovirus,*** 151
Cassia mild mosaic, ***Carlavirus,*** 343
Cassia yellow blotch, ***Bromovirus,*** 384
Cassia yellow blotch, ***Potyvirus,*** 354
Cattle rotavirus, ***Reoviridae,*** 191
Catu, ***Bunyaviridae,*** 276
Cauliflower mosaic, ***Caulimovirus,*** 150
Caviid herpesvirus 1, ***Herpesviridae***, 108
CbaAr426, ***Bunyaviridae,*** 275
CbVd, Viroid, 404
CbVd1, Viroid, 404
CbVd2, Viroid, 404
CbVd3, Viroid, 404
CCCVd, Viroid, 404
CEβ, ***Myoviridae,*** 162
Celery mosaic, ***Potyvirus,*** 352
Celery yellow mosaic, ***Potyvirus,*** 354
Celery yellow spot, ***Luteovirus,*** 310
Centrosema mosaic, ***Potexvirus,*** 349
Cercopithecid herpesvirus 1, ***Herpesviridae***, 106
Cercopithecine herpesvirus 2, ***Herpesviridae***, 109

Cymbidium ringspot, ***Tombusvirus,*** 334
Cynara, ***Rhabdoviridae,*** *260*
Cynodon chlorotic streak, ***Rhabdoviridae,*** 260
Cynodon mosaic, ***Carlavirus,*** 343
Cynosurus mottle, ***Sobemovirus,*** 329
Cypripedium calceolus, ***Potyvirus,*** 354
Cytoplasmic polyhedrosis from *Bombyx mori,* ***Reoviridae,*** 193
δ1, ***Microviridae,*** 179
D3, ***Siphoviridae,*** 164
δA, ***Inoviridae,*** 182
Dabakala, ***Bunyaviridae,*** 276
Dahlia mosaic, ***Caulimovirus,*** 151
Dakar bat, ***Flaviviridae,*** 226
DakArK 7292, ***Rhabdoviridae,*** 256
Dandelion latent, ***Carlavirus,*** 342
Dandelion yellow mosaic, Parsnip yellow fleck virus, 319
Danish plum line pattern, ***Ilarvirus,*** 390
Daphne S, ***Carlavirus,*** 343
Daphne X, ***Potexvirus,*** 349
Daphne Y, ***Potyvirus,*** 354
Darna trima, ***Tetraviridae,*** 331
Dasheen mosaic, ***Potyvirus,*** 352
Dasychira pudibunda, ***Tetraviridae,*** 331
Datura 437, ***Potyvirus,*** 354
Datura mosaic, ***Potyvirus,*** 354
Datura shoestring, ***Potyvirus,*** 352
Datura yellow vein, ***Rhabdoviridae,*** 258
DdVI, ***Myoviridae,*** 162
Deer papilloma, ***Papovaviridae,*** 147
Demodema boranensis, ***Poxviridae***, 99
Dendrobium leaf streak, ***Rhabdoviridae,*** 261
Dendrobium mosaic, ***Potyvirus,*** 352
Dengue types 1, 2, 3, 4, ***Flaviviridae,*** 227
Dera Ghazi Khan, ***Bunyaviridae,*** 279
Dermolepida albohirtum, ***Poxviridae***, 99
Desmodium mosaic, ***Potyvirus,*** 354
Desmodium yellow mottle, ***Tymovirus,*** 337
Desulfurolobus, ***SSV1***, 126
dϕ3, ***Microviridae,*** 179
dϕ4, ***Microviridae,*** 179
dϕ5, ***Microviridae,*** 179
DHBV, ***Hepadnaviridae***, 115
Dhori, ***Orthomyxoviridae,*** 271
Diatraea Densovirus, ***Parvoviridae,*** 171
Digitaria streak, ***Geminivirus,*** 174
Digitaria striate mosaic, ***Geminivirus,*** 174
Digitaria striate, ***Rhabdoviridae,*** 259
Diodia yellow vein, ***Closterovirus,*** 346
Dioscorea alata ring mottle, ***Potyvirus,*** 354
Dioscorea bacilliform, Commelina yellow mottle, 154
Dioscorea green banding, ***Potyvirus,*** 353
Dioscorea latent, ***Potexvirus,*** 349
Dioscorea trifida, ***Potyvirus,*** 354
Diplocarpon rosae, ***Partitiviridae,*** 209
Dock mottling mosaic, ***Potyvirus,*** 354
Dog papilloma, ***Papovaviridae,*** 147
Douglas, ***Bunyaviridae,*** 276
Drosophila C, ***Picornaviridae,*** 324
Drosophila P and A, ***Picornaviridae,*** 324
Drosophila X, ***Birnaviridae,*** 201
DrV, ***Partitiviridae,*** 209
Duck hepatitis B, ***Hepadnaviridae***, 115
Dugbe, ***Bunyaviridae,*** 279
Dulcamara A and B, ***Carlavirus,*** 343
Dulcamara mottle, ***Tymovirus,*** 337
Duvenhage, ***Rhabdoviridae,*** 255
DXV, ***Birnaviridae,*** 201
Eastern equine encephalitis, ***Togaviridae,*** 218
Ebola, ***Filoviridae,*** 248
Ec9, ***Inoviridae,*** 182
Echinochloa hoja blanca, ***Tenuivirus,*** 399
Echinochloa ragged stunt, ***Reoviridae,*** 197
Echovirus 9, ***Picornaviridae,*** 322
Ectromelia, ***Poxviridae***, 93
Edge Hill, ***Flaviviridae,*** 227
Eel American, ***Rhabdoviridae,*** 253
Eel B12, ***Rhabdoviridae,*** 255
Eel European, ***Rhabdoviridae,*** 253
EgAn1825-61, ***Bunyaviridae,*** 278
Eggplant carlavirus, ***Carlavirus,*** 343
Eggplant mild mottle, ***Carlavirus,*** 343
Eggplant mosaic, ***Tymovirus,*** 337
Eggplant mottled crinkle, ***Tombusvirus,*** 334
Eggplant mottled dwarf, ***Rhabdoviridae,*** 259
Eggplant yellow mosaic, ***Geminivirus,*** 176
Egtved, ***Rhabdoviridae,*** 256
Elderberry A, ***Carlavirus,*** 342
Elderberry carla, ***Carlavirus,*** 342
Elderberry latent, ***Carmovirus,*** 304
Elephant papilloma, ***Papovaviridae,*** 147
Elk papilloma, ***Papovaviridae,*** 147
Elm mottle, ***Ilarvirus,*** 390
Embu, ***Poxviridae***, 100
EMC, ***Picornaviridae,*** 323
Encephalomyocarditis, ***Picornaviridae,*** 323
Endive, ***Rhabdoviridae,*** 260
Enseada, ***Bunyaviridae,*** 282
Entebbe bat, ***Flaviviridae,*** 226
Epizootic hemorrhagic disease, ***Reoviridae,*** 189
Epstein-barr, ***Herpesviridae***, 109
equ1, ***Adenoviridae,*** 142
Equid (alpha) herpesvirus 1, ***Herpesviridae***, 106

Equid (alpha) herpesvirus 4, ***Herpesviridae***, 106
Equid herpesvirus 2, ***Herpesviridae***, 108
Equid herpesvirus 3, ***Herpesviridae***, 106
Equine AAV, ***Parvoviridae,*** 170
Equine abortion, ***Herpesviridae***, 106
Equine arteritis, ***Togaviridae,*** 220
Equine encephalosis, ***Reoviridae,*** 189
Equine infectious anemia, ***Reoviridae,*** 298
Equine rhinoviruses types 1 and 2, ***Picornaviridae,*** 324
Eret-147, ***Bunyaviridae,*** 276
Erve, ***Bunyaviridae,*** 280
Erysimum latent, ***Tymovirus,*** 337
Esc-7-11, ***Podoviridae,*** 166
Estero Real, ***Bunyaviridae,*** 276
Eubenangee, ***Reoviridae,*** 189
Eucocytis meeki, ***Tetraviridae,*** 331
Euonymus fasciation, ***Rhabdoviridae,*** 260
Eupatorium yellow vein, ***Geminivirus,*** 176
Euphorbia mosaic, ***Geminivirus,*** 176
Euphorbia ringspot, ***Potyvirus,*** 354
Euploea corea, ***Tetraviridae,*** 331
European bat type 1, ***Rhabdoviridae,*** 255
European bat type 2, ***Rhabdoviridae,*** 255
European harvest mice papilloma, ***Papovaviridae,*** 147
European swine fever, ***Flaviviridae,*** 229
European wheat striate mosaic, ***Tenuivirus,*** 399
Euxoa Densovirus, ***Parvoviridae,*** 171
Everglades, ***Togaviridae,*** 218
Evonymus mosaic, ***Carlavirus,*** 343
Eyach, ***Reoviridae,*** 190
F1, ***Siphoviridae,*** 164
f1, ***Inoviridae,*** 182
ϕ1, ***Myoviridae,*** 162
f2, ***Leviviridae,*** 307
ϕ17, ***Podoviridae,*** 166
ϕ29, ***Podoviridae,*** 166
ϕ31C, ***Siphoviridae,*** 164
ϕ80, ***Siphoviridae,*** 163
ϕ92, ***Myoviridae,*** 162
ϕ105, ***Siphoviridae,*** 164
ϕ115-A, ***Siphoviridae,*** 164
F116, ***Podoviridae,*** 166
ϕ150-A, ***Siphoviridae,*** 164
ϕ2037/1, ***Siphoviridae,*** 164
ϕ2042, 2, ***Podoviridae,*** 166
ϕA, ***Microviridae,*** 179
ϕA8010, ***Siphoviridae,*** 164
Facey's Paddock, ***Bunyaviridae,*** 276
Farallon, ***Bunyaviridae,*** 279
ϕC, ***Siphoviridae,*** 164
FC3-9, ***Myoviridae,*** 162
fcan, ***Leviviridae,*** 308
ϕCb2, ***Leviviridae,*** 308
ϕCb4, ***Leviviridae,*** 308
ϕCb5, ***Leviviridae,*** 308
ϕCb8r, ***Leviviridae,*** 308
ϕCb9, ***Leviviridae,*** 308
ϕCb12r, ***Leviviridae,*** 308
ϕCb23r, ***Leviviridae,*** 308
ϕCP2, ***Leviviridae,*** 308
ϕCP18, ***Leviviridae,*** 308
ϕCr14, ***Leviviridae,*** 308
ϕCr28, ***Leviviridae,*** 308
ϕD328, ***Siphoviridae,*** 163
FDV, ***Reoviridae,*** 195
Felid herpesvirus 1, ***Herpesviridae***, 106
Feline calicivirus, ***Caliciviridae,*** 301
Feline herpesvirus, ***Herpesviridae***, 106
Feline immunodeficiency, ***Reoviridae,*** 298
Feline infectious peritonitis, ***Coronaviridae,*** 235
Feline parvovirus, ***Parvoviridae,*** 168
Feline sarcoma and leukemia, ***Reoviridae,*** 294
Feline syncytial, ***Reoviridae,*** 296
feline panleukopenia, ***Parvoviridae,*** 168
Fer-de-Lance, ***Paramyxoviridae,*** 245
Fescue cryptic, ***Cryptovirus,*** 213
Festuca leaf streak, ***Rhabdoviridae,*** 258
Festuca necrosis, ***Closterovirus,*** 346
ϕFSW, ***Siphoviridae,*** 164
ϕgal-1-R, ***Myoviridae,*** 162
FH5, ***Leviviridae,*** 307
ϕI, ***Podoviridae,*** 165
Ficus carica, ***Potyvirus,*** 354
Fig S, ***Carlavirus,*** 343
Figulus sublaevis, ***Poxviridae***, 99
Figwort mosaic, ***Caulimovirus,*** 151
ϕII, ***Podoviridae,*** 165
Fiji disease , ***Reoviridae,*** 195
Filareè red leaf, ***Luteovirus,*** 310
Fin V-707, ***Bunyaviridae,*** 278
Finger millet mosaic, ***Rhabdoviridae,*** 259
ϕKZ, ***Myoviridae,*** 162
Flanders, ***Rhabdoviridae,*** 256
Flexal, ***Arenaviridae,*** 288
Flock house, ***Nodaviridae,*** 373
Flounder isolate, ***Iridoviridae,*** 135
ϕNS11, ***Tectiviridae,*** 155
Folac, ***Leviviridae,*** 308
Forecariah, ***Bunyaviridae,*** 282
Fort Morgan, ***Togaviridae,*** 218
Fort Sherman, ***Bunyaviridae,*** 275
Fowl adenovirus 1 (CELO), ***Adenoviridae,*** 143
Fowlpox, ***Poxviridae***, 94
Foxtail mosaic, ***Potexvirus,*** 349
FPLV, ***Parvoviridae,*** 168

ϕR, ***Microviridae,*** 179
fr, ***Leviviridae,*** 307
Frangipani mosaic, ***Tobamovirus,*** 358
Fraser Point, ***Bunyaviridae,*** 279
Freesia mosaic, ***Potyvirus,*** 354
fri, ***Myoviridae,*** 162
Frijoles, ***Bunyaviridae,*** 278
Frog 3, , ***Iridoviridae,*** 134
Frog viruses 1, 2, 5 - 24, L2, L4 and L5, , ***Iridoviridae,*** 134
Fukuoka, ***Rhabdoviridae,*** 256
Fuschia latent, ***Carlavirus,*** 343
ϕUW21, ***Siphoviridae,*** 164
FV3, ***Iridoviridae,*** 134
ϕW-14, ***Myoviridae,*** 162
G, ***Myoviridae,*** 162
G4, ***Microviridae,*** 179
G6, ***Microviridae,*** 179
G13, ***Microviridae,*** 179
G14, ***Microviridae,*** 179
GA, ***Leviviridae,*** 307
GA-1, ***Podoviridae,*** 166
Gabek Forest, ***Bunyaviridae,*** 278
Gadget's Gully, ***Flaviviridae,*** 227
Gaeumannomyces graminis 019/6-A, ***Partitiviridae,*** 208
Gaeumannomyces graminis 87-1-H, ***Totiviridae,*** 204
Gaeumannomyces graminis T1-A, ***Partitiviridae,*** 209
GAL 1, ***Adenoviridae,*** 143
gal1-gal2, ***Adenoviridae,*** 143
Galinsoga mosaic, ***Carmovirus,*** 304
Galleria densovirus , ***Parvoviridae,*** 171
Gamboa, ***Bunyaviridae,*** 276
Gan Gan, ***Bunyaviridae,*** 282
Garland chrysanthemum temperate, ***Cryptovirus,*** 213
Garlic mosaic, ***Carlavirus,*** 343
Garlic mosaic, ***Potyvirus,*** 352
Garlic yellow streak, ***Potyvirus,*** 354
Gentiana carlavirus, ***Carlavirus,*** 343
Geotrupes sylvaticus, ***Poxviridae***, 99
Gerbera symptomless, ***Rhabdoviridae,*** 260
Germiston, ***Bunyaviridae,*** 275
Getah, ***Togaviridae,*** 218
GF-2, ***Iridoviridae,*** 135
GFV-1, ***Iridoviridae,*** 135
GgV-019/6A, ***Partitiviridae,*** 208
Ggv-87-1-H, ***Totiviridae,*** 204
GgV-T1-A, ***Partitiviridae,*** 209
gh-1, ***Podoviridae,*** 166
Giardia Lamblia, ***Giardiavirus,*** 206
Gibbon ape leukemia, ***Reoviridae,*** 294
Ginger chlorotic fleck, ***Sobemovirus,*** 329
Gladiolus ringspot, ***Carlavirus,*** 342
Gloriosa fleck, ***Rhabdoviridae,*** 260
Gloriosa stripe mosaic, ***Potyvirus,*** 352
GLV, ***Giardiavirus,*** 206
Glycine mosaic, ***Comovirus,*** 362
Glycine mottle, ***Carmovirus,*** 304
Goatpox, ***Poxviridae***, 96
Goeldichironomus holoprasimus, ***Poxviridae***, 100
Golden shiner, ***Reoviridae,*** 192
Goldfish 1, ***Iridoviridae,*** 135
Goldfish 2, ***Iridoviridae,*** 135
Gomphrena, ***Rhabdoviridae,*** *260*
Gonometa, ***Picornaviridae,*** 324
Goose parvovirus, ***Parvoviridae,*** 168
Gordil, ***Bunyaviridae,*** 278
Gossas, ***Rhabdoviridae,*** 256
Grand Arbaud, ***Bunyaviridae,*** 278
Grapevine A, ***Closterovirus,*** 346
Grapevine Algerian latent, ***Tombusvirus,*** 334
Grapevine Bulgarian latent , Satellite, 401
Grapevine Bulgarian latent, ***Nepovirus,*** 370
Grapevine chrome mosaic, ***Nepovirus,*** 370
Grapevine fanleaf, ***Nepovirus,*** 370
Grapevine fanleaf, Satellite, 401
Grapevine leafroll-associated I, ***Closterovirus,*** 346
Grapevine leafroll-associated II, ***Closterovirus,*** 346
Grapevine leafroll-associated III, ***Closterovirus,*** 346
Grapevine leafroll-associated IV, ***Closterovirus,*** 346
Grapevine leafroll-associated V, ***Closterovirus,*** 346
Grapevine Tunisian ringspot, ***Nepovirus,*** 370
Grapevine yellow speckle 1, Viroid, 404
Grapevine yellow speckle 2, Viroid, 404
Grass carp reovirus, ***Reoviridae,*** 192
Grasscarp rhabdovirus, ***Rhabdoviridae,*** 253
Gray Lodge, ***Rhabdoviridae,*** 253
Great Saltee, ***Bunyaviridae,*** 279
Ground squirrel hepatitis B, ***Hepadnaviridae***, 115
Groundnut crinkle, ***Carlavirus,*** 343
Groundnut eyespot, ***Potyvirus,*** 352
Groundnut rosette assistor, ***Luteovirus,*** 310
Groundnut rosette, Satellite, 401
GSV, ***Reoviridae,*** 192
GU71u344, ***Bunyaviridae,*** 276
GU71u350, ***Bunyaviridae,*** 276
Guajara, ***Bunyaviridae,*** 276
Guama, ***Bunyaviridae,*** 276
Guar symptomless, ***Potyvirus,*** 354
Guaratuba, ***Bunyaviridae,*** 276

Guaroa, ***Bunyaviridae,*** 275
Guinea grass mosaic, ***Potyvirus,*** 352
Guinea pig cytomegalovirus, ***Herpesviridae,*** 108
Guinea pig type C, ***Reoviridae,*** 294
Gumbo Limbo, ***Bunyaviridae,*** 275
Gynura latent, ***Carlavirus,*** 343
Gypsy moth, ***Nodaviridae,*** 373
GYSVd 1, Viroid, 404
GYSVd 2+, Viroid, 404
H 2, ***Adenoviridae,*** 142
H-1, ***Parvoviridae,*** 168
H-19J, ***Siphoviridae,*** 164
h1-h47, ***Adenoviridae,*** 142
H1N1***Orthomyxoviridae,*** 263
η8, ***Microviridae,*** 179
H39, ***Podoviridae,*** 166
H387, ***Siphoviridae,*** 164
H32580, ***Bunyaviridae,*** 275
Habenaria mosaic, ***Potyvirus,*** 354
Hantaan, ***Bunyaviridae,*** 280
Hard clam reovirus, ***Reoviridae,*** 192
Hare fibroma, ***Poxviridae***, 96
Hart Park, ***Rhabdoviridae,*** 256
Hazara, ***Bunyaviridae,*** 279
HB, ***Parvoviridae,*** 168
HBV, ***Hepadnaviridae***, 115
HCV, ***Flaviviridae,*** 230
Helenium S, ***Carlavirus,*** 342
Helenium Y, ***Potyvirus,*** 352
Heliothis zea NOB, ***Baculoviridae***, 121
Heliothis zea SNPV, ***Baculoviridae***, 119
Helleborus mosaic, ***Carlavirus,*** 343
Helminthosporium victoriae 145S, ***Partitiviridae***, 210
Helminthosporium victoriae 190S, ***Totiviridae,*** 204
Henbane mosaic, ***Potyvirus,*** 353
Hepatitis B, ***Hepadnaviridae***, 115
Hepatitis C , ***Flaviviridae,*** 230
Hepatopancreatic parvo-like, ***Parvoviridae,*** 171
Heracleum latent, ***Closterovirus,*** 346
Heron hepatitis B, ***Hepadnaviridae***, 115
Herpes simplex 1, ***Herpesviridae***, 105
Herpes simplex 2, ***Herpesviridae***, 106
Herpesvirus ateles, ***Herpesviridae***, 109
HHBV, ***Hepadnaviridae***, 115
Hibiscus chlorotic ringspsot, ***Carmovirus,*** 304
Hibiscus latent ringspot, ***Nepovirus,*** 370
Highlands J, ***Togaviridae,*** 218
Hippeastrum latent, ***Carlavirus,*** 342
Hippeastrum mosaic, ***Potyvirus,*** 353
Hirame rhabdovirus, ***Rhabdoviridae,*** 255
HK022, ***Siphoviridae,*** 163
HK97, ***Siphoviridae,*** 163
HLVd, Viroid, 404
HM2, ***Podoviridae,*** 166
HM3, ***Myoviridae***, 162
HM7, ***Siphoviridae,*** 164
HN59, ***Reoviridae,*** 190
HN131, ***Reoviridae,*** 190
HN199, ***Reoviridae,*** 190
HN295, ***Reoviridae,*** 190
Hog cholera, ***Flaviviridae,*** 229
Holcus lanatus yellowing, ***Rhabdoviridae,*** 260
Holcus streak, ***Potyvirus,*** 354
Honeysuckle latent, ***Carlavirus,*** 342
Honeysuckle vein chlorosis, ***Rhabdoviridae,*** 260
Honeysuckle yellow vein mosaic, ***Geminivirus,*** 176
Hop (American) latent, ***Carlavirus,*** 342
Hop A, ***Ilarvirus,*** 390
Hop C, ***Ilarvirus,*** 390
Hop latent, ***Carlavirus,*** 342
Hop latent, Viroid, 404
Hop mosaic, ***Carlavirus,*** 342
Hop stunt, Viroid, 404
Hop trefoil cryptic 1, ***Cryptovirus,*** 213
Hop trefoil cryptic 2, ***Cryptovirus,*** 214
Hop trefoil cryptic 3, ***Cryptovirus,*** 213
Hordeum mosaic, ***Potyvirus,*** 355
Horse papilloma, ***Papovaviridae,*** 147
Horsegram yellow mosaic, ***Geminivirus,*** 176
Horseradish latent, ***Caulimovirus,*** 151
HR, ***Inoviridae,*** 132
HSVd, Viroid, 404
Hughes, ***Bunyaviridae,*** 279
Human (alpha) herpesvirus 1, ***Herpesviridae***, 105
Human (alpha) herpesvirus 2, ***Herpesviridae***, 106
Human (beta) herpesvirus 5, ***Herpesviridae***, 107
Human (gamma) herpesvirus 4, ***Herpesviridae***, 109
Human adenovirus 2, ***Adenoviridae,*** 142
Human coronavirus , ***Coronaviridae,*** 235
Human coxsackieviruses A1-22, 24, ***Picornaviridae,*** 322
Human coxsackieviruses B1-6, ***Picornaviridae,*** 322
Human cytomegalovirus, ***Herpesviridae***, 107
Human echoviruses 1-9, 11-27, 29-34, ***Picornaviridae,*** 322
Human enteroviruses 68-71, ***Picornaviridae,*** 322

Human foamy, ***Reoviridae,*** 296
Human hepatitis A, ***Picornaviridae,*** 322
Human immunodeficiency type 1, ***Reoviridae,*** 298
Human immunodeficiency type 2, ***Reoviridae,*** 298
Human immunodeficiency, ***Reoviridae,*** 297
Human poliovirus 1, ***Picornaviridae,*** 322
Human polioviruses 2-3, ***Picornaviridae,*** 322
Human respiratory syncytial, ***Paramyxoviridae,*** 245
Human rhinovirus 1a, ***Picornaviridae,*** 323
Human rhinoviruses 1B-100, ***Picornaviridae,*** 323
Human rotavirus, ***Reoviridae,*** 190, 191
Human T-cell lymphotropic type 1, ***Reoviridae,*** 297
Human T-cell lymphotropic type 2, ***Reoviridae,*** 297
Humans papilloma, ***Papovaviridae,*** 147
Humpty Doo, ***Rhabdoviridae,*** 255
Humulus japonicus, ***Ilarvirus,*** 390
Hungarian *Datura innoxia*, ***Potyvirus,*** 354
hv, ***Myoviridae,*** 162
HVCV-1, ***Phycodnaviridae,*** 138
HVCV-2, ***Phycodnaviridae,*** 138
HVCV-3, ***Phycodnaviridae,*** 138
hw, ***Myoviridae,*** 162
Hyacinth mosaic, ***Potyvirus,*** 354
Hydra viridis Chlorella viruses, ***Phycodnaviridae,*** 138
Hydrangea latent, ***Carlavirus,*** 342
Hydrangea mosaic, ***Ilarvirus,*** 390
Hydrangea ringspot, ***Potexvirus,*** 349
Hypochoeris mosaic, ***Furovirus,*** 378
Hypochoeris mosaic, ***Tobamovirus,*** 358
Hypocrita jacobeae, ***Tetraviridae,*** 331
HzNOB, ***Baculoviridae***, 121
HzSnpv, ***Baculoviridae***, 119
I, ***Podoviridae,*** 166
I_2-2, ***Inoviridae,*** 182
I3, ***Myoviridae,*** 162
Iα, ***Leviviridae,*** 308
Iaco, ***Bunyaviridae,*** 275
IBDV, ***Birnaviridae,*** 201
IBV, ***Coronaviridae,*** 234
Icoaraci, ***Bunyaviridae,*** 278
If1, ***Inoviridae,*** 182
If2, ***Inoviridae,*** 182
II, ***Siphoviridae,*** 164
III, ***Podoviridae,*** 166
Ike, ***Inoviridae,*** 182
Ilesha, ***Bunyaviridae,*** 275
Ilheus, ***Flaviviridae,*** 227
Impatiens latent, ***Carlavirus,*** 343
Indian cassava mosaic, ***Geminivirus,*** 176
Indonesian soybean dwarf, ***Luteovirus,*** 310
Infectious bovine rhino-tracheitis, ***Herpesviridae***, 106
Infectious bursal disease, ***Birnaviridae,*** 201
Infectious haematopoietic necrosis, ***Rhabdoviridae,*** 255
Infectious pancreatic necrosis, ***Birnaviridae,*** 200
Influenzavirus A/PR/8/34, ***Orthomyxoviridae,*** 263
Influenzavirus B, ***Orthomyxoviridae,*** 271
Influenzavirus C/Taylor/1233/47, ***Orthomyxoviridae,*** 271
Ingwavuma, ***Bunyaviridae,*** 276
Inini, ***Bunyaviridae,*** 276
Inkoo, ***Bunyaviridae,*** 276
Insect iridescent 1, 2, 6, 9, 10, 16-32, ***Iridoviridae,*** 133
Insect iridescent 3 - 5, 7, 8, 11-15, ***Iridoviridae,*** 134
IPNV, ***Birnaviridae,*** 200
Ippy, ***Arenaviridae,*** 288
IPy-1, ***Siphoviridae,*** 164
Iridescent , ***Iridoviridae,*** 134
Iris fulva mosaic, ***Potyvirus,*** 353
Iris germanica leaf stripe, ***Rhabdoviridae,*** 260
Iris mild mosaic, ***Potyvirus,*** 353
Iris severe mosaic, ***Potyvirus,*** 353
Isachne mosaic, ***Potyvirus,*** 354
Isfahan, ***Rhabdoviridae,*** 253
Israel turkey meningoencephalitis, ***Flaviviridae,*** 227
ISS Ph1 116, ***Rhabdoviridae,*** 253
Issyk-Kul (Keterah), ***Bunyaviridae,*** 282
Itaituba, ***Bunyaviridae,*** 278
Itaporanga, ***Bunyaviridae,*** 278
Itaqui, ***Bunyaviridae,*** 275
Itimirim, ***Bunyaviridae,*** 276
IV, ***Siphoviridae,*** 164
Ivy vein clearing, ***Rhabdoviridae,*** 260
Jamestown Canyon, ***Bunyaviridae,*** 276
Japanese encephalitis, ***Flaviviridae,*** 226
Jatropha mosaic, ***Geminivirus,*** 176
JC, ***Papovaviridae,*** 148
Jersey, ***Siphoviridae,*** 164
JKT-6423, ***Reoviridae,*** 190
JKT-6969, ***Reoviridae,*** 190
JKT-7041, ***Reoviridae,*** 190
JKT-7075, ***Reoviridae,*** 190
Joa, ***Bunyaviridae,*** 278
Johnsongrass mosaic, ***Potyvirus,*** 353
Joinjakaka, ***Rhabdoviridae,*** 256
jonquil mild mosaic, ***Potyvirus,*** 354
Juan Diaz, ***Bunyaviridae,*** 276

Jugra, ***Flaviviridae,*** 227
Juncopox, ***Poxviridae***, 95
Junin, ***Arenaviridae,*** 288
Junonia Densovirus, ***Parvoviridae,*** 171
Jurona, ***Rhabdoviridae,*** 253
Jutiapa, ***Flaviviridae,*** 227
K, ***Papovaviridae,*** 148
K19, ***Myoviridae,*** 162
Kachemak Bay, ***Bunyaviridae,*** 280
Kadam, ***Flaviviridae,*** 227
Kaeng Khoi, ***Bunyaviridae,*** 276
Kaikalur, ***Bunyaviridae,*** 276
Kairi, ***Bunyaviridae,*** 275
Kaisodi, ***Bunyaviridae,*** 282
Kalanchoe latent, ***Carlavirus,*** 342
Kalanchoe top-spotting, Commelina yellow mottle, 154
Kamese, ***Rhabdoviridae,*** 256
Kannamangalam, ***Rhabdoviridae,*** 256
Kao Shuan, ***Bunyaviridae,*** 279
kappa, ***Myoviridae,*** 162
Karimabad, ***Bunyaviridae,*** 278
Karshi, ***Flaviviridae,*** 226
Kasokero, ***Bunyaviridae,*** 282
Kemerovo, ***Reoviridae,*** 189
Kennedya Y, ***Potyvirus,*** 354
Kennedya yellow mosaic, ***Tymovirus,*** 337
Kern Canyon, ***Rhabdoviridae,*** 256
Ketapang, ***Bunyaviridae,*** 275
Keuraliba, ***Rhabdoviridae,*** 256
Keystone, ***Bunyaviridae,*** 276
Kf1, ***Siphoviridae,*** 164
Khasan, ***Bunyaviridae,*** 279
Kimberley, ***Rhabdoviridae,*** 255
Kismayo, ***Bunyaviridae,*** 282
Klamath, ***Rhabdoviridae,*** 253
Kokobera, ***Flaviviridae,*** 226
Kolongo, ***Rhabdoviridae,*** 255
Koongol, ***Bunyaviridae,*** 276
Kotonkan, ***Rhabdoviridae,*** 255
Koutango, ***Flaviviridae,*** 226
Kowanyama, ***Bunyaviridae,*** 282
KSY1, ***Podoviridae,*** 166
Kunjin, ***Flaviviridae,*** 226
Kwatta, ***Rhabdoviridae,*** 253
Kyasanur forest disease, ***Flaviviridae,*** 226
Kyuri green mottle mosaic, ***Tobamovirus,*** 358
Kyzylagach, ***Togaviridae,*** 218
L17, ***Tectiviridae,*** 155
La Crosse, ***Bunyaviridae,*** 276
La Joya, ***Rhabdoviridae,*** 253
La-Piedad-Michoacan-Mexico, ***Paramyxoviridae,*** 245
Laburnum yellow vein, ***Rhabdoviridae,*** 260
Lactic dehydrogenase, ***Togaviridae,*** 221
Lacticola, ***Siphoviridae,*** 164
Laelia red leafspot, ***Rhabdoviridae,*** 260
Lagos bat, ***Rhabdoviridae,*** 255
Lamium mild mosaic, ***Fabavirus,*** 367
Landjia, ***Rhabdoviridae,*** 256
Langat, ***Flaviviridae,*** 226
Langur (PO-1-Lu), ***Reoviridae,*** 295
Lanjan, ***Bunyaviridae,*** 282
Lapine parvovirus, ***Parvoviridae,*** 168
Las Maloyas, ***Bunyaviridae,*** 275
Lassa, ***Arenaviridae,*** 288
Latino, ***Arenaviridae,*** 288
Lato river, ***Tombusvirus,*** 334
Launea arborescens stunt, ***Rhabdoviridae,*** 260
LCDV-1, ***Iridoviridae,*** 135
LCDV-2, ***Iridoviridae,*** 135
LCM, ***Arenaviridae,*** 288
Le Dantec, ***Rhabdoviridae,*** 256
Leaky, ***Bunyaviridae,*** 280
Leanyer, ***Bunyaviridae,*** 276
Lednice, ***Bunyaviridae,*** 276
Leek yellow stripe, ***Potyvirus,*** 353
Legume yellows, ***Luteovirus,*** 310
Lemon scented thyme leaf chlorosis, ***Rhabdoviridae,*** 260
Leo, ***Siphoviridae,*** 164
Lettuce infectious yellows, ***Closterovirus,*** 346
Lettuce mosaic, ***Potyvirus,*** 353
Lettuce necrotic yellows, ***Rhabdoviridae,*** 258
Leucorrhinia Densovirus, ***Parvoviridae,*** 171
Lilac chlorotic leaf spot, ***Capillovirus,*** 340
Lilac mottle, ***Carlavirus,*** 342
Lilac ring mottle, ***Ilarvirus,*** 390
Lilac ringspot, ***Carlavirus,*** 343
Lily mild mottle, ***Potyvirus,*** 354
Lily symptomless, ***Carlavirus,*** 342
Lily X, ***Potexvirus,*** 349
Limabean golden mosaic, ***Geminivirus,*** 176
Lisianthus necrosis, ***Necrovirus,*** 317
Locusta migratoria, ***Poxviridae***, 100
Lokern, ***Bunyaviridae,*** 275
Lolium (ryegrass), ***Rhabdoviridae,*** 260
Lolium enation, ***Reoviridae,*** 196
Lone Star, ***Bunyaviridae,*** 282
Lotus streak, ***Rhabdoviridae,*** 260
Louping ill, ***Flaviviridae,*** 226
LPMV, ***Paramyxoviridae,*** 245
LPP-1, ***Podoviridae,*** 166
Lucerne Australian latent, ***Nepovirus,*** 370
Lucerne Australian symptomless, ***Nepovirus,*** 371
Lucerne enation, ***Rhabdoviridae,*** 259
Lucerne transient streak, Satellite, 401

Lucerne transient streak, ***Sobemovirus,*** 328
LuIII, ***Parvoviridae,*** 168
Lukuni, ***Bunyaviridae,*** 275
Lumpy skin disease, ***Poxviridae***, 96
Lupin leaf curl, ***Geminivirus,*** 176
Lupin yellow vein, ***Rhabdoviridae,*** 260
Lychnis potexvirus, ***Potexvirus,*** 349
Lychnis ringspot, ***Hordeivirus,*** 396
Lymantria ninayi, ***Tetraviridae,*** 331
Lymphocystis disease dab isolate, ***Iridoviridae,*** 135
Lymphocytic choriomeningitis, ***Arenaviridae,*** 284
M, ***Leviviridae,*** 308
m, ***Myoviridae,*** 162
M_1, ***Siphoviridae,*** 164
M6, ***Siphoviridae,*** 164
M12, ***Leviviridae,*** 307
M13, ***Inoviridae,*** 182
M14, ***Reoviridae,*** 190
M20, ***Microviridae,*** 179
MAC-1', ***Microviridae,*** 180
MAC-1, ***Microviridae,*** 180
MAC-2, ***Microviridae,*** 180
MAC-4', ***Microviridae,*** 180
MAC-4, ***Microviridae,*** 180
MAC-5, ***Microviridae,*** 180
MAC-7, ***Microviridae,*** 180
Macaua, ***Bunyaviridae,*** 275
Maclura mosaic, ***Potyvirus,*** 354
Macupo, ***Arenaviridae,*** 288
Madrid, ***Bunyaviridae,*** 275
Maguari, ***Bunyaviridae,*** 275
Mahogany Hammock, ***Bunyaviridae,*** 276
Main Drain, ***Bunyaviridae,*** 275
Maize chlorotic dwarf, Maize chlorotic dwarf virus, 312
Maize chlorotic mottle, ***Sobemovirus,*** 329
Maize dwarf mosaic, ***Potyvirus,*** 353
Maize mosaic, ***Rhabdoviridae,*** 258
Maize rayado fino, ***Marafivirus,*** 314
Maize rough dwarf , ***Reoviridae,*** 196
Maize sterile stunt, ***Rhabdoviridae,*** 259
Maize streak, ***Geminivirus,*** 174
Maize stripe, ***Tenuivirus,*** 399
Maize white line mosaic, Satellite, 401
Mal de Rio Cuarto disease, ***Reoviridae,*** 196
Malakal, ***Rhabdoviridae,*** 255
Malignant rabbit fibroma, ***Poxviridae***, 96
Malpais Spring, ***Rhabdoviridae,*** 253
Malva silvestris, ***Rhabdoviridae,*** *260*
Malva vein clearing, ***Potyvirus,*** 354
Malva veinal necrosis, ***Potexvirus,*** 349
Malva yellows, ***Luteovirus,*** 310
Malvaceous chlorosis, ***Geminivirus,*** 176
Mamestra brassicae MNPV, ***Baculoviridae***, 119
Mammalian orthoreoviruses, ***Reoviridae,*** 187
Mammalian rotavirus, ***Reoviridae,*** 191
Mammalian type B, ***Reoviridae,*** 293
Mammalian type C, ***Reoviridae,*** 294
Manawa, ***Bunyaviridae,*** 278
Manawatu, ***Nodaviridae,*** 373
Manzanilla, ***Bunyaviridae,*** 276
Mapputta, ***Bunyaviridae,*** 282
Maprik, ***Bunyaviridae,*** 282
Mapuera, ***Paramyxoviridae,*** 245
Maraba, ***Rhabdoviridae,*** 253
Marburg, ***Filoviridae,*** 247
Marco, ***Rhabdoviridae,*** 256
Marigold mottle, ***Potyvirus,*** 354
Marituba, ***Bunyaviridae,*** 275
Marmosetpox, ***Poxviridae***, 100
Marsupialpox, ***Poxviridae***, 100
Mason-Pfizer monkey, ***Reoviridae,*** 295
Matruh, ***Bunyaviridae,*** 276
MAV, ***Luteovirus,*** 310
Mayaro, ***Togaviridae,*** 218
MbMNPV, ***Baculoviridae***, 119
Mboke, ***Bunyaviridae,*** 275
MCDV, Maize chlorotic dwarf virus, 312
Meaban, ***Flaviviridae,*** 227
Measles , ***Paramyxoviridae,*** 244
mel1-mel3, ***Adenoviridae,*** 143
Melandrium yellow fleck, ***Bromovirus,*** 384
Melanoplus sanguinipes, ***Poxviridae***, 100
Melao, ***Bunyaviridae,*** 276
Melilotus latent, ***Rhabdoviridae,*** 260
Melilotus mosaic, ***Potyvirus,*** 354
Melolontha melolontha, ***Poxviridae***, 99
Melon leaf curl, ***Geminivirus,*** 176
Melon leaf variegation, ***Rhabdoviridae,*** 260
Melon necrotic spot, ***Carmovirus,*** 304
Mengovirus, ***Picornaviridae,*** 323
Mentha piperita latent, ***Rhabdoviridae,*** 260
Mermet, ***Bunyaviridae,*** 276
Methanococcus, ***SSV1***, 126
MEV, ***Parvoviridae,*** 168
Mibuna temperate, ***Cryptovirus,*** 213
Michigan alfalfa, ***Luteovirus,*** 310
Middelburg, ***Togaviridae,*** 218
Milk vetch dwarf, ***Luteovirus,*** 310
Milkers' nodule, ***Poxviridae***, 94
Millet red leaf, ***Luteovirus,*** 310
Mimosa bacilliform, Commelina yellow mottle, 154
Minatitlan, ***Bunyaviridae,*** 276
Mink enteritis, ***Parvoviridae,*** 168
Minute of canines, ***Parvoviridae,*** 168
Minute of mice, ***Parvoviridae,*** 168

Mirabilis mosaic, ***Caulimovirus,*** 151
Mirim, ***Bunyaviridae,*** 276
Miscanthus streak, ***Geminivirus,*** 174
MM, ***Picornaviridae,*** 323
Mn 936-77, ***Rhabdoviridae,*** 256
Mobala, ***Arenaviridae,*** 288
Modoc, ***Flaviviridae,*** 227
Moju, ***Bunyaviridae,*** 276
Mojui dos Campos, ***Bunyaviridae,*** 276
Mokola, ***Rhabdoviridae,*** 255
Molluscum contagiosum, ***Poxviridae***, 97
Monkeypox, ***Poxviridae***, 93
Montana myotis leukoencephalitis, ***Flaviviridae,*** 227
Mopeia, ***Arenaviridae,*** 288
mor1, ***Siphoviridae,*** 164
Moriche, ***Bunyaviridae,*** 276
Moroccan pepper, ***Tombusvirus,*** 334
Mosqueiro, ***Rhabdoviridae,*** 256
Mosquito iridescent , ***Iridoviridae,*** 134
Mossuril ((I, V), ***Rhabdoviridae,*** 256
Mount Elgon bat, ***Rhabdoviridae,*** 253
Mouse cytomegalovirus, ***Herpesviridae***, 108
Mouse mammary oncovirus, ***Reoviridae,*** 294
Mousepox, ***Poxviridae***, 93
MP13, ***Myoviridae,*** 162
MP15, ***Siphoviridae,*** 164
MpV, ***Totiviridae,*** 204
MRFV, ***Marafivirus,*** 314
MS2, ***Leviviridae,*** 307
MSP8, ***Siphoviridae,*** 164
MSV, ***Geminivirus,*** 174
MT, ***Siphoviridae,*** 164
Mu, ***Myoviridae,*** 162
Mucambo, ***Togaviridae,*** 218
Mucosal disease, ***Flaviviridae,*** 228
Muir Springs, ***Rhabdoviridae,*** 256
Mulberry latent, ***Carlavirus,*** 342
Mulberry ringspot, ***Nepovirus,*** 370
Mule deer *poxvirus* , ***Poxviridae***, 100
Multimammate papilloma, ***Papovaviridae,*** 147
Mumps, ***Paramyxoviridae,*** 244
Mungbean mosaic, ***Potyvirus,*** 354
Mungbean mottle, ***Potyvirus,*** 354
Mungbean yellow mosaic, ***Geminivirus,*** 176
Munguba, ***Bunyaviridae,*** 278
Murid (beta) herpesvirus 1, ***Herpesviridae,*** 108
Murid herpesvirus 2, ***Herpesviridae***, 108
Murine encephalomyelitis (ME), ***Picornaviridae,*** 323
Murine hepatitis , ***Coronaviridae,*** 235
Murine poliovirus, ***Picornaviridae,*** 322
Murine sarcoma and leukemia, ***Reoviridae,*** 294
Murray Valley encephalitisJ, ***Flaviviridae,*** 226
Murre, ***Bunyaviridae,*** 278
Murutucu, ***Bunyaviridae,*** 275
mus1-mus2, ***Adenoviridae,*** 142
Muskmelon vein necrosis, ***Carlavirus,*** 342
MV-L1, ***Inoviridae,*** 183
MVC, ***Parvoviridae,*** 168
MVG51, ***Inoviridae,*** 183
MVM, ***Parvoviridae,*** 168
Mycogone perniciosa, ***Totiviridae,*** 204
Mynahpox, ***Poxviridae***, 95
Myrobalan latent ringspot , Satellite, 401
Myrobalan latent ringspot, ***Nepovirus,*** 370
Myxoma, ***Poxviridae***, 96
N1, ***Myoviridae,*** 162
N1, ***Siphoviridae,*** 164
N4, ***Podoviridae,*** 166
N5, ***Siphoviridae,*** 164
Nairobi sheep disease, ***Bunyaviridae,*** 279
Nandina mosaic, ***Potexvirus,*** 349
Nandina stem pitting, ***Capillovirus,*** 340
Naranjal, ***Flaviviridae,*** 227
Narcissus degeneration, ***Potyvirus,*** 353
Narcissus late season yellows , ***Potyvirus,*** 354
Narcissus latent, ***Carlavirus,*** 342
Narcissus mosaic, ***Potexvirus,*** 349
Narcissus tip necrosis, ***Carmovirus,*** 304
Narcissus yellow stripe, ***Potyvirus,*** 353
Nariva, ***Paramyxoviridae***245
Nasoule, ***Rhabdoviridae,*** 255
Nasturtium mosaic. ***Carlavirus,*** 343
Naudurelia β , ***Tetraviridae,*** 330
Navarro, ***Rhabdoviridae,*** 256
Ndumu, ***Togaviridae,*** 218
NE-8D, ***Phycodnaviridae***, 138
Neckar river, ***Tombusvirus,*** 334
Neethling, ***Poxviridae***, 96
Negishi, ***Flaviviridae,*** 226
Negro coffee mosaic, ***Potexvirus,*** 349
Nelson Bay, ***Reoviridae,*** 187
Nepuyo, ***Bunyaviridae,*** 275
Nerine latent, ***Carlavirus,*** 342
Nerine X, ***Potexvirus,*** 349
Nerine, ***Potyvirus,*** 354
New Minto, ***Rhabdoviridae,*** 256
Newcastle disease, ***Paramyxoviridae,*** 244
Ngaingan, ***Rhabdoviridae,*** 255
Ngari, ***Bunyaviridae,*** 275
Nicotiana velutina mosaic, ***Furovirus,*** 378
Nique, ***Bunyaviridae,*** 278
Nkolbisson, ***Rhabdoviridae,*** 256
NM1, ***Siphoviridae,*** 164

Nodamura, ***Nodaviridae,*** 372
Nola, ***Bunyaviridae,*** 275
Northern cereal mosaic, ***Rhabdoviridae,*** 258
Northway, ***Bunyaviridae,*** 275
Nothoscordum mosaic, ***Potyvirus,*** 353
nt-1, ***Myoviridae,*** 162
NT2, ***Siphoviridae,*** 164
Ntaya, ***Flaviviridae,*** 227
Nyando, ***Bunyaviridae,*** 276
O'nyong-nyong, ***Togaviridae,*** 218
O1, ***Myoviridae,*** 162
o6, ***Microviridae,*** 179
Oak-Vale, ***Rhabdoviridae,*** 255
Oat blue dwarf, ***Marafivirus,*** 315
Oat chlorotic stripe, ***Geminivirus,*** 174
Oat golden stripe, ***Furovirus,*** 378
Oat mosaic, ***Potyvirus,*** 355
Oat necrotic mottle, ***Potyvirus,*** 355
Oat sterile dwarf, ***Reoviridae,*** 196
Oat striate mosaic, ***Rhabdoviridae,*** 259
Obodhiang, ***Rhabdoviridae,*** 255
Oceanside, ***Bunyaviridae,*** 278
Ockelbo, ***Togaviridae,*** 218
Octopus vulgaris disease, ***Iridoviridae,*** 135
Odontoglossum ringspot, ***Tobamovirus,*** 358
Odrenisrou, ***Bunyaviridae,*** 278
Oedaleus senugalensis, ***Poxviridae***, 100
Oita 296, ***Rhabdoviridae,*** 256
Okola, ***Bunyaviridae,*** 282
Okra mosaic, ***Tymovirus,*** 337
Olifantsvlei, ***Bunyaviridae,*** 276
Olive latent ringspot, ***Nepovirus,*** 370
Olive latent-1, ***Sobemovirus,*** 329
Omo, ***Bunyaviridae,*** 279
Omsk hemorrhagic fever, ***Flaviviridae,*** 226
Onion yellow dwarf, ***Potyvirus,*** 353
Ononis yellow mosaic, ***Tymovirus,*** 337
Operophtera brumata, ***Poxviridae***, 100
OpMNPV, ***Baculoviridae***, 119
Opossum papilloma, ***Papovaviridae,*** 147
Orchid fleck, ***Rhabdoviridae,*** 261
Orf, ***Poxviridae***, 94
Orgyia pseudotsugata MNPV, ***Baculoviridae***, 119
Oriboca, ***Bunyaviridae,*** 275
Oriximina, ***Bunyaviridae,*** 278
Ornithogalum mosaic, ***Potyvirus,*** 353
Oropouche, ***Bunyaviridae,*** 276
Oryctes rhinoceros, ***Baculoviridae***, 121
Ossa, ***Bunyaviridae,*** 275
Ouango, ***Rhabdoviridae,*** 256
Oubi, ***Bunyaviridae,*** 276
OV, ***Birnaviridae,*** 201
ovi1-ovi6, ***Adenoviridae,*** 142
Ovine AAV, ***Parvoviridae,*** 170
OXN-52P, ***Siphoviridae,*** 164
OXN-100P, ***Podoviridae,*** 166
Oyster, ***Birnaviridae,*** 201
P-a-1, ***Siphoviridae,*** 164
P1, ***Myoviridae,*** 162
P2, ***Myoviridae,*** 162
P22, ***Podoviridae,*** 166
P034, ***Podoviridae,*** 166
P087, ***Siphoviridae,*** 164
P107, ***Siphoviridae,*** 164
P335, ***Siphoviridae,*** 164
PA-2, ***Siphoviridae,*** 163
Pacora, ***Bunyaviridae,*** 282
Pacui, ***Bunyaviridae,*** 278
Pahayokee, ***Bunyaviridae,*** 276
Palestina, ***Bunyaviridae,*** 276
Palm mosaic, ***Potyvirus,*** 355
Palyam, ***Reoviridae,*** 189
Pangola stunt, ***Reoviridae,*** 196
Panicum mosaic, Satellite, 401
Panicum mosaic, ***Sobemovirus,*** 329
Papaya apical necrosis, ***Rhabdoviridae,*** 259
Papaya leaf distortion, ***Potyvirus,*** 355
Papaya mosaic, ***Potexvirus,*** 349
Papaya ringspot, ***Potyvirus,*** 353
Papillomavirus sylvilagi papilloma, ***Papovaviridae,*** 147
Para, ***Bunyaviridae,*** 276
Parainfluenza type 1, ***Paramyxoviridae,*** 244
Parainfluenza type 2, ***Paramyxoviridae,*** 244
Parainfluenza type 3, ***Paramyxoviridae,*** 244
Parainfluenza type 4, ***Paramyxoviridae,*** 244
Paramecium bursaria chlorella - 1, ***Phycodnaviridae,*** 137
Paramushir, ***Bunyaviridae,*** 280
Parana, ***Arenaviridae,*** 288
Parapoxvirus of New Zealand red deer ***Poxviridae***, 94
Paravaccinia, ***Poxviridae***, 94
Parietaria mottle, ***Ilarvirus,*** 390
Parrot. papilloma, ***Papovaviridae,*** 147
Parry Creek, ***Rhabdoviridae,*** 255
Parsley 5, ***Potexvirus,*** 349
Parsley latent, ***Rhabdoviridae,*** 260
Parsnip 3, ***Potexvirus,*** 349
Parsnip 5, ***Potexvirus,*** 349
Parsnip mosaic, ***Potyvirus,*** 353
Parsnip yellow fleck, Parsnip yellow fleck virus, 318
Parvo-like from the crab Carcinus mediterraneus, ***Parvoviridae,*** 171
Paspalum striate mosaic, ***Geminivirus,*** 174

Passiflora latent, ***Carlavirus,*** 342
Passion fruit yellow mosaic, ***Tymovirus,*** 337
Passionfruit ringspot, ***Potyvirus,*** 355
Passionfruit vein clearing, ***Rhabdoviridae,*** 260
Passionfruit woodiness, ***Potyvirus,*** 353
Patchouli (*Pogostemon patchouli*) mottle, ***Rhabdoviridae,*** 260
Patchouli mottle, ***Potyvirus,*** 355
Pathum Thani, ***Bunyaviridae,*** 279
Patois, ***Bunyaviridae,*** 276
PAV, ***Luteovirus,*** 310
PB-1, ***Myoviridae,*** 162
PBCV - 1, ***Phycodnaviridae,*** 137
PbGV, ***Baculoviridae,*** 120
PBP1, ***Siphoviridae,*** 164
PBS1, ***Myoviridae,*** 162
PbV, ***Partitiviridae,*** 210
PC 84, ***Parvoviridae,*** 171
Pc-fV, ***Partitiviridae,*** 210
PcV, ***Partitiviridae,*** 209
PE1, ***Siphoviridae,*** 164
Pea early-browning, ***Tobravirus,*** 381
Pea enation mosaic, Pea enation mosaic, 375
Pea enation mosaic, Satellite, 401
Pea leaf roll, ***Luteovirus,*** 310
Pea mild mosaic, ***Comovirus,*** 362
Pea mosaic, crocus tomasinianus, ***Potyvirus,*** 352
Pea necrosis, ***Potyvirus,*** 352
Pea seed-borne mosaic, ***Potyvirus,*** 353
Pea streak, ***Carlavirus,*** 342
Peach latent mosaic, Viroid, 404
Peach rosette mosaic, ***Nepovirus,*** 370
Peacockpox, ***Poxviridae,*** 95
Peanut chlorotic ring mottle, ***Potyvirus,*** 353
Peanut chlorotic streak, ***Caulimovirus,*** 151
Peanut clump, ***Furovirus,*** 378
Peanut green mosaic, ***Potyvirus,*** 355
Peanut mild mottle, ***Potyvirus,*** 353
Peanut mosaic, ***Potyvirus,*** 355
Peanut mottle, ***Potyvirus,*** 353
Peanut stripe , ***Potyvirus,*** 353
Peanut stunt, ***Cucumovirus,*** 387
Peanut stunt, Satellite, 401
Peanut veinal chlorosis, ***Rhabdoviridae,*** 260
Peanut yellow mosaic, ***Tymovirus,*** 337
Peaton, ***Bunyaviridae,*** 276
Pecteilis mosaic, ***Potyvirus,*** 355
Pelargonium flower break, ***Carmovirus,*** 304
Pelargonium leaf curl, Satellite, 401
Pelargonium leaf curl, ***Tombusvirus,*** 334
Pelargonium vein clearing, ***Rhabdoviridae,*** 260
PEMV, Pea enation mosaic virus, 375
Penguinpox, ***Poxviridae,*** 95
Penicillium brevi compactum, ***Partitiviridae,*** 210
Penicillium cyaneo-fulvum , ***Partitiviridae,*** 210
Penicillium stoloniferum F, ***Partitiviridae,*** 209
Penicillium stoloniferum S, ***Partitiviridae,*** 209
Pepino latent, ***Carlavirus,*** 342
Pepino mosaic, ***Potexvirus,*** 349
Pepper mild mosaic, ***Potyvirus,*** 355
Pepper mild mottle, ***Tobamovirus,*** 358
Pepper mottle, ***Potyvirus,*** 355
Pepper ringspot, ***Tobravirus,*** 381
Pepper severe mosaic, ***Potyvirus,*** 353
Pepper veinal mottle, ***Potyvirus,*** 353
Perch rhabdovirus, ***Rhabdoviridae,*** 256
Perilla mottle, ***Potyvirus,*** 355
Perinet, ***Rhabdoviridae,*** 253
Periplanata Densovirus, ***Parvoviridae,*** 171
Peste-des-petits-ruminants, ***Paramyxoviridae,*** 245
Petunia asteroid mosaic, Satellite, 401
Petunia asteroid mosaic, ***Tombusvirus,*** 334
Petunia vein clearing, ***Caulimovirus,*** 151
Pf1, ***Inoviridae,*** 182
Pf2, ***Inoviridae,*** 182
Pf3, ***Inoviridae,*** 182
pha1, ***Adenoviridae,*** 143
Phage PRD1 group, ***Tectiviridae,*** 155
Phalaenopsis chlorotic spot, ***Rhabdoviridae,*** 261
Phialophora radicicola 2-2-A, ***Partitiviridae,*** 209
Phialophora sp. (lobed hyphopodia) 2-2-A, ***Partitiviridae,*** 209
Philosamia ricini, ***Tetraviridae,*** 331
Phnom Penh bat , ***Flaviviridae,*** 226
Phocine distemper, ***Paramyxoviridae,*** 244
Physalis mild chlorosis, ***Luteovirus,*** 310
Physalis mosaic, ***Tymovirus,*** 337
Physalis vein blotch, ***Luteovirus,*** 310
Pichinde, ***Arenaviridae,*** 288
Pieris brassicae granulosis, ***Baculoviridae,*** 120
Pieris Densovirus, ***Parvoviridae,*** 171
Pig cytomegalovirus, ***Herpesviridae,*** 108
Pig rotavirus, ***Reoviridae,*** 191
Pigeon pea (*Cajanus cajan*) proliferation, ***Rhabdoviridae,*** 260
Pigeonpox, ***Poxviridae,*** 95
PiGV, ***Baculoviridae,*** 120

PIIBNV6, ***Myoviridae,*** 162
PIIBNV6-C, ***Podoviridae,*** 166
Pike fry rhabdovirus, ***Rhabdoviridae,*** 253
PilHα, ***Leviviridae,*** 308
Pineapple chlorotic leaf streak, ***Rhabdoviridae,*** 260
Pineapple mealybug wilt-associated, ***Closterovirus,*** 346
Piry, ***Rhabdoviridae,*** 253
Pisum, ***Rhabdoviridae,*** *260*
Pittosporum vein yellowing and tomato vein yellowing, ***Rhabdoviridae,*** 259
Pittosporum vein yellowing, ***Rhabdoviridae,*** 260
Pixuna, ***Togaviridae,*** 218
PL-1, ***Siphoviridae,*** 164
Plantago 4, **Caulimovirus,** 151
Plantago mottle, ***Tymovirus,*** 337
Plantago severe mottle, ***Potexvirus,*** 349
Plantain (*Plantago lanceolata*) mottle, ***Rhabdoviridae,*** 261
Plantain 6, ***Carmovirus,*** 304
Plantain 7, ***Potyvirus,*** 355
Plantain 8, ***Carlavirus,*** 343
Plantain X, ***Potexvirus,*** 349
Playas, ***Bunyaviridae,*** 275
Pleioblastus mosaic, ***Potyvirus,*** 355
PLMVd, Viroid, 404
Plodia interpunctella granulosis, ***Baculoviridae***, 120
Plum pox, ***Potyvirus,*** 353
PMV-1, ***Paramyxoviridae,*** 244
PMV-2, ***Paramyxoviridae,*** 244
PMV-3, ***Paramyxoviridae,*** 244
PMV-4, ***Paramyxoviridae,*** 244
PMV-5, ***Paramyxoviridae,*** 244
PMV-6, ***Paramyxoviridae,*** 244
PMV-7, ***Paramyxoviridae,*** 244
PMV-8, ***Paramyxoviridae,*** 244
PMV-9, ***Paramyxoviridae,*** 244
Pneumonia of mice, ***Paramyxoviridae,*** 245
Poa semilatent, ***Hordeivirus,*** 396
Poinsettia cryptic, ***Cryptovirus,*** 213
Poinsettia mosaic, ***Tymovirus,*** 337
Pokeweed mosaic, ***Potyvirus,*** 353
Polyoma, ***Papovaviridae,*** 148
Polyomavirus bovis , ***Papovaviridae,*** 148
Polyomavirus cercopitheci (lymphotropic), ***Papovaviridae,*** 148
Polyomavirus hominis 1, ***Papovaviridae,*** 148
Polyomavirus hominis 2, ***Papovaviridae,*** 148
Polyomavirus maccacae 1, ***Papovaviridae,*** 148
Polyomavirus muris 1, ***Papovaviridae,*** 148
Polyomavirus muris 2, ***Papovaviridae,*** 148
Polyomavirus papionis 1, ***Papovaviridae,*** 148
Polyomavirus papionis 2, ***Papovaviridae,*** 148
Polyomavirus sylvilagi , ***Papovaviridae,*** 148
Pongine herpesvirus 1, ***Herpesviridae***, 109
Pongola, ***Bunyaviridae,*** 275
Ponteves, ***Bunyaviridae,*** 278
Poplar mosaic, ***Carlavirus,*** 342
Populus, ***Potyvirus,*** *355*
Porcine enteroviruses 1-8, ***Picornaviridae,*** 322
Porcine epidemic diarrhea, ***Coronaviridae,*** 236
Porcine hemagglutinating encephalomyelitis, ***Coronaviridae,*** 235
Porcine parvovirus, ***Parvoviridae,*** 168
Porcine transmissible gastroenteritis , ***Coronaviridae,*** 235
Porcine type C, ***Reoviridae,*** 294
Porton, ***Rhabdoviridae,*** 253
Potato A, ***Potyvirus,*** 353
Potato aucuba mosaic, ***Potexvirus,*** 349
Potato black ringspot, ***Nepovirus,*** 370
Potato leaf roll, ***Luteovirus,*** 310
Potato M, ***Carlavirus,*** 342
Potato mop-top, ***Furovirus,*** 378
Potato S, ***Carlavirus,*** 342
Potato spindle tuber, Viroid, 403, 404
Potato T, ***Capillovirus,*** 340
Potato U, ***Nepovirus,*** 370
Potato V, ***Potyvirus,*** 353
Potato X, ***Potexvirus,*** 348
Potato Y, ***Potyvirus,*** 351
Potato yellow dwarf, ***Rhabdoviridae,*** 259
Potato yellow mosaic, ***Geminivirus,*** 176
Poultry rotavirus, ***Reoviridae,*** 191
Powassan, ***Flaviviridae,*** 226
PP7, ***Leviviridae,*** 308
PP8, ***Myoviridae,*** 162
PR3, ***Tectiviridae,*** 155
PR4, ***Tectiviridae,*** 155
PR5, ***Tectiviridae,*** 155
PR64FS, ***Inoviridae,*** 182
PR772, ***Tectiviridae,*** 155
Precarious Point, ***Bunyaviridae,*** 278
Pretoria, ***Bunyaviridae,*** 279
Primula mosaic, ***Potyvirus,*** 355
pro2, ***Siphoviridae,*** 164
Prospect Hill, ***Bunyaviridae,*** 280
PRR1, ***Leviviridae,*** 308
Prune dwarf, ***Ilarvirus,*** 390
Prunus necrotic ringspot, ***Ilarvirus,*** 390
Prunus S, ***Carlavirus,*** 343

PrV-2-2-A, ***Partitiviridae***, 209
Ps V-F, ***Partitiviridae***, 209
Ps V-S, ***Partitiviridae***, 209
PS4, ***Siphoviridae***, 164
PS8, ***Siphoviridae***, 164
PS17, ***Myoviridae***, 162
Pseudocowpox, ***Poxviridae***, 94
Pseudomonas Phage φ6, ***Cystoviridae***, 184
Pseudorabies, ***Herpesviridae***, 106
Psittacinepox, ***Poxviridae***, 95
Psophocarpus necrotic mosaic, ***Carlavirus***, 343
PST, ***Myoviridae***, 162
PSTVd, Viroid, 403, 404
PT11, ***Siphoviridae***, 164
PTB, ***Podoviridae***, 165
PTFV, Parsnip yellow fleck virus 318
Puchong, ***Rhabdoviridae***, 255
Pueblo Viejo, ***Bunyaviridae***, 276
Puffin Island, ***Bunyaviridae***, 279
Punta Salinas, ***Bunyaviridae***, 279
Punta Toro, ***Bunyaviridae***, 278
Puumala, ***Bunyaviridae***, 280
PVX, ***Potexvirus***, 348
PVY, ***Potyvirus***, 351
Qalyub, ***Bunyaviridae***, 279
Qβ, ***Leviviridae***, 308
Quail pea mosaic, ***Comovirus***, 362
Quailpox, ***Poxviridae***, 95
Queensland fruitfly, ***Picornaviridae***, 324
R, ***Podoviridae***, 165
R_1, ***Siphoviridae***, 164
R1-Myb, ***Siphoviridae***, 164
R_2, ***Siphoviridae***, 164
R17, ***Leviviridae***, 307
R23, ***Leviviridae***, 308
R34, ***Leviviridae***, 308
RA-1, ***Parvoviridae***, 168
Rabbit (Shope) fibroma, ***Poxviridae***, 96
Rabbit (Shope) papilloma, ***Papovaviridae***, 147
Rabbit coronavirus, ***Coronaviridae***, 236
Rabbit kidney vacuolating, ***Papovaviridae***, 148
Rabbitpox, ***Poxviridae***, 93
Rabies, ***Rhabdoviridae***, 254
Raccoon parvovirus, ***Parvoviridae***, 168
Raccoonpox, ***Poxviridae***, 93
Radi, ***Rhabdoviridae***, 253
Radish mosaic, ***Comovirus***, 362
Radish yellow edge, ***Cryptovirus***, 213
Rana catesbriana LT 1 - 4, ***Iridoviridae***, 134
Ranunculus repens symptomless, ***Rhabdoviridae***, 261
Raphanus, ***Rhabdoviridae***, *260*
Raspberry leaf curl, ***Luteovirus***, 310
Raspberry ringspot, ***Nepovirus***, 370
Raspberry vein chlorosis, ***Rhabdoviridae***, 259
Rat coronavirus , ***Coronaviridae***, 236
Rat cytomegalovirus, ***Herpesviridae***, 108
Rat rotavirus, ***Reoviridae***, 191
Rat, ***Parvoviridae***, 168
Raza, ***Bunyaviridae***, 279
Razdan, ***Bunyaviridae***, 282
Red clover cryptic 2, ***Cryptovirus***, 214
Red clover mosaic, ***Rhabdoviridae***, 261
Red clover mottle , ***Comovirus***, 362
Red clover necrotic mosaic, ***Dianthovirus***, 365
Red clover vein mosaic, ***Carlavirus***, 342
Red pepper cryptic 1, ***Cryptovirus***, 214
Red pepper cryptic 2, ***Cryptovirus***, 214
Reed canary mosaic, ***Potyvirus***, 355
Reed Ranch, ***Rhabdoviridae***, 256
Reovirus type 1, ***Reoviridae***, 187
Reptilian type C, ***Reoviridae***, 295
Resistencia, ***Bunyaviridae***, 282
Respiratory infection, ***Herpesviridae***, 106
Restan, ***Bunyaviridae***, 275
Reticuloendotheliosis, ***Reoviridae***, 294
RGV, ***Luteovirus***, 310
Rhabdovirus carpia, ***Rhabdoviridae***, 253
Rhabdovirus of blue crab, ***Rhabdoviridae***, 256
Rhabdovirus of entamoeba, ***Rhabdoviridae***, 257
Rhabdovirus salmonis, ***Rhabdoviridae***, 257
Rhesus monkeys papilloma, ***Papovaviridae***, 147
Rhizidiomyces, ***Rhizidiovirus***, 145
Rhizoctonia solani, ***Partitiviridae***, 209
Rhododendron necrotic ringspot, ***Potexvirus***, 349
Rhubarb 1, ***Potexvirus***, 349
Rhubarb temperate, ***Cryptovirus***, 214
Rhynochosia mosaic, ***Geminivirus***, 176
Ribgrass mosaic, ***Tobamovirus***, 358
Rice black streaked dwarf, ***Reoviridae***, 196
Rice dwarf, ***Reoviridae***, 195
Rice gall dwarf, ***Reoviridae***, 195
Rice giallume, ***Luteovirus***, 310
Rice grassy stunt, ***Tenuivirus***, 399
Rice hoja blanca, ***Tenuivirus***, 399
Rice necrosis mosaic, ***Potyvirus***, 355
Rice ragged stunt, ***Reoviridae***, 197
Rice stripe necrosis, ***Furovirus***, 378
Rice stripe, ***Tenuivirus***, 398
Rice transitory yellowing, ***Rhabdoviridae***, 259
Rice tungro bacilliform, Commelina yellow mottle, 154

SV5, ***Paramyxoviridae,*** 244
SV40), ***Papovaviridae,*** 148
SV41, ***Paramyxoviridae,*** 244
Sweet clover necrotic mosaic, ***Dianthovirus,*** 365
Sweet potato feathery mottle, ***Potyvirus,*** 353
Sweet potato mild mottle, ***Potyvirus,*** 355
Sweet potato russett crack, sweet potato A, ***Potyvirus,*** 353
Sweetwater Branch, ***Rhabdoviridae,*** 255
Swine vesicular disease, ***Picornaviridae,*** 322
Swinepox, ***Poxviridae***, 97
τ, ***Leviviridae,*** 308
T1, ***Siphoviridae,*** 164
T2, ***Myoviridae,*** 162
T3, ***Podoviridae,*** 165
T4, ***Myoviridae,*** 162
T5, , ***Siphoviridae,*** 164
T6, ***Myoviridae,*** 162
T6-T20, ***Iridoviridae,*** 134
T21 from Xenopus, ***Iridoviridae,*** 134
Ta1, ***Podoviridae,*** 165
Tacaiuma, ***Bunyaviridae,*** 275
Tacaribe, ***Arenaviridae,*** 288
Tadpole edema, ***Iridoviridae,*** 134
Taggert, ***Bunyaviridae,*** 280
Tahyna, ***Bunyaviridae,*** 276
Tai, ***Bunyaviridae,*** 282
Taiassui, ***Bunyaviridae,*** 275
Tamana bat , ***Flaviviridae,*** 227
Tamarillo mosaic, ***Potyvirus,*** 353
Tamdy, ***Bunyaviridae,*** 282
Tamiami, ***Arenaviridae,*** 288
Tanapox, ***Poxviridae***, 98
Tanga, ***Bunyaviridae,*** 282
Tanjong Rabok, ***Bunyaviridae,*** 275
TASVd, Viroid, 404
Tataguine, ***Bunyaviridae,*** 282
Taterapox, ***Poxviridae***, 93
Tb, ***Podoviridae,*** 166
TBSV, ***Tombusvirus,*** 333
TBTVd, Viroid, 404
Teasel mosaic, ***Potyvirus,*** 355
Tehran, ***Bunyaviridae,*** 278
Telfairia mosaic, ***Potyvirus,*** 353
Tellina, ***Birnaviridae,*** 201
Telok Forest, ***Bunyaviridae,*** 275
Tembusu, ***Flaviviridae,*** 227
Tench reovirus, ***Reoviridae,*** 192
Tensaw, ***Bunyaviridae,*** 275
Tephrosia symptomless, ***Carmovirus,*** 304
Termeil, ***Bunyaviridae,*** 276
Tete, ***Bunyaviridae,*** 276
tf-1, **Inoviridae**, 182
Theiler's encephalomyelitis, TO, FA, GD7, ***Picornaviridae,*** 322
Thermoproteus Phage TTV1, ***Lipothrixviridae***, 127
Thermoproteus Phage TTV2, ***Lipothrixviridae***, 127
Thiafora, ***Bunyaviridae,*** 280
Thielaviopsis basicola, ***Totiviridae,*** 204
Thimiri, ***Bunyaviridae,*** 276
Thistle mottle, ***Caulimovirus,*** 151
Thogot, ***Orthomyxoviridae,*** 271
Thosea asiona, ***Tetraviridae,*** 331
Thottapalayam, ***Bunyaviridae,*** 280
Tibrogargan, ***Rhabdoviridae,*** 256
Tick-borne encephalitis, ***Flaviviridae,*** 226
Tigre disease, ***Geminivirus,*** 176
Tillamook., ***Bunyaviridae,*** 280
Timbo, ***Rhabdoviridae,*** 256
Timboteua, ***Bunyaviridae,*** 276
Tinaroo, ***Bunyaviridae,*** 276
Tlacotalpan, ***Bunyaviridae,*** 275
TMV, ***Tobamovirus,*** 357
TnGV, ***Baculoviridae***, 120
TnSnpv, ***Baculoviridae***, 119
TNV, ***Necrovirus,*** 316
Tobacco etch, ***Potyvirus,*** 353
Tobacco leaf curl, ***Geminivirus,*** 176
Tobacco mild green mosaic, ***Tobamovirus,*** 358
Tobacco mosaic common or U1 strain, ***Tobamovirus,*** 357
Tobacco mosaic, Satellite, 401
Tobacco necrosis, ***Necrovirus,*** 316
Tobacco necrosis, Satellite, 401
Tobacco necrotic dwarf, ***Luteovirus,*** 310
Tobacco rattle, ***Tobravirus,*** 380
Tobacco ringspot, ***Nepovirus,*** 368
Tobacco ringspot, Satellite, 401
Tobacco streak, ***Ilarvirus,*** 389
Tobacco vein banding mosaic, ***Potyvirus,*** 355
Tobacco vein distorting, ***Luteovirus,*** 310
Tobacco vein mottling, ***Potyvirus,*** 353
Tobacco yellow dwarf, ***Geminivirus,*** 175
Tobacco yellow net, ***Luteovirus,*** 310
Tobacco yellow vein assistor,***Luteovirus,*** 310
TobRV, ***Nepovirus,*** 368
Tomato (Peru) mosaic, ***Potyvirus,*** 353
Tomato apical stunt, Viroid, 404
Tomato aspermy, ***Cucumovirus,*** 387
Tomato black ring, ***Nepovirus,*** 370
Tomato black ring, Satellite, 401
Tomato bunchy top, Viroid, 404
Tomato bushy stunt , Satellite, 401
Tomato bushy stunt, ***Tombusvirus,*** 333

Tomato golden mosaic, ***Geminivirus,*** 176
Tomato leaf curl, ***Geminivirus,*** 176
Tomato leafroll, ***Geminivirus,*** 175
Tomato mosaic, ***Tobamovirus,*** 358
Tomato pale chlorosis, ***Carlavirus,*** 343
Tomato planto macho, Viroid, 404
Tomato pseudo-curly top, ***Geminivirus,*** 175
Tomato ringspot, ***Nepovirus,*** 370
Tomato spotted wilt, ***Bunyaviridae,*** 281
Tomato top necrosis, ***Nepovirus,*** 371
Tomato vein clearing, ***Rhabdoviridae,*** *261*
Tomato yellow dwarf, ***Geminivirus,*** 176
Tomato yellow leaf curl, ***Geminivirus,*** 176
Tomato yellow mosaic, ***Geminivirus,*** 176
Toscana, ***Bunyaviridae,*** 278
TPMVd, Viroid, 404
Tradescantia/Zebrina, ***Potyvirus,*** 355
Triatoma lethal paralysis, ***Picornaviridae,*** 324
Trichomonas vaginalis, ***Giardiavirus,*** 206
Trichoplusia ni granulosis, ***Baculoviridae***, 120
Trichoplusia ni SNPV, ***Baculoviridae***, 119
Trichoplusia ni, ***Tetraviridae,*** 331
Triticum aestivum chlorotic spot, ***Rhabdoviridae,*** 261
Trivittatus, ***Bunyaviridae,*** 276
Trombetas, ***Bunyaviridae,*** 275
Tropaeolum 1, ***Potyvirus,*** 355
Tropaeolum 2, ***Potyvirus,*** 355
Trubanaman, ***Bunyaviridae,*** 282
TRV, ***Tobravirus,*** 380
Tsuruse, ***Bunyaviridae,*** 276
TSV, ***Ilarvirus,*** 389
TSWV, ***Bunyaviridae,*** 281
Tucunduba, ***Bunyaviridae,*** 275
Tulare apple mosaic, ***Ilarvirus,*** 390
Tulip breaking, ***Potyvirus,*** 353
Tulip chlorotic blotch, ***Potyvirus,*** 353
Tulip X, ***Potexvirus,*** 349
Tupaia, ***Rhabdoviridae,*** 254
Turbot reovirus, ***Reoviridae,*** 192
Turkey coronavirus, ***Coronaviridae,*** 236
Turkey rhinotracheitis, ***Paramyxoviridae,*** 245
Turkeypox, ***Poxviridae***, 95
Turlock, ***Bunyaviridae,*** 276
Turnip crinkle, ***Carmovirus,*** 304
Turnip crinkle, Satellite, 401
Turnip mosaic, ***Potyvirus,*** 353
Turnip rosette, ***Sobemovirus,*** 328
Turnip yellow mosaic, ***Tymovirus,*** 336
Turnip yellows, ***Luteovirus,*** 310
Turtle papilloma, ***Papovaviridae,*** 147
Turuna, ***Bunyaviridae,*** 278
TV, ***Birnaviridae,*** 201
TVV, ***Giardiavirus,*** 206
TVX, ***Parvoviridae,*** 168
Twort, ***Myoviridae,*** 162
TYMV, ***Tymovirus,*** 336
Type 2 from *Inachis io*, ***Reoviridae,*** 194
Type 3 from *Spodoptera exempta*, ***Reoviridae,*** 194
Type 3, regular strain, ***Iridoviridae,*** 134
Type 4 from *Actias selene*, ***Reoviridae,*** 194
Type 5 from *Trichoplusia ni*, ***Reoviridae,*** 194
Type 6 from *Biston betularia*, ***Reoviridae,*** 194
Type 7 from *Triphena pronuba*, ***Reoviridae,*** 194
Type 8 from *Abraxas grossulariata*, ***Reoviridae,*** 194
Type 9 from *Agrotis segetum*, ***Reoviridae,*** 194
Type 10 from *Aporophylla lutulenta*, ***Reoviridae,*** 194
Type 11 from *Spodoptera exigua*, ***Reoviridae,*** 194
Type 12 from *Spodoptera exempta*, ***Reoviridae,*** *194*
Tyuleniy, ***Flaviviridae,*** 227
U3, ***Microviridae,*** 179
Uasin Gishu disease, ***Poxviridae***, 93
Uganda S, ***Flaviviridae,*** 227
Ulcerative disease rhabdovirus, ***Rhabdoviridae,*** 254
Ullucus C, ***Comovirus,*** 362
Ullucus mild mottle, ***Tobamovirus,*** 358
Ullucus mosaic, ***Potyvirus,*** 355
Umbre, ***Bunyaviridae,*** 276
Una, ***Togaviridae,*** 218
Upolu, ***Bunyaviridae,*** 282
Urucuri, ***Bunyaviridae,*** 278
Ustilago maydis P1, P4 and P6, ***Totiviridae,*** 204
Usutu, ***Flaviviridae,*** 226
Utinga, ***Bunyaviridae,*** 276
Utive, ***Bunyaviridae,*** 276
Uukuniemi, ***Bunyaviridae,*** 278
v1, ***Plasmaviridae***, 124
v2, ***Plasmaviridae***, 124
v4, ***Plasmaviridae***, 124
v5, ***Plasmaviridae***, 124
v6, ***Inoviridae,*** 182
v7, ***Plasmaviridae***, 124
Vaccinia, ***Poxviridae***, 92
Vallota mosaic, ***Potyvirus,*** 355
Vanilla mosaic, ***Potyvirus,*** 355
Vanilla necrosis, ***Potyvirus,*** 355
Variola, ***Poxviridae***, 93

VD13, ***Siphoviridae,*** 164
Velvet tobacco mottle, Satellite, 401
Velvet tobacco mottle, ***Sobemovirus,*** 328
Venezuelan equine encephalitis, ***Togaviridae,*** 219
Vesicular exanthema of swine, ***Caliciviridae,*** 300
Vesicular stomatitis -indiana, ***Rhabdoviridae,*** 253
Vesicular stomatitis Alagoas, ***Rhabdoviridae,*** 254
Vesicular stomatitis New Jersey, ***Rhabdoviridae,*** 254
VESV, ***Caliciviridae,*** 300
Vf12, ***Inoviridae,*** 182
Vf33, ***Inoviridae,*** 182
Vicia cryptic, ***Cryptovirus,*** 213
Vigna sinensis mosaic, ***Rhabdoviridae,*** 261
ViII, , ***Siphoviridae,*** 164
Vil, ***Myoviridae,*** 162
Vilyuisk, ***Picornaviridae,*** 322
Vinces, ***Bunyaviridae,*** 275
Viola mottle, ***Potexvirus,*** 349
Viper retrovirus, ***Reoviridae,*** 295
Viral haemorrhagic septicemia, ***Rhabdoviridae,*** 256
Virgin River, ***Bunyaviridae,*** 275
Visna/Maedi, ***Reoviridae,*** 298
Voandzeia mosaic, ***Carlavirus,*** 343
Voandzeia necrotic mosaic, ***Tymovirus,*** 337
Volepox, ***Poxviridae***, 93
Volepox, ***Poxviridae***, 101
VP1, ***Myoviridae,*** 162
VP3, ***Siphoviridae,*** 164
VP5, ***Siphoviridae,*** 164
VP5, ***Siphoviridae,*** 164
VP11, ***Siphoviridae,*** 164
ω3, 24, ***Siphoviridae,*** 164
Ω8, ***Podoviridae,*** 166
W31, ***Podoviridae,*** 165
WA/1, ***Microviridae,*** 179
Wallal, ***Reoviridae,*** 189
Wanowrie, ***Bunyaviridae,*** 282
Warrego, ***Reoviridae,*** 189
Watermelon chlorotic stunt, ***Geminivirus,*** 176
Watermelon curly mottle, ***Geminivirus,*** 176
Watermelon mosaic 1, ***Potyvirus,*** 353
Watermelon mosaic 2, ***Potyvirus,*** 353
WCCV-1, ***Cryptovirus,*** 213
WCCV-2, ***Cryptovirus,*** 214
Weldona, ***Bunyaviridae,*** 276
Wesselsbron, ***Flaviviridae,*** 227
West Nile, ***Flaviviridae,*** 226
Western equine encephalitis, ***Togaviridae,*** 219
WF/1, ***Microviridae,*** 179
Whataroa, ***Togaviridae,*** 219
Wheat American striate mosaic, ***Rhabdoviridae,*** 258
Wheat chlorotic streak, ***Rhabdoviridae,*** 259
Wheat dwarf, ***Geminivirus,*** 174
Wheat rosette stunt, ***Rhabdoviridae,*** 259
Wheat spindle streak mosaic, ***Potyvirus,*** 355
Wheat streak mosaic, ***Potyvirus,*** 355
Wheat yellow leaf, ***Closterovirus,*** 346
Wheat yellow mosaic, ***Potyvirus,*** 355
White bryony mosaic, ***Carlavirus,*** 343
White bryony mosaic, ***Potyvirus,*** 355
White clover cryptic 1, ***Cryptovirus,*** 213
White clover cryptic 3, ***Cryptovirus,*** 213
White clover crypticvirus 2, ***Cryptovirus,*** 214
White clover mosaic, ***Potexvirus,*** 349
Wild cucumber mosaic, ***Tymovirus,*** 337
Wild potato mosaic, ***Potyvirus,*** 355
Wineberry latent, ***Potexvirus,*** 349
Winter wheat mosaic, ***Tenuivirus,*** 399
Winter wheat Russian mosaic, ***Rhabdoviridae,*** 259
Wissadula mosaic, ***Geminivirus,*** 176
Wisteria vein mosaic, ***Potyvirus,*** 353
Witwatersrand, ***Bunyaviridae,*** 282
Wongal, ***Bunyaviridae,*** 276
Woodchuck hepatitis B, ***Hepadnaviridae,*** 115
Woolly monkey sarcoma, ***Reoviridae,*** 294
Wound tumor, ***Reoviridae,*** 194
WRSV, ***Papovaviridae,*** 148
WT1, ***Myoviridae,*** 162
WTV, ***Reoviridae,*** 194
WW/1, ***Microviridae,*** 179
Wyeomyia, ***Bunyaviridae,*** 275
X, ***Inoviridae,*** 182
X29, ***Myoviridae,*** 162
Xf, ***Inoviridae,*** 182
Xf2, ***Inoviridae,*** 182
Xiburema, ***Rhabdoviridae,*** 257
Xingu, ***Bunyaviridae,*** 275
XP5, ***Myoviridae,*** 162
ψ, ***Siphoviridae,*** 164
ψ5, ***Siphoviridae,*** 164
Yaba monkey tumor, ***Poxviridae***, 93
Yaba-1, ***Bunyaviridae,*** 276
Yaba-7, ***Bunyaviridae,*** 276
Yaba-like disease, ***Poxviridae***, 98
Yacaaba., ***Bunyaviridae,*** 282
Yam mosaic, ***Potyvirus,*** 353
Yarrowia lipolytica, ***Totiviridae,*** 204

Yata , ***Rhabdoviridae,*** 257
Yellow fever, ***Flaviviridae,*** 223
Yellow fever, ***Flaviviridae,*** 227
YIV, ***Totiviridae,*** 204
Yogue, ***Bunyaviridae,*** 282
Yokase, ***Flaviviridae,*** 227
ψpe F, ***Siphoviridae,*** 164
Yucca bacilliform, Commelina yellow mottle, 154
Yug Bogdanovac, ***Rhabdoviridae,*** 254
ζ3, ***Microviridae,*** 179
Zaliv Terpeniya, ***Bunyaviridae,*** 278
Zea mays, ***Rhabdoviridae,*** *261*
Zegla, ***Bunyaviridae,*** 276
ZG/1, ***Leviviridae,*** 308
ZG/2, ***Inoviridae,*** 182
ZG/3A, ***Siphoviridae,*** 164
ZIK/1, ***Leviviridae,*** 308
Zika, ***Flaviviridae,*** 227
Zirqa, ***Bunyaviridae,*** 279
ZJ/1, ***Leviviridae,*** 308
ZJ/2, ***Inoviridae,*** 182
ZL/3, ***Leviviridae,*** 308
Zoysia mosaic, ***Potyvirus,*** 355
ZS/3, ***Leviviridae,*** 308
Zucchini yellow fleck, ***Potyvirus,*** 353
Zucchini yellow mosaic, ***Potyvirus,*** 353
Zygocactus, ***Potexvirus,*** 349
μ2, ***Leviviridae,*** 307, 308

Orders, Families, Groups and Genera Index

Adeno-associated virus group, 170
ADENOVIRIDAE, 140
ADENOVIRUS FAMILY, 140
ALFALFA MOSAIC VIRUS GROUP, 392
Allolevivirus, 307
Alphaherpesvirinae, 105
Alphavirus, 217
Aphthovirus, 324
APPLE STEM GROOVING VIRUS GROUP, 339
Aquareovirus, 192
Arbovirus group A, 217
Arbovirus group B, 223
ARENAVIRIDAE, 284
ARENAVIRUS GROUP, 284
Arenavirus, 284
Arterivirus, 220
Ateline herpesvirus group, 143
Aviadenovirus, 143
Avian adenoviruses, 143
Avian type C Retrovirus group (ALV-related viruses), 295
Avihepadnavirus, 115
Avipoxvirus, 94
BACULOVIRIDAE, 117
BACULOVIRUSES, 117
BARLEY STRIPE MOSAIC VIRUS GROUP, 395
BARLEY YELLOW DWARF VIRUS GROUP, 309
Bean golden mosaic virus sub-group, 175
BEET YELLOWS VIRUS GROUP, 345
Beet curly top virus sub-group, 175
Betaherpesvirinae, 107
Bipartite dsRNA mycovirus group, 208
BIRNAVIRIDAE, 200
Birnavirus, 200
BISEGMENTED dsRNA VIRUS GROUP, 200
Bracovirus, 130
BROAD BEAN WILT VIRUS GROUP, 366
BROME MOSAIC VIRUS GROUP, 382
BROMOVIRUS, 382
BULLET-SHAPED VIRUS GROUP, 250
Bunyamwera supergroup, 274
BUNYAVIRIDAE, 273
Bunyavirus, 274
CALICIVIRIDAE, 300
CALICIVIRUS FAMILY, 300
Calicivirus group, 300
Calicivirus, 300
CAPILLOVIRUS, 339
Capripoxvirus, 95
Cardiovirus, 323
CARLAVIRUS, 341
CARMOVIRUS, 303
CARNATION LATENT VIRUS GROUP, 341
CARNATION MOTTLE VIRUS GROUP, 303
CARNATION RINGSPOT VIRUS GROUP, 364
CAULIFLOWER MOSAIC VIRUS GROUP, 150
CAULIMOVIRUS, 150
Chloriridovirus, 134
Chordopoxvirinae, 92
CLOSTEROVIRUS, 345
Coliphage λ group, 163
Coliphage MS2-GA group, 307
Coliphage Qβ-SP group, 307
Coliphage T4 Group, 161
Coliphage T7 group, 165
Coltivirus, 189
COMMELINA YELLOW MOTTLE VIRUS GROUP, 152
Common cold virus group, 322
COMOVIRUS, 360
CORONAVIRIDAE, 234
CORONAVIRUS GROUP, 234
Coronavirus, 234
CORTICOVIRIDAE, 157
Corticovirus, 157
COWPEA MOSAIC VIRUS GROUP, 360
CRYPTIC VIRUS GROUP, 212
CRYPTOVIRUS, 212
CUCUMBER MOSAIC VIRUS GROUP, 386
CUCUMOVIRUS, 386
Cypovirus, 193
CYSTOVIRIDAE, 184
Cystovirus, 184
Cytomegalovirus group, 107
Cytomegalovirus, 107
Cytoplasmic polyhedrosis virus group, 193
Densovirus, 171
Dependovirus, 170
DIANTHOVIRUS, 364
dsDNA ALGAL VIRUSES, 137
dsDNA Phycovirus group, 137
dsRNA MYCOVIRUSES WITH DIVIDED GENOMES, 208
Duck hepatitis B virus group, 115
EMC virus group, 323
Enterovirus group, 322
Enterovirus, 322
Entomopoxvirinae, 98
Entomopoxvirus A, 99
Entomopoxvirus B, 99
Entomopoxvirus C, 100
Equine arteritis virus, 220

Eubaculovirinae, 118
φ6 PHAGE GROUP, 184
φ6 Phage Group, 184
FABAVIRUS, 366
Fijivirus, 195
Filamentous phages, 181
FILOVIRIDAE, 247
Filovirus, 247
FLAVIVIRIDAE, 223
Flavivirus, 223
Foamy virus group, 296
Foot-and-mouth disease virus group, 324
Fowlpox subgroup, 94
Frog virus group, 134
FUROVIRUS, 377
φX-type phages, 178
Gammaherpesvirinae, 108
GEMINIVIRUS, 173
Giardia virus group, 206
Giardiavirus, 206
Goldfish virus group (proposed), 135
Granulosis virus (GV), 120
Hantaan group, 280
Hantavirus, 280
HEPADNAVIRIDAE, 111
Hepatitis A virus group, 322
Hepatitis B virus group, 115
Hepatitis C virus group, 230
Hepatovirus, 322
Herpes simplex virus group, 105
HERPESVIRIDAE, 103
HERPESVIRUS GROUP, 103
HORDEIVIRUS, 395
HTLV-BLV group, 297
Human cytomegalovirus group, 107
Human herpesvirus 1 group, 105
Human herpesvirus 3 group, 106
Human herpesvirus 4 group, 109
Ichnovirus, 130
ICOSAHEDRAL CYTOPLASMIC DEOXYRIBOVIRUSES, 132
ILARVIRUS, 389
Influenza virus A and B, 263
Influenza virus C, 270
INOVIRIDAE, 181
Inovirus, 181
Insect parvovirus group, 171
IRIDOVIRIDAE, 132
Iridovirus, 133
ISOMETRIC PHAGES WITH ssDNA, 178
Large iridescent insect virus group, 134
Lentivirus group, 297
Lentivirus, 297
Leporipoxvirus, 96
LEVIVIRIDAE, 306
Levivirus, 307
LIPOTHRIXVIRIDAE, 127
Lipothrixvirus, 127
LUTEOVIRUS, 309
Lymphocryptovirus, 109
Lymphocystis disease virus group, 135
Lymphocystivirus, 135
Lymphocytic choriomeningitis virus group, 284
Lymphoproliferative virus group, 108
Lyssavirus, 254
MAC-1-type phages, 180
MAIZE CHLOROTIC DWARF VIRUS GROUP, 312
MAIZE RAYADO FINO VIRUS GROUP, 314
Maize streak virus sub-group, 174
Mammalian adenoviruses, 142
Mammalian Type B Oncovirus group, 293
Mammalian Type D Retrovirus group, 295
MARAFIVIRUS, 314
MARBURG VIRUS GROUP, 247
Mastadenovirus, 142
Measles-rinderpest-distemper virus group, 244
MICROVIRIDAE, 178
Microvirus, 178
MLV-related viruses (Mammalian type C Retrovirus group), 294
Molluscipoxvirus, 97
Molluscum subgroup, 97
MONONEGAVIRALES, 239
MONOPARTITIE dsRNA MYCOVIRUS GROUP, 203
Monopartitie dsRNA mycovirus group, 203
Morbillivirus, 244
Mucosal disease virus group, 228
Multiple nucleocapsid viruses (MNPV), 119
Murine cytomegalovirus group, 108
Muromegalovirus, 108
MYOVIRIDAE, 161
Myxoma subgroup, 96
Nairobi sheep disease group, 279
Nairovirus, 279
NECROVIRUS, 316
NEPOVIRUS, 368
NODAMURA VIRUS GROUP, 372
NODAVIRIDAE, 372
Nodavirus, 372
Nonoccluded baculoviruses (NOB), 120
Nonoccluded baculoviruses (NOB), 121
Nuclear polyhedrosis virus (NPV), 118
NUDAURELIA ß VIRUS GROUP, 330
Nudibaculovirinae, 120
Occluded baculoviruses, 118
Orbivirus, 188
Orf subgroup, 94
Orthohepadnavirus, 115

ORTHOMYXOVIRIDAE, 263
Orthopoxvirus, 92
Orthoreovirus, 187
Papillomavirus, 147
PAPOVAVIRIDAE, 146
PAPOVAVIRUS GROUP, 146
Parainfluenza virus group, 244
PARAMYXOVIRIDAE, 242
Paramyxovirinae, 243
Paramyxovirus, 244
Parapoxvirus, 94
PARSNIP YELLOW FLECK VIRUS GROUP, 318
PARTITIVIRIDAE, 208
Partitivirus, 208
PARVOVIRIDAE, 167
Parvovirus group, 168
Parvovirus, 168
PEA ENATION MOSAIC VIRUS GROUP, 375
Penicillium chrysogenum virus group, 209
Pestivirus, 228
PHAGES WITH CONTRACTILE TAILS, 161
PHAGES WITH LONG NON-CONTRACTILE TAILS, 163
PHAGES WITH SHORT TAILS, 165
PHAGES WITH TAILED PHAGES, 155
Phages with double capsids, 155
Phlebovirus, 277
PHYCODNAVIRIDAE, 137
Phycodnavirus, 137
Phytoreovirus, 194
PICORNAVIRIDAE, 320
PICORNAVIRUS GROUP, 320
Plant reovirus subgroup 1, 194
Plant reovirus subgroup 2, 195
Plant reovirus subgroup 3, 197
Plant rhabdovirus A, 258
Plant rhabdovirus B, 259
Plant rhabdovirus group, 257
PLASMAVIRIDAE, 124
Plasmavirus, 124
Plectrovirus, 182
PLEOMORPHIC PHAGES, 125
Pleomorphic phages, 125
PM2 PHAGE GROUP, 157
PM2 Phage group, 157
Pneumovirinae, 245
Pneumovirus, 245
PODOVIRIDAE, 165
POLYDNAVIRIDAE, 129
POLYDNAVIRUS GROUP, 129
Polyomavirus, 148
POTATO VIRUS X GROUP, 348
POTATO VIRUS Y GROUP, 351
POTEXVIRUS, 348
POTYVIRUS, 351
POXVIRIDAE, 91
POXVIRUS GROUP, 91
Poxvirus of *Coleoptera*, 99
Poxvirus of *Diptera*, 100
Poxvirus of *Lepidoptera* and *Orthoptera*, 99
Poxviruses of insects, 98
Poxviruses of vertebrates, 92
Rabies virus group, 254
Ranavirus, 134
REOVIRIDAE, 186
Reovirus subgroup, 187
Respiratory syncytial virus group, 245
RETROVIRIDAE, 290
RHABDOVIRIDAE, 250
Rhadinovirus, 109
Rhinovirus, 323
Rhizidiomyces virus, 145
Rhizidiovirus, 145
RICE STRIPE VIRUS GROUP, 398
RNA TUMOR VIRUSES, 290
ROD-SHAPED PHAGES, 181
Rod-shaped phages, 182
Rotavirus, 190
Rubella virus, 219
Rubivirus, 219
Sandfly fever and Uukuniemi group, 277
SATELLITES, 400
Sheep pox subgroup, 95
Simplexvirus, 105
Single nucleocapsid viruses (SNPV), 119
SIPHOVIRIDAE, 163
Small iridescent insect virus group, 133
SOBEMOVIRUS, 327
SOIL-BORNE WHEAT MOSAIC VIRUS GROUP, 377
SOUTHERN BEAN MOSAIC VIRUS GROUP, 327
Spiromicrovirus, 179
Spumavirus, 296
SpV4-type phages, 179
ssRNA PHAGES, 306
SSV1 Group, 126
SSV1-TYPE PHAGES, 126
Suipoxvirus, 97
Swinepox subgroup, 97
TAILED PHAGES, 159
TECTIVIRIDAE, 155
Tectivirus, 155
TENUIVIRUS, 398
TETRAVIRIDAE, 330
TOBACCO MOSAIC VIRUS GROUP, 357
TOBACCO NECROSIS VIRUS GROUP, 316
TOBACCO RATTLE VIRUS GROUP, 380
TOBACCO RINGSPOT VIRUS GROUP, 368
TOBACCO STREAK VIRUS GROUP, 387
TOBAMOVIRUS, 357

TOBRAVIRUS, 380
TOGAVIRIDAE, 216
TOMATO BUSHY STUNT VIRUS GROUP, 322
Tomato spotted wilt group, 281
TOMBUSVIRUS, 332
Torovirus, 237
Tospovirus, 281
TOTIVIRIDAE, 203
Totivirus, 203
TTV1 FAMILY, 127
TTV1 Group, 127
TURNIP YELLOW MOSAIC VIRUS GROUP, 336
TYMOVIRUS, 336
Vaccinia subgroup, 92
Varicellovirus, 106
Vesicular stomatitis virus group, 252
Vesiculovirus, 252
VIROIDS, 403
White clover cryptic virus I, 213
White clover cryptic virus II, 214
Yaba/Tanapox subgroup, 98
Yatapoxvirus, 98

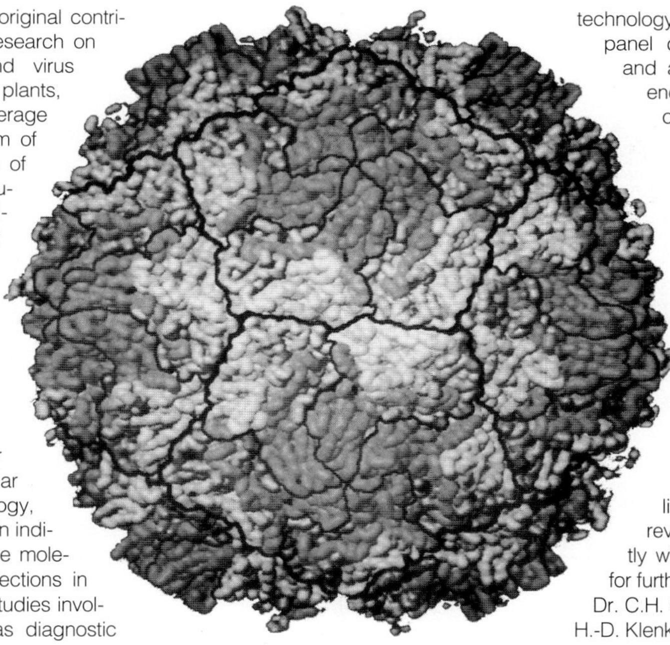